6"

8"

MASK FOR HALF-FRAME FILMSTRIP FRAME

The area enclosed within the black rule gives the outline of the mask
for a 35mm half-frame filmstrip (proportion 3 × 4).

PLANNING AND PRODUCING INSTRUCTIONAL MEDIA

FIFTH EDITION

JERROLD E. KEMP
San Jose State University

DEANE K. DAYTON
Educational Technologies, Inc. Charlotte, NC

With the assistance of
RON CARRAHER
Art
University of Washington
and
RICHARD F. SZUMSKI
Photography
San Jose State University

1817

HARPER & ROW, PUBLISHERS, New York
Cambridge, Philadelphia, San Francisco,
London, Mexico City, São Paulo, Singapore, Sydney

Sponsoring Editor: Louise H. Waller
Project Editor: Brigitte Pelner
Production Assistant: Debra Forrest Bochner
Compositor: ComCom Division of Haddon Craftsmen, Inc.
Printer and Binder: The Murray Printing Company
Cover Design: Ron Carraher

This book was formerly published as *PLANNING AND PRODUCING AUDIOVISUAL MATERIALS.*

PLANNING AND PRODUCING INSTRUCTIONAL MEDIA, Fifth Edition
Copyright © 1985 by Harper & Row, Publishers, Inc.

Library of Congress Cataloging in Publication Data

Kemp, Jerrold E.
    Planning and producing instructional media.

    Rev. ed. of: Planning and producing audiovisual materials. 4th ed. c1980.
    Includes bibliographies and index.
    1. Audiovisual materials. 2. Audiovisual equipment.
I. Dayton, Deane K. II. Kemp, Jerrold E. Planning
and producing audiovisual materials. III. Title.
LB1043.K4 1985   371.3′3   84-10909
ISBN 0-06-043588-7

85 86 87 88 9 8 7 6 5 4 3 2 1

BRIEF CONTENTS

CONTENTS

Part Two   PLANNING INSTRUCTIONAL MEDIA

Contents

Part Five   PRODUCING INSTRUCTIONAL MEDIA

Part Five   MANAGEMENT

MANAGING MEDIA PRODUCTION SERVICES
263

Contents

Appendixes

PREFACE

The title of previous editions of this book contained the term *audiovisual materials*. With this edition, the title is changed to *Planning and Producing **Instructional Media***. It is a more inclusive title since attention is now given to printed media forms and to computer-based instruction as well as to audiovisual media—display materials, audio recordings, overhead transparencies, slides, filmstrips, multi-image presentations, motion pictures, and video recordings.

The overall goal of this book continues to emphasize effective communication through the development of media to accomplish specific ideas or identified needs. The purposes served may be instructional, informational, or motivational. In order to satisfy this goal and serve the purposes stated, the reader will gain competencies relative to:

- Recognizing the contributions of instructional media to the learning process
- Identifying the role of media in systematically planned programs
- Using information about perception, communications, and learning theory when planning the media
- Using evidence from media research when designing and preparing the media
- Selecting the most appropriate medium for use with groups and individuals to serve the ideas and needs
- Applying necessary planning steps prior to production
- Using fundamental skills in photography, graphic arts, and sound recording for preparing materials
- Applying techniques for producing ten types of instructional media
- Managing media production services within an organization

While your initial interest in using the book may be to learn certain production skills, it is hoped that you will also explore information in chapters relating to the more advanced objectives stated above. Cross references throughout the book will help you to make use of the fundamental skills in your specific area of interest.

In addition to the descriptive information and procedures comprising the words and pictures of the 21 chapters in this book, review questions are included as self-check exercises for the reader. They are found at the ends of chapters and at the ends of many sections within the longer chapters. These questions are designed not only to test your recall of information presented but also to encourage your application of concepts and principles as you study them. Check your answers with those in the Appendix.

Planning and producing any type of instructional media can be a creative and intellectually satisfying experience. To ensure quality for the resulting product, give careful attention to the information in this book. Good luck!

Jerrold E. Kemp
Deane K. Dayton

ACKNOWLEDGMENTS

The continued success of this book is due in large measure to the "team members" whose names appear on the title page. Each one of us has contributed not only specific media skills, but the ability to deal with ideas in a creative fashion. We've had many spirited conversations and communications concerning the selection and arrangements of topics and then their verbal and visual treatment. It has been a superb professional experience working with Deane, Ron, and Dick on this fifth edition. A word about each of my colleagues . . .

I welcome Deane Dayton as co-author. He is a graduate of the Instructional Systems Technology program at Indiana University, where he taught AV graphics, supervised media production, and took a leading role in introducing computer-based instruction and computer graphics into the program. He prepared the computer graphics section of Chapter 11 and wrote Chapter 20, Computer-Based Instruction. His suggestions and carefully worded criticisms strengthened many other chapters.

Ron Carraher has developed the art since the first edition of this book. He teaches photography at the University of Washington and does free-lance art consultation. He is the type of person who can take a "two word" idea and visualize it in *three* exciting ways! Then we have the difficult task of making a choice!

Dick Szumski translates ideas into photographs. He does this in all visual forms—print, slide, film, and video—to which he adds further talent with the audio medium. As media production specialist at San Jose State University, he designs effective media products for the faculty and administration. The results of his efforts typify what this book attempts to help you do!

Other persons provided us with professional and technical assistance during the preparation of this edition. Our thanks to the following:

- Bob Reynolds—instructional television producer/director at San Jose State for his guidance with the content of Chapter 19, Video and Film. I appreciate his patience while explaining the complex newer developments in video.
- Barney Hazarian—professor of graphic arts at San Jose State. He reviewed the textual content on graphic reproduction processes and assisted in the preparation of special graphics examples.
- Daryle Webb, Dick Mills, Romaldo Lopez, and Tom Tutt—the talented staff members of our photographic and graphic media production department at San Jose State. These fellows were most helpful in offering advice and assistance as visual materials were being prepared.
- Jessica Klein—an artist in Seattle who assisted Ron Carraher in the preparation of the final art work.
- Ann Roomel—a graphics specialist in San Jose who was responsible for doing the final paste-up of the camera-ready copy for each page.

The frequent remarks, "I'll be done in a minute. Yes, I'll be right with you," will no longer be heard on this project! For their understanding and patience, more than thanks to Dee, Carol, Laurel, and Joan—our wives.

Yes dear, it *is* finished!

Jerrold E. Kemp

Part One

BACKGROUND FOR PLANNING AND PRODUCING INSTRUCTIONAL MEDIA

1. Instructional Media for Education and Training

2. Perception, Communication, and Learning Theory

3. Research in the Design and Production of Instructional Media

Chapter 1

INSTRUCTIONAL MEDIA FOR EDUCATION AND TRAINING

- Contributions of Media to the Learning Process
- Designing for Instruction
- Patterns for Teaching and Learning
- Media in the Instructional-Design Process
- Levels of Instructional Media Production
- Students and Instructional Media Production

"George, you always make everything seem so simple."

When skillfully combined, pictures, words, and sounds have the power to evoke emotions, change attitudes, and motivate actions. Examples of this power can be seen every day on television: the commercial that motivates the viewer to buy a product, the political spot that attempts to sway a voter's choice of candidate, or the emotional appeal for donations to a charity. We have become accustomed to living in a world of audiovisual impressions. The impressions that are created by combinations of pictures, words, and sounds have been shown to be retained by viewers significantly longer than when they are only heard or read (Wilkinson).

Instructional media also make use of the power of pictures, words, and sounds to compel attention, to help an audience understand ideas and acquire information too complex for verbal explanation alone, and to help overcome the limitations of time, size, and space.

CONTRIBUTIONS OF MEDIA TO THE LEARNING PROCESS

While the advantages of using instructional media have been recognized for a long time, their acceptance and integration within instructional programs have been slow. Recently, there has been increasing evidence that positive results take place when carefully designed, high quality instructional media are used either as an integral part of classroom instruction or as the principal means of direct instruction. The outcomes often realized are:

- **The delivery of instruction can be more standardized.** Each student seeing and hearing a media presentation receives the same message. Instructors may interpret subject content in different ways but by using media the variations can be reduced and the same information can be communicated to all students as the bases for further study, practice, and application.
- **The instruction can be more interesting.** There is an attention-getting factor associated with instructional media that keeps members of an audience alert. The clarity and coherence of a message, the attractiveness of changing images, the use of special effects, and the impact of ideas that can arouse curiosity cause an audience to laugh or be thoughtful; all contribute to the motivational and interest-creating aspects of instructional media.
- **Learning becomes more interactive through applying accepted learning theory.** The content of instructional media can be organized and presented in a manner that represents good instruction. In the design of media and related materials for student use, attention should be given to such psychological principles as

learner participation, feedback, and reinforcement if the student is to become actively engaged in learning experiences. There is a continual interaction between the learner and the media, resulting in effective instruction and learning.

- **The length of time required for instruction can be reduced.** Most media presentations require a short time to transmit their messages. But during this brief period, a large amount of information can be communicated to and absorbed by the learner. This can lead to greater efficiency in the use of time for both the instructor and the student during an instructional session.
- **The quality of learning can be improved.** When there is a careful integration of pictures and words, instructional media can communicate elements of knowledge in a well-organized, specific, and clearly defined manner. As a result, with suitable study effort on the part of the student and appropriate follow-up activities, learning can be expected to reach an acceptable competency level.
- **The instruction can be provided when and where desired or necessary.** When instructional media are designed for individualized use, then a student can study at a time and place that is personally convenient. This flexibility is particularly important when individuals must integrate study activities with vocational and personal responsibilities.
- **The positive attitude of students toward what they are learning and to the learning process itself can be enhanced.** Students frequently express preference for using media as a means of studying. This may be due to both the motivational aspect and the contributions that media can make to a person's success in learning. Students find learning with instructional media both enjoyable and satisfying.
- **The role of the instructor can be appreciably changed in positive directions.** While most benefits for the use of instructional media are directed to the student and to his or her accomplishment in learning, there are advantages also for the instructor. First, much of the burden for repeated explanations of content and skills can be eliminated. Second, by not having to present as much information verbally, other, possibly more important aspects of a course can be given attention. Third, the instructor has increased opportunity to fulfill the role of being a consultant and advisor to students.

Taken together, these eight outcomes indicate that through the use of media both the efficiency of learning and positive attitudes toward learning may be enhanced. Each person must decide which of these contributions should receive attention when planning and producing media for instruction.

To satisfy the outcomes specified above, instructional media have to be not only of high quality but should also be selected or designed and produced as an integral part of an instructional program. They must make a definite contribution to the achievement of the program's objectives. For this reason, a person interested in planning and producing instructional media should also become acquainted with the process of systematic instructional planning within which instructional media are to function.

DESIGNING FOR INSTRUCTION

Traditionally, plans for instruction most often are made in intuitive fashion and may be based on ambiguous purposes. Subject content is the basis for planning, and only casual attention is given to other details. It is now recognized that the instructional process is complex and that attention must be given to many interrelated factors if outcomes are to be successful.

For an instructional program to be successful, the following should occur:

- Satisfactory learning takes place so that students have acquired necessary knowledge, skills, and attitudinal behavior patterns, and after training, perform productively in their assignments.
- The learning is accomplished with due regard for reasonable expenditures of money and time.
- The learning experiences are meaningful and interesting so that students are encouraged to continue with their studies.
- The planning and implementation of an instructional program prove to be a satisfactory set of experiences for the instructor and support staff.

The term **instructional development** applies to the broad process of designing an instructional program—whether a single module, a complete unit, or a total course—using an objective, systematic procedure. This method starts with answers to four questions:

1. For whom is the program being developed? (nature of the **students** or **trainees**)
2. What do you want the students or trainees to learn or be able to do? **(objectives)**
3. How is the subject content or skills best learned? (teaching/learning **methods** and **activities** with **resources**)
4. How do you determine the extent to which the learning has been achieved? **(evaluation)**

Four elements—learners, objectives, methods, and evaluation—form the framework of instructional development procedures. In addition, there are other factors that either support or relate to these four elements. Taking all these pieces together, we can develop an **instructional-design** *plan,* which consists of these interrelated components (Kemp):

1. Assess **learning needs** for designing an instructional program; state **goals, constraints,** and **priorities** that must be recognized.
2. Select **topics** or **job tasks** to be treated and indicate **general purposes** to be served.
3. Examine **characteristics** of **students** or **trainees** which should receive attention during planning.
4. Identify **subject content** and **analyze tasks** relating to the topic and job.
5. State **learning objectives** to be accomplished by students or trainees in terms of subject content and job tasks.

6. Through **pretesting,** determine preparation of students or trainees for studying the topic.
7. Select a **teaching/learning method** and design **activities** to accomplish learning objectives.
8. Select **resources** (including **media**) to support the activities.
9. Specify **support services** required for developing and implementing activities and acquiring or producing materials.
10. Prepare to **evaluate student learning** and **program outcomes** in terms of the accomplishment of objectives, with a view to revising and reevaluating any phases of the instructional plan that need improvements.

The ten elements of this instructional-design plan can be illustrated by a diagram:

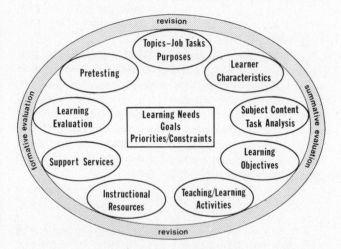

The diagram is circular in format because not all designers will start their planning with the same element. One person might start with a consideration of the learner and another with one of content or objectives. Note the location and relationship of the instructional resources item to other elements of the instructional-design plan. You will discover in Part II of this book that a design procedure similar to that shown above is also applied when planning for the production of instructional media.

As used here, the term **instructional media** refers to **audiovisual** and **related materials** that serve instructional functions for education and training. As you will see in Chapter 4 (page 28), media forms are also useful for *informational* and *motivational* purposes. Thus, their value extends beyond the "instructional" intent of the label *instructional media.* If your interest is in developing a type of media for informational or motivational purposes, much of this background and the planning procedures in Part II will still merit careful attention.

One important consideration when planning and producing instructional media is to take into account the audience mode in which the media might be used. You should recognize the possible alternatives that are available within instructional situations along with their advantages and limitations.

PATTERNS FOR TEACHING AND LEARNING

There are three broad methods within which most learning takes place: (1) **presentation** of information to groups of students, (2) **self-paced study,** or individualized learning, with each student working on his or her own, and (3) small-group **interaction** between instructor and students or among students. There is a gradual shift through innovative instructional methods away from complete dependence on group teaching to more emphasis on self-paced learning procedures. While instructional media do provide many necessary learning experiences within each of the three patterns, the media have essential functions in the newer, nontraditional instructional formats.

By recognizing the features of these three methods, you should be better able to decide on appropriate instructional media for use in any of them, and then to design the materials to fit the requirements of the pattern.

Instructor Presentation to Groups

This method is typified by one-way communications from instructor to students, as in a lecture. Information is presented at the instructor's rate of delivery. Students are physically passive, although listening, taking notes, or completing related worksheets. In this pattern, the students have no flexibility of individual pacing and choice of study methods and materials. With the current trend to reduce the amount of time spent in the conventional presentation of subject content by the teacher in preference to self-paced study of content by students, the purposes served by this pattern are changing. Often, for efficiency, essential information may be transmitted to numbers of students, in regular classes or in large groups, to serve these needs:

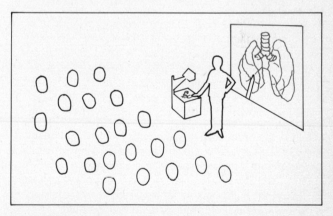

- Introduce new topics and provide orientation to activities in a unit of study.
- Provide motivation for studying a subject or topic, possibly through a videotape recording or a multi-image presentation.
- Illustrate relations or integration of one topic with another.
- Point out special applications or new developments in a topic that may be too recent for inclusion in the self-paced study materials.
- Provide special enrichment materials and experiences,

like a film, or a guest speaker who cannot be available to small groups or individual students.

To complement or possibly replace an instructor's usual verbal presentation, instructional media, such as overhead transparencies, slides, motion pictures, videotape recordings, or multi-image presentations, may be selected to serve one or more of these instructional needs relative to a topic. The usual film or other media formats, consisting of 10- to 30-minute presentations, can be modified to more succinct structures in terms of specific objectives to be treated. Also, the value of learner participation during a presentation can be increased by providing activities for students, like responding periodically to questions on an exercise sheet or allowing the selection of items for followup work.

Individualized or Self-paced Learning

The individualization of learning may take many forms and is given numerous labels. See the references on page 10. Its main attributes include the students' assuming responsibility for their learning, proceeding with activities and materials at their own level, and studying at their own pace in school, home, office, or elsewhere. In a basic program all students may follow the same track, using the same materials, with only their individual pace of study being different. In more advanced programs, alternative methods for accomplishing the objectives are provided along with a correlated variety of materials. Choice of learning experiences is made by the student.

When the procedure of systematic instructional development is applied to the design of individualized learning programs, these elements are often included:

- Learning objectives and required levels of student knowledge or performance are clearly stated.
- Pretesting permits the student to skip study of one or more objectives if competency is demonstrated.
- Alternative procedures for accomplishing the objectives are specified.
- Participation activities and required responses for the learner are included.
- Confirmation or correction of performance or response is immediately available to the learner.
- Opportunities are provided for the learner to self-check understanding, progress, and performance against the objectives.

- The learner decides when ready to have knowledge or performance evaluated by the teacher.
- If the results of tests or other evaluation methods do not indicate a satisfactory competency level, the student may restudy and be retested.
- Followup applications or projects permit the student to use the knowledge or skill learned.

An individualized or self-paced learning program consists of study units often called **modules.** Each module treats a topic (examples: *The Tablesaw, Cardiogenic Shock, The Cold War*). The supporting resources, such as learning aids, slides, a filmstrip, audiocassette recording, a short videotape recording, or a computer-assisted program, are specific to the topic. This means that media selected must be in a form suitable for self-paced study, should be brief in serving one or only a few objectives, and should be carefully integrated with other activities.

Instructor–Student Interaction

The third pattern for teaching and learning provides opportunities for instructors and students to work together in small groups to discuss, to question, to report, to perform, to be evaluated, or to engage in other forms of personalized interchange. In light of the shift toward individualized learning, with the student spending more time working on his or her own, it is necessary to provide opportunities for direct contact with instructors and with other students. This pattern provides such experiences.

Some of the same resources used in presentations and for individualized learning may be available for reference during small-group discussion. Also, special materials can be prepared for the purposes of motivating discussion, illustrating concepts, presenting problem situations for group consideration, and evaluating learning. A real need exists for imaginatively designed media materials for use in group activities.

A prime advantage of small-group interaction is the encouragement of student participation in socialization, leadership development, and peer recognition purposes. One of the best methods of providing experiences that lead to accomplishing one or more of these important purposes is to encourage and assist students to plan and produce their own instructional media, then present the results to their

An Example of
Elements Within an
Instructional Design

Subject:
 Solar energy

Topics:
 Collecting solar energy
 Storing the energy
 Using the energy for heating

General Purpose:
 To understand how solar energy can be used

Objectives:
 1. To explain how four scientific principles are applied
    for creating heat in a solar collector
 2. To identify four parts of a solar collector and the
    functions of each
 3. To assemble a solar collector when provided with the
    four essential parts

Teaching/Learning Activities: Media Resources:
A. Presentation: Teacher to
   student group
   Scientific principles   Overhead transparencies
   applied to collecting solar   16mm commercial film clips
   energy   Printed worksheets

B. Self-paced Learning
   Parts, placement, and use of   Slides/tape recording
   solar collectors   Worksheet review
                                  Self-check test
C. Group Activities
   1. Visit local solar
      collector manufacturer
      and installer
   2. Make plans for projects

[Note: The slides and tape recording for B above are planned in
detail in the chapters of Part II.]

group. This can be a very practical experience for students in elementary through college level academic programs.

The three patterns we have been examining—instructor presentation to groups, self-paced learning methods, and instructor–student interaction activities—provide the framework within which experiences for learning can be planned. Many of these activities rely on the use of appropriate instructional media.

MEDIA IN THE INSTRUCTIONAL-DESIGN PROCESS

Instructional media of any type should be carefully planned and produced, whether they will be part of an application of an instructional-design plan or individual entities of their own. The techniques described in subsequent chapters apply to materials being prepared for any purposes.

Media to be used within the instructional design are determined by the requirements of students, objectives, content, and instructional methods. Media are *not* supplementary to, or in support of, instruction, but *are* the instructional

input itself. In this light, the old concept of audiovisual *aids* as supplements to teaching can no longer be accepted. Determination must be made of which media, in what form, and at what time, will most effectively and efficiently provide the most relevant experiences for learners.

Just as various instructional objectives require different kinds of learning, appropriate instructional resources should be matched to required tasks. Each separate concept to be taught should require a separate consideration of resources. Certain media can best serve certain purposes (sound or print, motion or still pictures). In other cases, available equipment, convenience, costs, and such factors may be the determiners of choice. See page 42 for further discussion of media-selection methods.

This approach to teaching and learning is developed around specificity—specificity in terms of learning objectives to serve the needs of particular students. Commercial materials will usually not be suitable, since in the main they are too generalized and too broad in treatment of subjects. On the other hand, dependence on local production for all necessary materials is unduly costly in time and money. Perhaps forward-looking producers will treat the most commonly taught subject topics and concepts by providing carefully designed and interrelated materials that may be of use in a variety of locally developed instructional systems. But openings will remain for the addition of materials having local applications or particular local emphases. For example, a unit on solar energy includes the study of how the sun's rays are collected and converted from light into heat. Commercial materials may be available to describe how scientific principles are applied in the process. The teaching of the design and use of local solar collectors should be taught with locally prepared materials. See page 28.

LEVELS OF INSTRUCTIONAL MEDIA PRODUCTION

The local production of instructional media can take place on any of three levels.

Mechanical Level: Preparation

First, there is the *mechanical* level; here the concern is solely with the techniques of preparation. Mounting pictures on cardboard or cloth, copying pictures on film for slides, recording a speech or other presentation on audio or videotape, and running a printed page or clipping through a copy machine to make a transparency are examples of the mechanical preparation of materials. Even though the individual has a purposeful use in mind, little planning is required and the actual preparation follows a routine procedure. Many persons start at this level in instructional media production and go on to other levels of activity.

Creative Level: Production

A step above the mechanical level is the *creative* level. Here, materials being considered for production require decisions; planning accordingly becomes an important forerunner of production. *Production* implies an order of activity beyond *preparation,* with its more routine connotations. The production of an instructional bulletin board, of a slide series with a recording, of a filmstrip for self-instruction, of a set of thoughtfully designed transparencies to teach a concept, of a videotape recording that illustrates a process, or a multi-image presentation as the motivational introduction to a topic—all are examples of materials produced on the creative level. The skills developed on the mechanical level become tools for use on this level.

Design Level: Conception

As previously explained, the production of instructional media that can be carefully integrated into learning activities to serve specific instructional objectives and to meet the needs of individuals or a specific group of students, may be part of a *design for instruction.* The design of a slide/tape presentation or of an interactive computer/video program, as part of a self-paced learning module, are examples of this third level. Now *instructional media* are conceived within a carefully designed instructional framework

for group or individual uses. During the planning for materials on this level attention needs to be given to such factors as *specificity* in serving objectives, *adaptability* for certain individuals or groups, *flexibility* in method of use, and *integration* with other experiences. The skills developed on both the mechanical and creative levels serve important functions here.

While your interest in using this book may start with the mechanical level, it is hoped that you will find potentials for developing materials on either the creative or design levels.

On each level of production you can prepare appropriate materials to serve personal or instructional purposes of your own. Or, you might be developing materials for other persons to use in satisfying their motivational, instructional, or informational communication needs.

But another group is showing increasing interest in activities involving instructional media. They are *students* on all levels of education—nursery school to graduate school. Many teachers plan specific activities that will involve students in the preparation of photographs, slides, motion pictures, videotape recordings, or multi-image presentations. Others find that students, on their own, are ready to engage in such enterprises. Whatever the base, this increasing interest and enthusiasm on the part of students for planning and producing instructional media should be recognized and encouraged. There are definite educational benefits for students who engage in such activities.

STUDENTS AND INSTRUCTIONAL MEDIA PRODUCTION

The recognition of the need to involve students with instructional media in an active and productive way is a recent thrust in education. Its purpose is not only to allow students to produce materials for projects and reports, but also to make students more visually literate. Developing the skills to understand and use visual communication techniques is especially important in our society since so much information is transmitted in nonverbal modes—graphic design, still photography, motion pictures, and television. Learners need opportunities to become perceptive and analytical of the visual world in which they live so as to make their own judgments and choices of what may be appropriate and

aesthetically pleasing in a situation. To do this, students must develop the skills needed for interpreting the messages they receive in visual form and must also become fluent in expressing their own ideas visually.

This visual awareness, comprehension, and expression can be obtained first by developing an intimate familiarity with design principles and visual tools (page 107) and elements of composition (page 94), and then by manipulating these items through involvement with a variety of graphic, pictorial, and other nonverbal communications media.

The skills an individual develops in interpreting, judging, responding to, and using visual representations of reality (i.e., a visual intelligence) result in what is called **visual literacy.** For students on all levels, the process of becoming visually literate requires experiences which allow them to:

● Recognize and "read" graphic and photographic illustrations that represent objects, events, places, and people.
● Sort and organize such visual representations into patterns and relationships that apply a *vocabulary* of nonverbal visual expression.
● Produce visual materials as their own interpretations of actual objects, events, places, and people.

One of the best ways for students to become literate in this visual sense is by actively selecting an idea, developing it by planning as described in Part II of this book, and then translating the written words and sketches into an audiovisual form. This can be done by students individually or in groups.

When students work together to plan and produce successfully a photographic picture display, a slide series, a motion picture, or a videotape recording of a school or community activity, they take part in a mentally vigorous process. The planning phase includes the assumption of responsibilities by individuals in the group, doing research work, expressing and organizing ideas, and structuring the visual presentation to communicate the intended ideas. Then follows the hard work and excitement of production, which brings the verbal thoughts to visual life in a logical sequence.

For students of any age, such an activity can be a stimulus to growth and toward visual intelligence, which means toward better interpretation and understanding of meanings and expressions that take visual forms and require visual decisions in their lives.

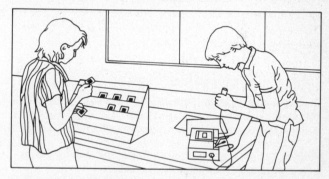

Review What You Have Learned in This Introduction to Instructional Media:

1. Return to page 3 and number in order the eight *effects of media on the learning process.* Use those numbers in answering this question: To which effect does each of the following relate?
   a. On a rating scale students complete at the end of the course, they overwhelmingly indicate a preference in using media for learning.
   b. After seeing the visual report on the benefits of the children's program, many people volunteered their services.
   c. Technicians using computer-based materials completed their recertification in less time than when using only the manual.
   d. While developing the media, the planning team came

up with ideas for additional teaching activities in the unit of instruction.

e. All students in each of the three classes learned the procedures for equipment assembly in the same way.

f. For upgrading skills, staff members were able to study, with the media materials, when they had free time while on duty.

g. The teacher found she had more time to spend with individual students when they used the media to study on their own.

2. What *four* elements form the basis of instructional design?

3. Why is a knowledge of instructional design important to a person interested in developing media for instructional uses?

4. What are the three patterns within which teaching and learning activities can be placed?

5. To which teaching and learning pattern does each situation apply?
   a. Student studying at a computer terminal
   b. Instructor showing a film to a class
   c. Team of students working together to edit their videotape

d. Instructor using transparencies before a training group

e. Salesman listening to an audiocassette tape containing examples of sales methods while driving in his car.

6. What are the *three* levels on which instructional media may be produced?

7. To which level of media production does each of the following relate?
   a. Audiotaping a discussion session
   b. Developing transparencies in terms of objectives for a lesson
   c. Making a multi-image slide presentation
   d. Mounting a set of magazine pictures on cardboard
   e. Scripting and then shooting a how-to-do-it videotape recording
   f. Applying the knowledge learned to problem situations after each student completes the computer program

8. Define *visual literacy*.

9. What are some benefits to students who engage in instructional media production activities?

REFERENCES

Media in the Instructional Process

Brown, James W., et al. *AV Instruction: Technology, Media, and Methods.* 6th ed. New York: McGraw-Hill, 1983.

Hartley, James, and Davies, Ivor, eds. *Contributions to Educational Technology,* volume 2. New York: Nichols, 1978.

Heinich, Robert, et al. *Instructional Media: The New Technologies of Instruction.* New York: Wiley, 1983.

Knapper, Christopher K. *Expanding Learning Through New Communications Technologies.* San Francisco: Jossey-Bass, 1982.

Wilkinson, Gene. *Media in Instruction: 60 Years of Research.* Washington, D.C.: Association for Educational Communications and Technology, 1980.

Instructional Development and Design

Dick, Walter, and Carey, Lou. *The Systematic Design of Instruction.* Glenview, IL: Scott Foresman, 1978.

Gagné, Robert M., and Briggs, Leslie. *Principles of Instructional Design.* Englewood Cliffs, NJ: Prentice-Hall, 1979.

Gerlach, Vernon, and Ely, Don. *Teaching and Media: A Systematic Approach.* 2nd ed. Englewood Cliffs, NJ: Prentice-Hall, 1980.

Kemp, Jerrold E. *The Instructional Design Process.* New York: Harper & Row, 1985.

Nadler, Leonard. *Designing Training Programs.* Reading, MA: Addison-Wesley, 1982.

Romiszowski, A. J. *Designing Instructional Systems.* New York: Nichols, 1981.

Individualized Learning

Bell, Normal T., and Abedor, Allan J. *Developing Audio-Visual Instructional Modules for Vocational and Technical Training.* Englewood Cliffs, NJ: Educational Technology Publications, 1973.

Russell, James D. "Audio-Tutorial Systems." In Danny Langdon, ed., *Instructional Design Library Series,* volume 3., Englewood Cliffs NJ: Educational Technology Publications, 1978.

Russell, James D., and Johanningsmeier, Kathleen A. *Improving Competence Through Modular Instruction.* Dubuque, IA: Kendall-Hunt, 1981.

Visual Literacy

Ashburn, Lynna, and Ausburn, Floyd. "Visual Literacy: Background, Theory, and Practice." *Programmed Learning and Educational Technology* 15 (November 1978): 291–297.

Dondis, Donis A. *A Primer of Visual Literacy.* Cambridge: MIT Press, 1973.

Dwyer, Frank M. "Communicative Potential of Visual Literacy: Research and Implications." *Educational Media International* 2 (1979): 19–25.

Journal of Visual/Verbal Literacy. Lida Cochran. Ed. International Visual Literacy Association, 35 Olive Court, Iowa City, IA. Semiannual.

Wileman, Ralph. *Exercises in Visual Thinking.* New York: Hastings House, 1980.

Media Production Periodicals and Journals

Audio-Visual Communications. United Business Publications, Inc., 475 Park Avenue South, New York, NY 10016.

AV Directions. Montage Publishing Inc., 25550 Hawthorne Blvd., Suite 314, Torrance, CA 90505.

Biomedical Communications. United Business Publications, Inc. (See above.)

Educational and Industrial Television. C. S. Tepfer Publishing Co., 51 Sugar Hollow Road, Danbury, CT 06810.

Law Enforcement Communications. United Business Publications, Inc. (See above.)

Moving Images. Sheptow Publishing, 609 Mission Street, San Francisco, CA 94105.

Video Systems. Intertec Publishing Corp., P.O. Box 12901, Overland Park, KS 66212

Chapter 2

PERCEPTION, COMMUNICATION, AND LEARNING THEORY

- Perception
- Communication
- Learning Theory
- Summary

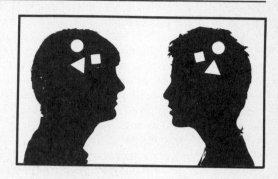

Slides, filmstrips, motion pictures, video recordings, and other instructional media have been in use for many years. While some of these materials do excellent jobs of informing or instructing, of teaching skills, of motivating, or of influencing attitudes, others are less effective, and are of poor quality or may even be detrimental to accomplishing the purposes they were made to serve.

Too often the production of a videotape recording or the planning for a multi-image presentation is based on intuition, subjective judgment, personal preferences for one's own way of doing things, or even on a committee decision. These, unfortunately, are ineffective bases for insuring satisfactory results.

How can you be more sure that the materials you plan and produce will be effective for the purposes you intend? Is there evidence from research and some general principles to guide you?

Three areas should be of particular concern. One is the logical steps of developing objectives, of planning, and of getting ready to take or draw pictures and to make recordings. These procedures will ensure some degree of success for your audiovisual materials. Part II of this book presents the planning steps you should consider using.

The second area from which you can obtain help in designing effective instructional media includes reports on experimental studies measuring the effectiveness of such materials. In these studies specific elements that affect production have been controlled, thus providing evidence for handling various elements in media productions. Summaries of these research findings are reported in the next chapter.

Third, and fundamental to both media research and careful planning for media production, is the need to know how people perceive things around them, how people communicate with each other, and how people learn. Therefore, our immediate concern is to examine evidence from the fields of psychology and communication.

The discussions that follow have one purpose—to make the reader aware of (or to review) some generalizations from the areas of perception, communication, and learning theory that can be applied to make instructional media effective. Admittedly the treatment of each topic is greatly simplified, and only the minimum essentials are presented. But even these can be useful to you as you plan your materials and consider the place of your materials for motivating, informing, or teaching in an instructional sequence.

PERCEPTION

Perception is the process whereby one becomes aware of the world around oneself. In perception we use our senses to apprehend objects and events. The eyes, ears, and nerve endings in the skin are primary means through which we maintain contact with our environment. These, and other senses, are the tools of perception; they collect data for the nervous system. Within the nervous system the impressions so received are changed into electrical impulses, which then trigger a chain of further electrical and chemical events in the brain. The result is an internal awareness of the object or event. Thus, perception precedes communication. Communication leads to learning.

Two things are of major importance about perception. First, any perceptual event consists of many sensory messages that do not occur in isolation, but are related and combined into complex patterns. These become the basis of a person's knowledge of the world. Second, an individual reacts to only a small part of all that is taking place at any one instance. The part of an event to be experienced is "selected" by a person on the basis of desire or what attracts his or her attention at any one time. Hence, one needs first to design materials that will attract the attention and hold the interest of the learner, and then to make certain that in this sampling procedure the learner gets the "right" sample, relevant to the learning task. The experience of perception is individual and unique. It is not exactly alike for any two people. A person perceives an event in terms of individual past experiences, present motivation, and present circumstances.

While any one perceptual experience is uniquely individual, a series of perceptions by different persons can be related to become nearly identical. If you walk around a statue, its shape will constantly change as you change the angles at which you look at it. If someone else then walks around the same statue and looks at it from the same angles, this other person will have different individual experiences, but the series will result much the same as it was for you. Thus a succession of individual experiences enables us to agree upon what we have experienced, even though the individual experiences are somewhat different.

The instructional media field rests on the assumptions that people learn primarily from what they perceive and that carefully designed visual experiences can be common experiences and thus influence behavior in a positive way.

A useful summary of research-based principles from the behavioral sciences that can be applied to the design of instructional media has been prepared by Fleming and Levie. In it over 200 principles and corollaries relate to areas of perception, memory, concept formation, and attitude change. Here is a selection of the major conclusions, concerning perception, that Fleming and Levie present:

- **Basic principles**
  1. Perception is relative, rather than absolute.
     a. Provide reference points to which unknown objects or events can be related.
     b. Present a difficult concept through small steps.
  2. Perception is selective.
     Limit the range of aspects being presented to essential factors, presented a step at a time.
  3. Perception is organized.
     Use numbering and verbal cues ("next," "either-or") to give order to a message.
  4. Perception is influenced by set.
     Give instructions that call attention to elements, or directions for finding an answer in an illustration.
- **Attention and preattention**
  Attention is drawn to changes in how relevant ideas in a message are presented (by means of brightness, movement, novelty, asking questions, posing problems).
- **Perceptual elements and processing**
  Such characteristics as brightness, color, texture, form, and size should be selected and arranged carefully because they have a positive influence on perceptions.
- **Perceiving pictures and words**
  Use the visual channel for presenting spatial (space) concepts and the auditory channel for representing temporal (time) concepts.

- **Perceptual capacity**
  1. For difficult material presented aurally, use short sentences, redundancy, and excellent technical quality.
  2. The most compatible modes that permit the highest information level are simultaneous auditory and visual presentation of a subject provided by slides and tape, sound film, and video recording.
- **Perceptual distinguishing, grouping, and organizing**
  1. Use lines around, under, and between, to cue groupings; accentuate and relate elements in a visual.
  2. Facilitate recognition of similarities and differences by presenting several related objects together.
  3. Make the organizational outline of a message apparent (subtitles, transitional statements).
- **Perception and cognition**
  The better an object or event is perceived (by means of applying the above-stated and other perception principles), the more feasible and reliable will be memory, concept formation, problem solving, creativity, and attitude change.

This is only a sampling of the perception principles important in the design of instructional media, as summarized by Fleming and Levie. For further explanations, examples, and illustrations, see the reference.

As you design instructional media, keep in mind the importance of providing carefully for desirable perceptual experiences in terms of the learner's experience background and of the present situation. Such production elements as methods of treating the topic (expository, dramatic, inquiry, or other), vocabulary level, kinds and number of examples, pacing of narration and visuals, and graphic techniques can each contribute to successful perception. In this way communication will be more effective and learning should be positive.

COMMUNICATION

Perception leads to communication. In all communication, however simple or complex, a sequence similar to this occurs:

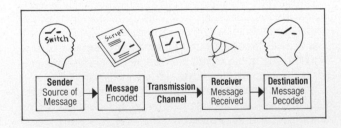

This model (originally developed by Shannon and Weaver) illustrates that a *message* (at the mental level), generally in the form of information, originated by a *source* or *sender* (the brain of an individual), is *encoded*—converted into transmittable form (a thought verbalized, as for conversation, by being turned into sound waves, or words written for a script). The message then passes through a *transmitter*

(print, slides, film, videotape) via a suitable *channel* (air, wire, paper, light) to the *receiver* (a person's senses—eyes, ears, other sensory nerve endings), then to the *destination* (brain of the receiver) where the message is *decoded* (converted into mental symbols).

Effective communication depends upon participation of the receiver. A person reacts by answering, questioning, or performing, mentally or physically. There is then a return, or response loop of this cycle, from receiver *to* sender. It is termed *feedback.* It happens through words, expressions, gestures, or other actions. This reverse communication advises the originator how satisfactorily the message was received. Feedback enables the originator to correct omissions and errors in the transmitted message, or to improve the encoding and transmission process, or even to assist the recipient in decoding the message.

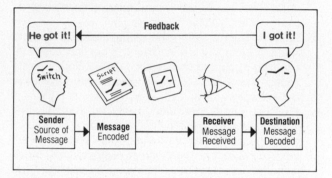

One additional element must be added to this communication model:

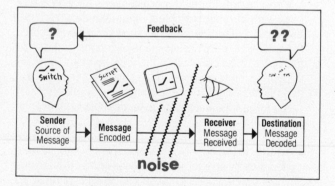

Noise is *any* disturbance that interferes with or distorts transmission of the message. The factor of *noise* can have serious impact on the success or failure of communication. Static on a radio broadcast is a simple example of noise. A flashing light can be a distracting "noise" when a person is reading a book. Ambiguous or misleading material in a film can be deemed noise. Noise can be created internally, within the receiver, to upset satisfactory communication—for example, a lack of attention. Even conflicting past experience can be an inhibiting noise source. Recall the importance of an individual's background experience in affecting perception. Noise clouds and masks information transmission to varying degrees and must be recognized as an obstacle to be overcome.

At times noise cannot be avoided, and in planning materials the factor of *redundancy* is often used to overcome the effect of evident or anticipated noise. Redundancy refers to the repeated transmission of a message, possibly in different channels, to overcome or bypass distracting noise. Some examples of redundancy are: showing and also explaining an activity, projecting a visual and distributing paper copies of the same material for study, and providing multiple applications of a principle in different contexts.

In working with instructional media you should understand where the materials, as channels of communication, fit within the framework and process of message movement between senders and receivers. Then, you should know how the various elements, along with factors of noise and redundancy, function to affect the success of your efforts to communicate effectively.

LEARNING THEORY

The process of learning is an individual experience for each person. Learning takes place whenever an individual's behavior is modified—when a person thinks or acts differently, when he or she has acquired new knowledge or a new skill.

Since a major purpose for preparing instructional media is to affect behaviors that serve objectives, it is appropriate to turn to the psychology of learning for help in locating principles that would guide the planning of effective instructional media.

Learning theories fall into three major families. One is the so-called *behaviorist,* or *connectionist,* group, which interprets human behavior as connections between stimuli and responses. This is the **stimulus-response** *(S-R)* pattern of learning. A stimulus is the **message,** containing content, that is transmitted to the **receiver** (the learner) during the communications process.

Each specific reaction is an exact *response* to a specific sensation, or *stimulus.* Spoken and written words, simple pictures, and all instructional media are examples of stimuli. Some stimuli are more effective than others for certain purposes.

Much instruction is of this stimulus-response type. This concept is implicit in the "programmed-instruction" approach introduced by B. F. Skinner. The emphasis here, as in most newer approaches to instruction, is on the learner and the correctness of his or her response to questions as the instruction proceeds.

In programmed instruction, each sequence of learning is broken into small steps, requiring an appropriate response to each item followed by immediate knowledge of results (known as **feedback**). If the response is correct, the knowledge is a **reinforcement,** a rewarding recognition of each correct response. Much of the attention being given to individualized learning follows this pattern.

The second group of theories is referred to variously as the *organismic, gestalt, field,* or *cognitive* theories. The common feature of these theories is that they assume that cognitive processes—meaningfulness, understanding, and organizational abilities—are the fundamental characteristics of human behavior. Human learning is seen as marked

13

by a quality of intelligence and the ability to create relationships.

A third category of psychological behavior, social psychology, is receiving increased attention. It is often called *social learning theory*. In it, attention is given to personality factors and the interactions among people. Whether we learn from direct experience or through vicarious experiences with instructional media, much learning involves other people in a social setting.

There are areas of agreement and similar emphasis among the learning theories from which generalizations can be made. The following psychological conditions and principles are important factors to consider in the design and use of instructional media.

1. **Motivation.** There must be a need, an interest, or a desire to learn on the part of the student before attention can be given to the task to be accomplished. Moreover, the experiences in which the learner will engage must be relevant and meaningful to him or her. Therefore, it may be necessary to create the interest by means of motivational treatment of the information presented in instructional media.

2. **Individual differences.** Students learn at various rates and in different ways. Such factors as intellectual ability, educational level, personality, and cognitive learning styles affect an individual's readiness and ability to engage in learning. The rate at which information is presented in instructional media should be considered in terms of the anticipated comprehension rates of students.

3. **Learning objectives.** When students are informed of what they can expect to learn through the use of instructional media, their chances for success are greater than when not so informed. Also, as we will see in Chapter 4, a statement of objectives to be accomplished with the materials is helpful to those who will plan the materials. The objectives indicate what content will receive attention in the media.

4. **Organization of content.** Learning is easier when content and procedures or physical skills to be learned are organized into meaningful sequences. Students will understand and remember material longer when it is logically structured and carefully sequenced. Also, the rate of information to be presented is established in terms of the complexity and difficulty of content. By employing these suggestions in the design of media, the student can be helped to better synthesize and integrate the knowledge to be learned.

5. **Prelearning preparation.** Students should have satisfactorily achieved the preparatory learning or have had the necessary experiences that may be prerequisite to their successful use of the instructional media to be studied. This means that, when planning materials, careful attention should be given to the nature and probable level of preparation of the group for which the materials are to be designed.

6. **Emotions.** Learning which involves the emotions and personal feelings, as well as the intellect, is influential and lasting. Instructional media are powerful means of generating emotional responses such as fear, anxiety, empathy, love, and excitement. Therefore, careful attention should be given to media design elements if emotional results are desired for learning or motivational purposes.

7. **Participation.** In order for learning to take place, a person must *internalize* the information, not simply be told it. Therefore, learning requires activity. Active participation by the student is preferable to lengthy periods of passive listening and viewing. Participation means engaging in mental or physical activity interspersed during an instructional presentation. Through participation, there will be a greater probability that students will understand and retain the information presented.

8. **Feedback.** Learning is increased when students are periodically informed of progress in their learning. Knowledge of successful results, a good performance, or the need for certain improvement will contribute to continued motivation for learning.

9. **Reinforcement.** When the student is successful in learning, he or she is encouraged to continue learning. Learning motivated by success is rewarding; it builds confidence, and it will affect subsequent behavior in positive ways.

10. **Practice and repetition.** Rarely is anything new learned effectively with only one exposure. For knowledge, or a skill, to become a confirmed part of an individual's intellectual repertoire or competencies, provision should be made for frequent practice and repetition, often in different contexts. This can lead to long-term retention.

11. **Application.** A desired outcome of learning is to increase the individual's ability to apply or transfer the learning to new problems or situations. Unless a student can do this, complete understanding has not taken place. First, the learner must have been helped to recognize or discover generalizations (concepts, principles, rules) relating to the topic or task. Then opportunities must be provided for the learner to reason and make decisions by applying the generalizations or procedures to a variety of new, realistic problems or tasks.

Each one of these conditions or principles of learning can be applied directly or indirectly in the design of the various instructional media. Many of these principles also relate to the manner in which the media are subsequently used in correlation with accompanying printed materials and activities. Therefore, the planning and production phases cannot be entirely separated from plans for utilization. When all factors are considered, the quality of the media and their resulting effectiveness for learning can be greatly enhanced.

Domains of Learning

Starting in 1948 an attempt was made to develop a system for classifying the goals of the educational process. Its purpose was to standardize the terminology used to appraise learning. Such goals, stated in behavioral terms, could represent most kinds of human behavior. The result has been the development of *taxonomies,* or classification systems, in three areas:

- **Cognitive domain**—knowledge, information, other intellectual skills
- **Affective domain**—attitudes, values, appreciations
- **Psychomotor domain**—skeletal-muscle use and coordination

For each of these domains, progressive levels of higher-order behavior have been identified. The cognitive domain includes six levels of intellectual activity (Bloom and others):

1. **Knowledge**—recalling information
2. **Comprehension**—interpreting information
3. **Application**—applying information
4. **Analysis**—breaking information into parts
5. **Synthesis**—bringing together elements of information to form a new whole
6. **Evaluation**—making judgments against agreed criteria

The affective domain consists of five levels of attitudes, interests, and/or personal involvement (Krathwohl and others):

1. **Receiving**—attracting the learner's attention
2. **Responding**—learner willing to reply or take action
3. **Valuing**—commiting oneself to take an attitudinal position
4. **Organization**—making adjustments or decisions from among several alternatives
5. **Characterization of a value complex**—integrating one's beliefs, ideas, and attitudes into a total philosophy

Although a taxonomy of the psychomotor domain has been developed, which includes six major classes of behavior from reflex motions to skilled movements and nondiscursive communication, its usefulness is difficult to interpret (Harrow). Another scale of physical activities may be of more value (Kibler).

1. **Gross body movements**—arms, shoulders, feet, and legs
2. **Finely coordinated movements**—hand and fingers; hand and eye; hand and ear; hand, eye, and foot
3. **Nonverbal communication**—facial expression, gestures, bodily movements
4. **Speech behaviors**—sound production and projection, sound–gesture coordination

These three taxonomies can be appropriate references as instructional media are planned and developed. The objectives to be served by any media form represent the organizing point for your planning. Since a majority of instructional media are designed to provide information, they serve objectives in the cognitive domain. The lowest level, knowledge, represents rote learning through memorization and recall of facts. The five other levels require higher intellec-

tual learning. These latter levels should receive greater attention when designing instructional media.

Kinds of Learning

Robert Gagné, another psychologist, classified observations about learning and decided that various educational objectives require different *conditions of learning.* He developed a hierarchy (a classification sequence similar to a taxonomy) that includes eight kinds of learning, cutting across all learning theories, ranging from simple involuntary learning to more complex and abstract levels. Of the eight categories of behavior, the four highest levels provide for the sequential organization of subject content and related learning activities:

1. **Factual foundations**—elements of subject matter which provide basic terminology and facts relating to a topic
2. **Conceptual understandings**—grouping facts with common features under a generalized name
3. **Principles and rules**—high-level generalizations involving statements that show relationships among two or more concepts
4. **Problem solving and content applications**—using the facts, concepts, and principles in various situations

When instructional media of any type are developed, attention within the topic should be given to the relationships of these four levels. In this way you can be more certain of establishing conditions for successful learning.

SUMMARY

It has been shown that there are principles and practices from the fields of perception, communication, and learning theory which can contribute to the design and development of all instructional media forms. In the words of Fleming and Levie, cautions are expressed for using this information:

Adherence to the procedures and principles offered will not automatically result in better learning and these ideas are not offered as substitutes for experience and creativity. It is hoped, however, that this information may guide the insightful designer to analyze problems from more than one point of view, and may suggest effective solutions which might otherwise have been overlooked.

Finally, a media psychologist (Witt) suggests some practical guidelines for designing media to present factual information:

1. Design the production for your specific audience.
2. Tell your viewers what is coming, and what they should learn from the media presentation.
3. Associate new facts and ideas with ones the viewer already knows.
4. Rely on visuals and mental imagery (associating words with pictures) to help viewers remember.

15

5. Don't overload your production with information.

6. Give the viewer time to "let the information sink in."

7. Use repetition to hammer in critical facts.

8. Present a closing review of the major points in an organized pyramidal structure.

Review What You Have Learned About Perception, Communication, and Learning Theory:

1. In your own words, what is "human perception"? How does perception relate to the design of instructional media?

2. If you wished to make an in-depth study of perception as background for designing instructional media, to what reference might you refer?

3. Starting on page 12, a number of major conclusions regarding perception are presented. To which heading does each of the following relate?

  a. When preparing to take pictures, place the tools on a background of contrasting color so they are easy to see.

  b. Include a person in the scene as size comparison for the equipment.

  c. Use a number of the recognized principles of perception when designing a slide/tape program.

  d. Ask this question: "In what way does this object differ from the one shown on the previous page?"

  e. An audiotape recording might be the best media form to use when time relationships are to be presented.

  f. As a procedure is shown, explain each action with a few words.

  g. For narration—"In frame 8 examine the two related diagrams."

  h. Introduce a new section of the program with a title and brief introductory statement.

4. Recall the seven elements of the communication process in the communications model presented in this chapter. Where do instructional media fit into the model?

5. Which learning theories support the shift in instruction from teacher-centered to self-paced learning?

6. To which psychological condition or principle of learning does each statement refer?

  a. Your decision to include a slide/tape presentation as part of a study unit.

  b. Make a list of what students should be able to do after viewing the videotape recording and your informing them of these anticipated outcomes in the introductory materials.

  c. Show the trainee the accepted answer to a problem after his or her work is completed.

  d. Interpose questions for students to answer as the transparencies are presented.

  e. Design materials for each person to use individually, at his or her own pace of study.

  f. Have each person demonstrate present skill with equipment before starting advanced training.

  g. Provide a kit of materials so each person can carry out the procedures illustrated, after studying a media presentation.

  h. Present new situations in which the student must use the information learned from the media.

  i. Answers to review questions, associated with the media presentation, reveal that students are consistently correct.

  j. Divide the topic into small sections and relate them sequentially.

  k. Show a brief media presentation, as an introduction to build interest in the subject to be studied.

  l. Divide content for a topic into small sections with participation activity for each section before student starts next section.

  m. Show a film to a class as introduction for a new topic.

  n. Follow Gagné's structure of content for designing activities.

7. What are the *three* domains of learning? Which domain is the most difficult for instructional media to serve?

8. To which domain of learning does each activity relate?

  a. Throw a football.

  b. List six important practices when applying for a job interview.

  c. Take blood pressure with a sphygmomanometer.

  d. Agree to contribute time to a community agency.

  e. Drill holes accurately in a sheet of metal.

  f. Judge the quality level of units as they are assembled.

  g. Be on time for all assignments.

9. The cognitive domain includes a low recall level (knowledge) and higher intellectual levels. Which of the following learning activities would be on a *low* level and which ones on a *higher* level?

  a. Decide which tool to use for correcting a malfunction.

  b. List the steps to follow in a procedure.

  c. Repeat a name after hearing it.

  d. Formulate a plan to solve a problem.

  e. Describe an event in your own words.

10. To which of the four levels for sequencing content, as classified by Gagné, does each item apply on the topic of solar energy?

  a. Insulation material

  b. Heated air rising

  c. 47 percent of energy from the sun reaching the earth's surface

  d. Building a solar collector

  e. A solar collector

REFERENCES

Perception

Fleming, Malcolm, and Levie, W. Howard. *Instructional Message Design: Principles from the Behavioral Sciences.* Englewood Cliffs, NJ: Educational Technology Publications, 1978.

Communication

Hill, Harold. "Communication and Educational Technology." In *Educational Media Yearbook* 1981, pp. 40–49. Eds. James W. Brown and Shirley N. Brown. Littleton, Co: Libraries Unlimited, 1981.

Randhawn, Bikkar, and Coffman, William E. *Visual Learning, Thinking, and Communication.* New York: Academic Press, 1978.

Shannon, Claude, and Weaver, W. *The Mathematical Theory of Communication.* Urbana, IL: University of Illinois Press, 1949.

Sless, David. *Learning and Visual Communication.* New York: Wiley, 1981.

Learning Theory

Bloom, Benjamin S., et al. *A Taxonomy of Educational Objectives. Handbook I: The Cognitive Domain.* New York: Longman, 1956.

Gagné, Robert. *The Conditions of Learning.* 3rd ed. New York: Holt, Rinehart and Winston, 1977.

Harrow, Anita J. *A Taxonomy of the Psychomotor Domain.* New York: Longman, 1972.

Hergenhahn, B. R. *An Introduction to Theories of Learning.* 2nd ed. Englewood Cliffs, NJ: Prentice-Hall, 1982.

Kibler, Robert J. *Objectives for Instruction and Evaluation.* Boston: Allyn and Bacon, 1981.

Krathwohl, David R., et al. *A Taxonomy of Educational Objectives. Handbook II: The Affective Domain.* New York: David McKay, 1964.

Skinner, B.F. "The Science of Learning and the Art of Teaching." *Harvard Educational Review* 24 (Spring 1954):86–97.

Witt, Gary A. *Media Psychology for Trainers.* Dr. Gary Witt, 9th and Brazos, Suite 800, Austin, TX.

Chapter 3

RESEARCH IN THE DESIGN AND PRODUCTION OF INSTRUCTIONAL MEDIA

- Treatment of Subject
- Learner Participation and Knowledge of Results
- Presentation Elements
- Camera Angles
- Color
- Special Effects
- Directing Attention
- Picture–Narration Relationship
- Narration
- Music
- Printed Media
- Intellectual Abilities of Learners

In Chapter 2 the available knowledge derived from perception, communication theory, and learning theory, which should be of value in the design of instructional media, was summarized. Now, we turn to the conclusions derived from research studies directly relating to the planning and production of instructional media. It is important to be informed of this evidence so that materials to be prepared will have the maximum effectiveness in accomplishing the stated objectives.

A large portion of the research on instructional media has treated comparison studies relating to the use of specific media as compared with conventional teaching methods. In a small number of studies a particular aspect of media presentation has been controlled or varied in order to determine the effect on learning of that particular factor. Results of the latter group have relevance for the planning and subsequent production of instructional media.

Many studies were carried out some years ago. The majority were concerned with variables relating to motion picture design and production factors. These included such topics as format, camera angles, use of color and special effects, narration, and music. Many of the results apply as well to video recordings, sound-slide and filmstrip presentations, and to multi-image programs. In addition, there has been a sizeable research effort concerning preparation aspects of printed media. Only a few of the conclusions reached have been questioned, or retested in recent years. The reports provide evidence that can serve as guidelines for most media design situations.

Summaries of research findings on production elements have been prepared by a number of writers. From the earlier summaries and the more recent ones, the findings relating to planning and production aspects of instructional media have been abstracted and are presented in this chapter. Following each statement, the name in parenthesis relates to the literature source listed as a reference at the end of the chapter. Readers who have access to the original reports should refer to them for more detailed information.

These findings are fairly numerous and probably cannot be remembered or applied easily. At the end of the chapter, to assist in your understanding and recall of these findings, applications are offered for appraisal in a review exercise. The number of each finding is referred to in the review exercise.

In addition to the conclusions stated in this chapter, principles and generalizations presented in other chapters also relate to the planning and production processes. Of particular importance are the following:

- Principles of perception and learning theory in Chapter 2
- Picture composition factors on page 94
- Design principles for graphic materials on page 107
- Legibility standards for projected and nonprojected media on page 121

As you read through the following statements, in some instances you will find one generalization reemphasizing or extending another one. This indicates the value and strength of the conclusion. On the other hand, there are a few situations in which contradictory statements are included. Obviously, there will be variations in testing situations and learner responses may also differ over a period of time. Thus, those summarizing the research might then have reached different conclusions.

TREATMENT OF SUBJECT

1. Present the relevant information in an introduction and tell the viewer what is expected to be learned. (Hoban and van Ormer)
2. Summarize the important points in a clear, concise manner. Summaries probably do not improve learning unless they are complete enough to serve as repetition and review. (Hoban and van Ormer)
3. Ideas and concepts should be presented at a rate appropriate to the comprehension ability of the audience. (Hoban and van Ormer)
4. Instructional content may be more completely learned if it is presented to the learner two or more times, in identical or varied forms. (Allen 1973)
5. Organize the media so that important sequences or concepts are repeated. Repetition is one of the most effective means for increasing learning. (Hoban and van Ormer)
6. The learning of performance skills from media will be increased if you show common errors and how to avoid them. (Hoban and van Ormer)
7. Learning may be enhanced by organizing instruction sequentially to permit establishing subordinate skills before teaching those of higher order. (Allen 1973)
8. A demonstration should include only the basic elements of what is to be learned, but oversimplification can have a deleterious effect. (Travers)
9. When a presentation involving a media form can be reduced in complexity so that only the factors that directly contribute to accomplishing the task are included, learning will be more predictable and replicable. (Levie and Dickie)

LEARNER PARTICIPATION AND KNOWLEDGE OF RESULTS

10. Learning will increase if the viewer practices a skill while it is presented, provided the explanation is slow enough, or provided periods of time are allowed in which the learner is permitted to practice without missing new material. (Hoban and van Ormer)
11. Participation, relative to what is learned through media, does not have to be overt. Mental practice is also effective. (Travers)
12. When a student participates frequently by responding actively to some stimulus, learning of the material will be increased. (Allen 1973)
13. Furnishing knowledge of results, as part of the participation process, also has positive effects upon learning. (Travers)

PRESENTATION ELEMENTS

14. Stimuli that are pleasing, interesting, and satisfying are positive reinforcers and will increase the probability that the learner will remember and can reproduce what was presented in the media. (May)
15. Motivators, such as color, dramatic presentations, humor and comic effects, and inserted printed questions cause the learner to pay close attention, to look or listen for relevant and crucial clues, to have a ''set'' or put forth effort to learn, and to respond or practice. (May)
16. Cue identifiers, like color, arrows and pointers, animation, and ''implosion'' techniques (having assembled parts fall into place without being handled by the demonstrator) help the learner identify and recognize the relevant cues in a media presentation. (May)
17. Simplifiers, such as improving the readability of narration, eliminating irrelevant pictorial materials, repeating illustrations or adding additional illustrations are procedures for making presentations more effective. (May)
18. Incorporating dramatic sequences, such as comedy, singing commercials, or realistic settings to teach factual information, have not been shown to improve learning effectiveness. (Hoban and van Ormer)
19. Because color, optical effects, and dramatic effects have little to do with increasing learning, it is possible to eliminate them. (Hoban and van Ormer)
20. A crude presentation (pencil sketches of visuals) may be at least equal in effectiveness to a polished media presentation. (May and Lumsdaine)
21. The rate of development or pacing should be slow enough to permit the learners to grasp the material as it is shown. (Hoban and van Ormer)
22. If the audience is familiar with the setting being pictured, learning may be improved. (Hoban and van Ormer)

CAMERA ANGLES

23. Show a performance on the screen the way the learner would see it if actually doing the job (subjective camera position). (Hoban and van Ormer)
24. When taking a picture, avoid excessive detail by moving the camera closer or changing the viewpoint. (Saul)

COLOR

25. The fact that color adds to the attractiveness of a training device does not necessarily mean that it improves learning. Black-and-white is as effective as color for instructional purposes except when the learning involves an actual color discrimination. Learners prefer color versions despite the fact that the addition of color does not generally contribute to learning. (Travers)
26. There is an increasing amount of empirical evidence to support the use of color in visual illustrations as evi-

denced by improved achievement of specific educational objectives. (Dwyer)

SPECIAL EFFECTS

27. Special effects used as attention-getting devices have no positive influence on learning. (Hoban and van Ormer)
28. A film or video recording in which straight cuts have replaced optical effects (such as fades, wipes, and dissolves) teaches just as effectively as media that use these effects. (Hoban and van Ormer)
29. The special effects (fades, dissolves) that are used to represent lapses in time and other events are not effective in conveying the intended meanings. Printed titles seem to be more effective. Special sound effects appear to provide much more challenge to the producer than aid to the learner. The same can be said of humor and of other special means intended to retain the interest of the learner. (Travers)

DIRECTING ATTENTION

30. Media that treat discrete factual material appear to be improved by the use of an organizational outline in titles and commentary. (Hoban and van Ormer)
31. Liberal use of titles, questions, and other printed words can improve teaching effectiveness. (May and Lumsdaine)
32. It is useful to direct the learner's attention to particular elements of instructional messages through visual cueing and other attention-attracting devices. (Allen 1973)

PICTURE–NARRATION RELATIONSHIP

33. The audio channel is much more capable of obtaining attention if it is used as an interjection on the pictorial channel rather than being continuously parallel with the pictorial. (Hartman)
34. While concepts and principles can be acquired solely on the basis of visual presentations, to rely *only* on visual lessons is inefficient. Words serve an important cueing role and should be incorporated, for this secondary purpose, into a visual presentation. (Gropper)
35. In media designed to teach performance skills, the pictures should carry the main teaching burden. (Hoban and van Ormer)

NARRATION

36. The number of words in the commentary has a definite effect on learning. Care should be taken not to "pack" the sound track. (Hoban and van Ormer)
37. Use direct forms of address (imperative or second per-

son) in media narration. Avoid the passive voice. (Hoban and van Ormer)
38. Media that provide opportunity for the audience to identify with the narrator will be more successful than that not providing such an opportunity. (Hoban and van Ormer)
39. Except where the use of live dialog can have marked superiority for meeting particular objectives, narration has great advantages. (May and Lumsdaine)
40. Verbal simplification in media commentaries increases teaching effectiveness. (Travers)

MUSIC

41. There is little evidence to support the opinion that background or mood music facilitates learning from media productions. (Seidman)
42. Musical accompaniment enhances the emotional impact of a media production. (Seidman)
43. Music can provide continuity by tying together various scenes in a script. (Seidman)

PRINTED MEDIA

44. A wide variation in design seems permissible without greatly affecting efficiency in reading. (Wilson et al.)
45. Words are identified most rapidly when composed of lower case characters. (Hartley and Burnhill)
46. Spacing characters very close together does not aid legibility. (Hartley and Burnhill)
47. The brightness contrast between letter and background is one factor determining perceptibility of letters. (Wilson et al.)
48. Very short lines slow perception, while very long lines increase the number of regressions and cause inaccuracy in locating the beginning of each new line. (Wilson et al.)
49. There is little evidence to support the effectiveness of printing all lines in equal lengths (justified). (Wilson et al.)
50. The use of headings and underlining serves to accentuate selected elements in printed text with the expectation of improving learner acquisition and retention. (Wilson et al.)
51. Inclusion of pictures in printed material can substantially improve learning. (Levin)
52. There is reader preference for double-column format on a page. (Wilson et al.)
53. The generous use of open space in printed instructional materials is a necessity for aiding comprehension. (Wilson et al.)

INTELLECTUAL ABILITIES OF LEARNERS

Materials designed for learners of low mental ability should employ these design techniques: (Allen 1975)

54. Preparatory or motivational procedures that establish a readiness to learn the material, with attention to overviews, verbal directions, and questions to answer.

55. Organizational outlines or internal structuring of the content.

56. Attention-directing devices that point out, emphasize, or refer to relevant cues in the communication.

57. Procedures that elicit active participation and response from the learner to the content of the communication.

58. Provisions for correcting or confirming feedback to responses made by learners.

59. A slow rate of development, or a slow pace of presentation of the content to be learned.

Materials designed for learners of high mental ability should employ these design techniques: (Allen 1975)

60. High information density; high pictorial and conceptual complexity; and richness of images, ideas, and relationships.

61. Rapid rate of development of information and concepts being communicated.

62. A format that places requirements on the learner to organize, hypothesize, abstract, and manipulate the stimuli mentally in order to extract meaning from it.

A person interested in planning and producing instructional media should review and weigh all the evidence from the research findings and theory reported in this and the preceding chapters. These findings, rather than intuition, should be considered as you design your own materials for instruction. Start with these results and recommendations, realizing that some may have been derived from situations far afield of the applications you plan to make. (Yet they are starting points with positive evidence for improved learning at lower costs in terms of time, materials, and services.) Then adapt and change as you gain experience and test the results of your efforts.

Review What You Have Learned About Research Findings for the Design and Production of Instructional Media

Following are statements concerning instructional media that apply one or more of the findings described in this chapter. Some make recommended applications, while others apply elements in nonrecommended fashion. For each example indicate your *agreement* or *disagreement* with the proposed plan. Then check your answer, using the reference numbers at the right to locate the relevant numbered finding, or findings, used as a basis for each example.

AGREE OR DISAGREE
EVIDENCE

1. When demonstrating the proper method to use in casting with a fishing rod, show errors commonly made and ways to avoid them. _____ 6

2. In demonstrating a skill, like fingering a musical instrument, color will add to the instructional value of the medium used. _____ 19, 25, 26

3. In explaining the operation of a machine, use arrows to indicate each part as it is referred to. _____ 16, 32

4. Dramatic-type music can create a mood for an audience at the beginning of a slide/tape program. _____ 42

5. To demonstrate how a woman sews an intricate stitch by hand, film the action from over her shoulder. _____ 23

6. In teaching a skill like welding, limit the amount of narration and depend on the visuals for the major instructional effect. In narration, use words in the present tense to direct attention ("hold the tool . . ."; "notice the color . . ."). _____ 33–37

7. A video recording that shows action in many locations is more effective if an optical effect like a dissolve (page 232) is used between scenes to bridge distance rather than abrupt cuts from one scene to the next. Also, background music will enhance the presentation. _____ 28, 29, 41

8. To explain a sophisticated security system to new employees, present only the essential facts without repetition of any of the concepts. A brief, general summary should be included. _____ 2–5

9. In describing an industrial process, quickly present only the essential information. The commentary should describe, at length, what cannot easily be visualized. If it is a lengthy subject, plan for the commentary to move along rapidly. _____ 3, 21, 36, 40

10. Picture sketches from storyboard cards (page 49), converted to video tape, may be as effective as a high-quality polished treatment for illustrating a farming procedure. _____ 17, 20

11. In treating the subject of animal life in a mountain area, color should be used and important concepts presented through multiple examples. _____ 5, 25, 26

12. Introduce the demonstration of a safety procedure with an explanation of the purpose of the demonstration and what the student is expected to learn from it. Include titles that indicate the sequence of steps in the procedure. Describe carefully and visualize new technical terms. _____ 1, 31, 32

13. For learning to operate a piece of equipment, like a table saw, a short video recording that can be viewed any number of times may have advantages. Directions for the learner to stop the video, answer questions, and practice steps in the skill should be included. Correct answers to questions should be provided immediately after the questions are answered. _____ 10–13, 31

14. A self-learning package treating a topic for an electronic lab course includes a cassette recording on which the instructor outlines the facts basic to the topic and leads the student to laboratory applications. _____ 1, 7

15. When it is important that information be remembered for a media presentation, make the viewers feel uncomfortable and dissatisfied while watching the program. _____ 14

16. The media designed for new trainees shown to have low intelligence should present ideas at the same pace as that designed to be used with other trainees. _____ 54–59

17. On a transparency, use capital letters for all words. _____ 45

18. It may be preferable to have the instructor narrate the script rather than to use a professional narrator. _____ 38

19. A media program is designed to start by having professional actors make fun of an important situation so the technicians attending would give their full attention to the message that followed. _____ 15

20. Even though it is more costly to have a study guide printed with all text lines of the same length (justified), the use of this technique is preferred. _____ 49

21. When setting up for an important step in a how-to-do-it picture, place the camera so that only the part being handled will be seen in the camera. _____ 24

22. Underline key words for attention in an instructional manual. _____ 50

REFERENCES

Allen, William H. "Intellectual Abilities and Instructional Media." *AV Communication Review* (Summer 1975):139–170.

———. "Research in Educational Media." In *Educational Media Yearbook* 1973. James W. Brown, ed. New York: R. R. Bowker, 1973.

Dwyer, Frances M. *Strategies for Improved Visual Learning.* State College, PA: Learning Services, Box 784, 1978.

Gropper, George L. "Learning from Visuals: Some Behavioral Considerations." *AV Communication Review* (Spring 1966):37–70.

Hartley, James, and Burnhill, Peter. "Fifty Guidelines for Improving Instructional Text." *Programmed Learning and Instructional Technology,* 4 (February 1977):65–73.

Hartman, Frank R. "Single and Multiple Channel Communication: A Review of Research and a Proposed Model." *AV Communication Review* (November–December 1961):235–262.

Hoban, Charles F., Jr., and van Ormer, Edward B. *Instructional Film Research 1918–1950.* Technical Report No. SDC 269-7–19.

Port Washington, NY: U.S. Naval Special Devices Center, 1950.

Levie, W. Howard, and Dickie, Kenneth E. "The Analysis and Application of Media." In *Second Handbook of Research on Teaching.* Robert M. W. Travers, ed. Chicago: Rand McNally, 1973.

Levin, Joel R. "Pictorial Strategies for School Learning." In *Cognitive Strategy Research—Educational Applications.* M. Pressley and J.R. Levin, eds. New York: Springer-Verlag, 1983, pp. 203–237.

Lumsdaine, A. A. "Controlled Variations of Specific Factors in Design and Use of Instructional Media." In *Handbook of Research on Teaching.* N. L. Gage, ed. Chicago: Rand McNally, 1963.

May, Mark A. *Enhancements and Simplifications of Motivational and Stimulus Variables in Audiovisual Instructional Materials.* Washington: U.S. Office of Education, contract No. OE-5-16-006, 1965.

May, Mark A., and Lumsdaine, A. A. *Learning from Films.* New Haven, CT: Yale University Press, 1958.

Saul, Ezra V., et al. *A Review of the Literature Pertinent to the Design and Use of Effective Graphic Training Aids.* Technical Report SPECDEVCEN 494-08-1. Port Washington, NY: U.S. Naval Special Devices Center, 1951.

Seidman, Steven A. "On the Contributions of Music to Media Productions." *Educational Communications and Technology Journal,* 29 (Spring 1981):49–61.

Travers, Robert M. W. *Research and Theory Related to Audio-Visual Information Transmission.* Kalamazoo: Western Michigan University, 1967.

Wilson, Thomas C., et al. *The Design of Printed Instructional Materials: Research on Illustrations and Typography.* Syracuse, NY: ERIC Clearinghouse on Information Resources, Syracuse University, 1981.

Part Two

PLANNING INSTRUCTIONAL MEDIA

4. Preliminary Planning

5. The Kinds of Media

6. Designing the Media

7. Producing the Media

8. Using and Evaluating Media

9. Planning and Production Summary

Chapter 4

PRELIMINARY PLANNING

- Planning and Creativity
- Start with a Purpose or an Idea
- Develop the Objectives
- Consider the Audience (the Learner)
- Use a Team Approach
- Find Related Materials
- Review What You Have Done
- Prepare the Content Outline

All too often someone says, "Let's make a videotape to train our salespeople." Or, "How about shooting some slides to impress new staff members with their value to the organization." Or, "We should have some transparencies to use with this report."

Unfortunately such statements are frequently the signal to start taking pictures prematurely and to produce instructional media that often are unorganized and ineffective. These proposals are no more than bare ideas. To assure that the product will fulfill the need, they require further consideration, such as decisions about the objectives to be served, the specific content to be treated and its organization into planned sequences of pictures.

Occasionally, it is necessary to make pictures without any prior planning. These situations arise either when an event happens unexpectedly or when activities take place over which you may have little, if any, control. Examples of this would be the need to document the visit of an official to an industrial plant or a ceremony at an institution. But most often, effective instructional media are carefully planned *before* production takes place.

PLANNING AND CREATIVITY

There are two related, although seemingly opposed processes that should receive attention when developing instructional media. One is the structured planning procedure which requires organization, attention to a logical sequence of components, and their integration into a unified message. The other is the unstructured free flow of ideas and expressions typified by creative thinking, leading to the solution of problems encountered during planning. The intermingling of both systematic planning and creative thinking is important if effective and interesting media productions are to result.

During the planning process, many problems will be faced and numerous decisions must be made. Some are procedural—"Should we start on the storyboard now, or work on the script?" Or, "Can we consider using captions or depend entirely on narration?" Others are analytical—"Are the steps for the equipment operating procedure in the correct sequence? Has anything been left out?" As you proceed, ideas may come to mind which can influence the planning—"Let's include a dramatic sequence showing treatment with the drug before going to the research lab." Or, "You might show a diagram of the whole process followed by a detailed examination of each component."

Some of these decisions should be based on the principles relating to perception and learning theory presented in Chapter 2 and the research evidence summarized in Chapter 3. But many times ideas come to mind unexpectedly, or through brainstorming and other creative thinking efforts. These original thoughts can contribute significantly to the structure, appeal, and effectiveness of any instructional media.

An analysis of how the creative thinking process is successfully used reveals that most often six steps are followed:

1. **Desire**—Have an initial motivation to want to solve a problem with which you are concerned.
2. **Preparation**—Gather information relative to the problem as revealed by the planning stages of audience identification, objective specification, and content listing.
3. **Manipulation**—Play with a number of ideas to devise one or more possible solutions, or to find a new pattern of treating the content which differs from that which has been familiar.
4. **Incubation**—Takeover of the thought process by the subconscious mind, especially if a solution has not been attained.

5. **Illumination**—Sudden revelation of a solution or a new pattern as the result of subconscious thought.
6. **Verification**—Examine the solution to evaluate its feasibility, and then to accept it as being appropriate to the problem.

Each of us has the capacity of applying this creative thinking process, some persons to a greater degree than others. Both logical, sequential processing skills and random or creative thinking abilities are essential mental activities when developing instructional media. Prepare yourself to handle both responsibilities as you move through the planning stages that follow.

In this chapter, and in the following ones in Part II, we will examine each of the important elements that comprise the planning process. For some media forms—slide series, filmstrips, video recordings, and multi-image presentations—attention should be given to all steps. For other media types—printed and display media, overhead transparencies, audiotape recordings, and computer-based instruction—consideration need be given only to certain steps. The necessary emphasis will be indicated for each medium at the beginning of its chapter in Part IV.

START WITH A PURPOSE OR AN IDEA

An idea, a problem situation, or a learning need identified within an instructional-design plan for a unit or a course should be the starting point for the development of your instructional media. An idea may indicate an area of interest you have, but the more useful ideas are those conceived in terms of a need relating to a specific group—an audience's need for certain information or for a skill, or the need to establish a desired attitude.

So here is the first step: Express your idea or purpose concisely. For example:

1. Security has been an important issue within our company. We need to alert all employees to the need for maintaining a high level of security for company equipment, supplies, and new products under development.
2. As a financial advisor, you frequently are asked to meet with groups who are interested in making investments for future security. You would like to acquaint middle-aged people with advantages and limitations in alternative ways of investing money.
3. I conduct an adult education class on ecology and the environment. Each year I suggest topics of potential interest to the group for study. One that is frequently selected is the use of solar energy. Therefore, I want to develop a better understanding of how solar energy can be used in heating homes and for making hot water. (Note: The example of using solar energy will serve as an illustration of the planning steps that follow.)

In the instructional-design procedure described in Chapter 1, one element was to identify general purposes to be served by the topic. The purposes, as underlined in the examples above, are the beginning **ideas** referred to here.

Take the time to state them clearly so that you, and other persons involved in planning, will understand the aim of the materials to be produced.

The expression "instructional media" is used in this book to represent audio, visual, and certain print materials appropriate for communication in instruction and training. The procedures described here also are useful for satisfying such needs as intraorganizational communications and communications for external public relations.

To Motivate, to Inform, or to Instruct

Any one of three intentions may be served by instructional media when used with individuals, groups, or large audiences: (1) to **motivate** an interest or a degree of action; (2) to present **information;** or (3) to provide **instruction.** The difference among these three intentions should be recognized by the reader because the treatment of content will vary for each one.

For **motivational** purposes, dramatic or entertainment techniques may be employed. The desired result is to generate interest or stimulate members of an audience to take action (assume responsibility, volunteer service, or contribute money). This involves accomplishing objectives that affect personal attitudes, values, and emotions.

For **informational** purposes, the instructional media would most likely be used in a presentation made before an audience or class group. The content and form of the presentation would be general in nature, serving as an introduction, an overview, a report, or background knowledge. It also might employ entertainment, dramatic, or motivational techniques in order to attract and hold attention. When viewing and listening to informational-type materials, individuals are passive viewers and listeners. The anticipated response by people most likely would be limited to degrees of mental agreement or disagreement and to emotionally pleasant, neutral, or unpleasant feelings. Materials that are designed for informational purposes can in turn lead a person to involvement with the idea or topic on an instructional level.

For **instructional** purposes, while the presentation of information is important, attention must also be given to involving the participants in mental or overt activities relating to the instructional media being used so that learning can take place. The materials themselves should be designed more systematically and psychologically sound in terms of learning principles in order to provide effective instruction. At the same time they should be enjoyable and provide pleasant experiences. Making provisions for individuals to use the instructional media on their own can be desirable. In this way, each person will interact with the materials by answering questions, by engaging in performance as directed by the materials, by checking understanding, and by making use of information presented.

In the examples cited above, the company's security awareness would be an illustration of a motivational purpose; the solar energy topic has an instructional purpose; and the financial advising situation represents an informational purpose.

Keep in mind the differences among instructional media to be used for motivation, for information, and those to be used for instruction. The planning and treatment for each type differ in certain aspects. Although major emphasis is given to planning media for instructional purposes in the following chapters, reference will be made, when necessary, to items that affect motivational and informational materials.

DEVELOP THE OBJECTIVES

Build upon the idea or generalized statement of purposes. Doing this means translating the general idea into a clearcut and specific statement of one or more objectives for the planned learning within an instructional context. For a motivational or informational presentation, the objectives often are more broadly stated than for instructional purposes.

Much attention is given in the literature to the topic of **learning objectives;** these have a key role in instructional design as well as in choosing learning activities.

To plan successful instructional media and other learning experiences, it is necessary to know specifically what must be learned. The purpose of formulating objectives is to provide clear guidance that permits an orderly presentation of content leading to learning.

When learning takes place, a person changes in some way. This may be a mental growth as the individual acquires new knowledge (definitions of terms, steps in a procedure, criteria for making a decision). Or, the learning can become evident by the way the individual now performs a skill, spells a word, argues a position, or treats other persons. In each situation, as the result of the learning experience, the individual behaves differently.

In order to provide for the desired learning, objectives are written. They indicate what should be the outcome of the learning. Objectives are grouped into three major categories as described in Chapter 2—the **psychomotor** area, represented by performance skills involving the use of muscles as a task or job is being carried out; the **cognitive** area, which includes knowledge and information, represented by thinking and other intellectual skills; and the **affective** area of attitudes, appreciations, and values. This last category requires the most care when objectives are being specified.

The difficult problem is to spell out the objectives so that (1) learning experiences can be developed to satisfy each objective, and (2) tests or performance measurements can be designed to find out whether the learning has taken place.

The general nonspecific words that are often used to describe instructional purposes—to *know*, to *understand*, to *become familiar with*, to *appreciate*, to *believe*, to *gain insight into*, to *accept*, to *enjoy*, and so forth—are unsatisfactory guide words for objectives. They do not permit verification through specific observable behavior and they are open to many interpretations of how their accomplishment may be measured. Such expressions as the foregoing may be acceptable as indications of generalized objectives for motivational- and informational-type presentations.

Useful statements of objectives for instructional purposes are made up of two grammatical parts. First: a specific ACTION VERB like one of these—to *identify*, to *name*, to

demonstrate, to *show*, to *make* or *build*, to *order* or *arrange*, to *distinguish between*, to *compare*, or to *apply*. Second: CONTENT REFERENCE that follows the verb, like—to name the *five steps in a process*, to assemble *all parts of a machine* properly, to write a 500-word *theme*, to apply a *rule*, to solve four of five *problems*.

Notice that in addition to the action verb and the content reference, we may add a STANDARD OF COMPETENCY (*5 steps, all parts, 500 words, four of five problems*). The standard further provides for setting an attainment level that can be measured.

With this awareness of measurable learning objectives, how can we specifically indicate the objectives for the general idea example on solar energy indicated previously?

It is *not* sufficiently specific to write:

● To understand how solar energy can be used
  It is specific to write:

1. To describe the planning necessary for using solar energy
2. To list at least two features of each of the three processes required in utilizing solar energy for heating
3. To assemble a solar collector with 100 percent accuracy when provided with four essential parts
4. To assume responsibility for engaging in a solar energy project

Statements 1 and 2 above specify cognitive objectives (1 on a low level—recall of information; 2 higher intellectually—comprehension level). Statement 3 is a psychomotor objective, requiring physical performance by the student. Statement 4 specifies an attitudinal behavior. It is much more difficult to indicate behaviors and measurements for attitudes and appreciations than to do so for either knowledge/intellectual skills or performance skills. To repeat, only when the objectives are stated in terms of an individual's mental learning or physical performance is there much guidance for the design of instruction. (For further explanation of and suggestions for developing objectives, see Chapter 7 in Kemp, *The Instructional Design Process.*)

Remember that each topic for instruction (like the solar energy topic) requires a number of objectives, each to be considered individually or as a related set, in designing learning experiences. Therefore, it takes time and careful thought to develop and state objectives. Also, as was indicated during the discussion of instructional design in Chapter 1, it is natural to move gradually from general to specific objectives. Finally, realize that some objectives may become clearly evident only when content is being selected or even when specific instructional media are in the planning stage. In such a case, return to this beginning point, the statement of objectives, and check how well the content and learning experiences fit the stated objective; you may want to revise the statement.

You might prepare media for many instructional purposes. Here are some major general purposes with specific examples. They are stated as objectives in terms of student or audience's behavioral changes, from the learner or audience viewpoint.

- **To learn about a subject;** for example, "to determine how a worker should be selected for a job when there are several qualified applicants"
- **To apply the steps in a process;** for example, "to measure a patient's blood pressure within $\pm$ 5mm Hg"
- **To practice a certain attitude;** for example, "to form the habit of using safe procedures in operating shop equipment"
- **To respond to a social need;** for example, "to offer your services in a youth recreation program"

In planning materials, limit yourself to no more than a few concisely stated achievable objectives. However much you feel it necessary *to cover the whole topic,* you will realize eventually that limitations should be set. If you do not set limits, your materials may become too complex and unmanageable. You can maintain limits by aiming at a series of related instructional media, each of which includes a single phase of a large topic.

For the solar energy topic a number of matters require attention—forms in which energy is used in the home, scientific principles applied in collecting solar energy, process of energy collection, storing the energy, and so forth. For the purpose of the planning stages that follow, *collecting solar energy* is selected as the topic. The objectives will be:

1. *To explain how four scientific principles are applied for creating heat in a solar collector*
2. *To identify four parts of a solar collector and the functions of each*
3. *To assemble a solar collector when provided with the four essential parts*

Finally, objectives do not stand alone. It is obvious that they are dependent on the subject content that will be treated and are influenced by the needs and dispositions of the learner or intended audience.

CONSIDER THE AUDIENCE (THE LEARNER)

The characteristics of the learner or audience—those who will be seeing, using, and learning from your materials—cannot be separated from your statement of objectives. One influences the other. Such audience characteristics as age and educational level; knowledge of the subject, skills relating to it, attitude toward it, cultural context; and individual differences within the group, all have bearing on your objectives and treatment of the topic. The audience is the determinant when you consider the complexity of ideas to be presented, the rate at which the topic is developed, the vocabulary level for captions and narration, the number of examples to use, the kinds of involvement and degree of participation of the learner. These and similar matters will influence the complexity of the objectives and the way you should handle the topic.

At times more than one audience may fit your plans, but generally it is advisable to plan for **one major audience**

group. Then consider other **secondary** ones which also might use your materials. Describe the major audience, explicitly.

For the example of solar energy:

The audience will be middle-class adults who have at least a high school education and likely some college experience. They are very inquisitive and particularly interested in environmental matters that affect their lives.

If you plan instructional media for use with a group of young children or with individuals from an ethnic minority culture, anticipate difficulty with understanding language and accepting situations based on experience or cultural factors that differ from their own.

Make sure that the subject and the activities selected are appropriate to their interests and their abilities. Your own enthusiasm for a topic may take you far beyond the limited amount of interest in it that others have. Also, give careful consideration to the complexity of the subject so that your group does not become burdened with too many details and lose interest as a result.

USE A TEAM APPROACH

You may be capable of planning and preparing your instructional media without the assistance of others. If you are, you have skills in three areas. First, you have a good knowledge of the subject. Second, you know how to plan instructional media and how to interpret the subject visually. Third, you have the necessary technical skills in photography, graphic arts, and sound recording.

But if you feel inadequate in any of these areas, you should obtain assistance or carefully use this book (in the second and third areas). Even so, there is value in getting reactions and suggestions from other people, so plan to involve others during some phases of the planning and preparation processes.

Three individuals or three groups might make up the **production team.** The **subject specialist** is the person or persons having broad knowledge of the content to be treated and most often is familiar with the potential audience. The **communications specialist** is the individual who knows how to handle the content (treatment, scriptwriting, camera angles, and related essential knowledge and skills) and knows the advantages, limitations, and uses of the various instructional media so that the resulting materials will achieve the anticipated purposes. This person may also serve as the **instructional designer** during the planning described in Chapter 1. Finally, the **technical staff** comprises those responsible for the photography, the videotaping, the art work, the lighting, and the sound recording.

These separately described areas naturally overlap: The communications person may also take the pictures; or, as was mentioned, you may fill all three jobs. The important thing to recognize is that all three jobs exist. Keep them in mind as you consider the stages of planning and preparation that follow. For example:

While planning media on collecting solar energy, I will consult with a friend who is an engineer with a local heating and air conditioning company that sells and installs solar collectors. He and I in combination will fill the role of subject specialist. I will request assistance also from the adult education center's media coordinator, who will thus function as communications specialist. Since I have a good skill in photography, I will prepare the visual materials, but I will be assisted by a technical staff of three students, from the class, who have abilities in photography and art.

FIND RELATED MATERIALS

Before carrying the planning of your instructional media to an advanced stage, locate and examine any materials already prepared on your general topic or on topics closely related to it. They may offer you some useful ideas, or you may find that all or part of such materials may fit one or more of your objectives.

Communicate with media specialists in school systems, universities, colleges, or business and industry for suggestions of other possible materials and also for their reactions to your plans. Check your library for these references:

- A. S. Barnes and Co., 11175 Flintkote Ave., San Diego, CA 92121
  Radio's Golden Years
- R. R. Bowker Co., 1180 Avenue of the Americas, New York, NY 10036
  Educational Film Locator: Consortium of University Film Centers
- British University Film Council, Ltd., 81 Dean St., London W1V 6AA England
  Researcher's Guide to British Film and Television Collections
- Cassette Information Services, Box 9559, Glendale, CA 91206
  Audio Cassette Directory
- Chicorel Library Publications Corp., 275 Central Park West, New York, NY 10024
  Chicorel Index to Video Tapes & Cassettes
- Choice, 100 Riverview Center, Middletown, CT 06457
  Vocational–Technical Audiovisual Materials for Learning Resources Centers
- Educators Progress Service, Randolph, WI 53956
  Educators Guide to Free Audio and Video Materials
  Educators Guide to Free Films
  Educators Guide to Free Filmstrips
  Educators Guide to Free Health, Physical Education, and Recreation Materials
  Educators Guide to Free Science Materials
  Educators Guide to Free Social Studies Materials
- ERIC Documents Reproduction Service, Box 190, Arlington, VA 22210
  Catalogs of Audiovisual Materials: A Guide to Government Sources (ED 198 822)
  National Library of Medicine: Audiovisual Catalog
  Reference List of Audiovisual Materials Produced by the United States Government

- Hendershot Bibliography, 4114 Ridgewood, Bay City, MI 48706
  Programmed Learning and Individually Paced Instruction
- Marquis Academic Media, 200 East Ohio St., Chicago, IL 60611
  Selective Guide to Audiovisuals for Mental Health and Family Life Education
- National Aeronautics and Space Administration, Lyndon B. Johnson Space Center, Film Distribution Library, Houston, TX 77058
  NASA Johnson Space Center Film Catalog
- National Audiovisual Center, National Archives and Records Service, Washington, DC 20409
  Audiovisual Resource List
- National Information Center for Educational Media (NICEM), University of Southern California, Los Angeles, CA 90007
  Index to Educational Audio Tapes
  Index to Environmental Studies—Multimedia
  Index to Health and Safety—Multimedia
  Index to Nonprint Special Education Materials
  Index to Educational Overhead Transparencies
  Index to Psychology—Multimedia
  Index to Educational Records
  Index to 16mm Educational Films
  Index to Educational Slide Sets
  Index to 35mm Educational Filmstrips
  Index to Vocational & Technical Education—Multimedia
  Index to Educational Video Tapes
- National Public Radio, 2025 M Street, N.W., Washington, DC 20036
  National Public Radio Educational Cassette Programs
- National Video Clearinghouse, 100 Lafayette Dr., Syosset, NY 11791
  Video Sourcebook
- PBS Video, 475 L'Enfant Plaza, S.W., Washington, DC 20024
  PBS Video
- Pharmaceutical Communications, Inc., 42-15 Crescent St., Long Island City, NY 11101
  Educators International Guide to Free and Low Cost Health Audio-Visual Teaching Aids
  Guide to Health A/V Teaching Aids
- Televised Higher Education, 546 14th St., Boulder, CO 80302
  The Catalog: College and Adult Level Video Courses in Science, Business, and Engineering
  The "Other" Catalog: College and Adult Level Video Courses in Agriculture, Education, Health Sciences, Humanities, Law, Trades, and Crafts
- U.S. Department of Energy, Washington, DC 20545
  Energy Films
- Video-Forum, Division of Jeffrey Norton Publications, 145 East 49th St., New York, NY 10017
  Business and Technology Videolog
  General Interest and Education Videolog
  Health Science Videolog
- Young Filmmakers/Video Arts, 4 Rivington St., New York, NY 10002
  Young Filmmakers/Video Arts Catalog

Table 4-1 **EXAMPLES OF INITIAL PLANNING STEPS**

IDEA	OBJECTIVE	AUDIENCE
(1) Learn about plants in the community	To identify characteristics of 25 most common species of seed-bearing plants in our county	a. High-school biology classes b. Nature clubs
(2) Understand anatomy and physiology of fetal circulation	A. To name the two blood vessels found in the umbilical cord B. To locate the two shunts which are normal in fetal circulation C. To label a diagram of fetal circulation	Junior level nursing students
(3) Introduce new employees to the operation of our company	To understand their role in the successful operation of our company	Employees of the ABC Insurance Company
(4) Acquire the skills of soldering	To solder three types of conductor splices with 95% accuracy	Electronic technicians
(5) Persuade a potential customer to become interested in our XYZ equipment	To request a demonstration of the XYZ equipment	Managers, electronic assembly departments
(6) Take part in the youth program of our church	A. To take part in youth group activities B. To know how youth activities help to develop sound character and religious understanding among our young people	a. Children and teenagers of church members b. Church sponsors and adult members

REVIEW WHAT YOU HAVE DONE

The planning steps thus far examined are: (1) start with an idea or a specific purpose; (2) from this statement develop your objectives, with due regard for the intended audience and its characteristics; (3) obtain assistance as necessary, from persons who have special knowledge and skills relative to the topic being developed or from already prepared materials.

Examine the examples in Table 4-1. Are the steps clearly stated and easy to follow? Are the objectives stated in behavioral terms? Which of them may not be so stated? The review questions at the end of the chapter help you to correct any improperly stated objectives.

PREPARE THE CONTENT OUTLINE

Now consider the subject matter that relates to the objectives. Consult with your subject specialist, or, if you are handling the content yourself, do any necessary research work. Facts about a subject and details of a task are often found through interviews, during visits to suitable facilities, and in the library. After this background work you can feel confident that your basic facts are correct and that you will include all pertinent information on your topic.

From the data you have gathered prepare a content outline. This outline becomes the framework for your instructional media. It consists of (1) basic topics that support your objectives and (2) factual information that explains each topic. If you are treating a specific task like assembling, operating, or trouble-shooting equipment, then list the details or steps to be performed.

A word of caution: Keep in mind the people who will be your audience—their interests and their limitations. Decide what information must be included in detail and what can be treated lightly; what you can suggest for additional study and what should be left out or considered for other instructional media.

In the sequence of the instructional design, the examination of content follows the objectives. At this stage you are not as yet concerned about specific materials. There is no gain in asking at this point whether a video recording, a film, a set of transparencies, an audio recording, a printed program, or a combination of media will best serve the objectives. You must find out what content is required to support the objectives. Then you can make decisions about specific kinds of media.

Use Storyboard Cards

A good way of relating content to objectives is to connect the two visually. In a few pages we will examine the method of preparing a **storyboard** to visualize the treatment of a topic for instructional media. This same technique has value now.

Write each objective on a 3×5 inch card or slip of paper. Tack or tape the cards to a wall for display. Then, from your notes on paper, make a second set of cards listing the

An Example of
a Content Outline

Objective 1: To demonstrate how four scientific principles are applied
for creating heat in a solar collector
A. Radiation
1. Energy is generated within the sun
2. Travels through space to the earth
B. Conduction
1. Molecules become active at heating point
2. Heated molecules vibrate
3. Vibration causes increased motion of adjacent molecules
C. Greenhouse effect
1. Container with transparent cover
2. Heat is trapped inside
D. Convection
1. Heated material becomes less dense and rises
2. Replaced by cooler material which is heated in turn

Objective 2: To identify four parts of a solar collector and their functions
A. Clear glass or plastic cover
1. Allows light and heat to enter
2. Important in greenhouse effect to trap heat
B. Flat plate with tubes
1. Metal or plastic
2. Tubes filled with fluid
3. Coated dark to absorb heat
4. Circulation of fluid after heating
C. Insulation
1. To prevent loss of heat
2. Consists of plastic foam, fiberglass, or similar material
D. Wooden or metal weatherproof container

Objective 3: To assemble a solar collector when provided with the four
essential parts
A. Construct wooden container
1. Mark dimensions for frame and backing
2. Cut 2×4 frame and $\frac{3}{4}$ inch plywood backing
3. Nail container together
4. Notch upper edge for cover glass
B. Place insulation in container
1. Cut 2″ styrofoam to size
2. Set in container
C. Assemble flat plate
1. Place flat plate on insulation
2. Cut holes in frame for tubing
3. Solder inlet and outlet tubes
D. Cover with clear glass
1. Caulk edge of container
2. Set glass in place
3. Attach molding

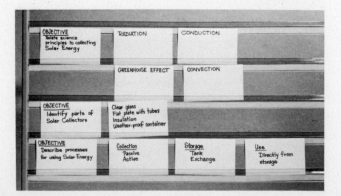

content—the factual information related to each objective —and display these cards under or beside each appropriate objective card. At this stage, list all the available content relating to the objectives, without considering what you may use and what will be discarded.

It is advisable to use cards of one color for objectives and of a second color for the content. Later you can add additional cards for specific materials that relate to single objectives and items of content, or to groups of either.

You will find that using cards frees you to experiment with the order of the ideas until they are in a logical sequence. What you start with as the first point may later become the last one. Additional objectives that occur as you organize can also be added easily at this stage, while anything that apparently disrupts the sequence can also, just as easily, be eliminated or relocated. Later, during the actual story-boarding and scripting, you may find need for further changes, but now you have a simple, natural guideline to follow.

It should be reemphasized that at this stage you include as much as possible about the content—facts, examples, locations, and special reminders. It will be easier to eliminate some points later than to search for them if needed. While you are listing content, visual ideas may come to mind. Note them also on cards.

A planning board for displaying cards consists of plastic or cardboard strips stapled to a stiff backing board. As illustrated, a card is held firmly in a strip, but it can be slipped in and out easily. (For further details on making and using a planning board see Chapter 13 in Kemp, *The Instructional Design Process.*)

If you do not have a planning board available, information can be written on pieces of paper and tacked to a wall. Or, consider using Scotch 3M product 655 (POST-IT Note Pad) which has a strip of adhesive on back of each slip of paper. This allows for positioning, removing, and resetting each slip of paper.

Once the content has been listed in relation to the objectives, you may consider this to be a **checkpoint.**

Although you may already have a team that includes subject consultants, it is a good idea to bring in one or more qualified persons to examine your work. This may be a colleague, a manager, or other individual who either initiated the project or has final approval for its acceptance. They may find something important left out, or offer a comment that strikes a spark to give a direction you had not considered.

Also, there is real benefit in asking one or more members of the potential audience group to review your planning as you proceed. The perceptions that they may have for your topic and its organization may differ appreciably from yours. Involve them at this checkpoint and others that will be indicated. Their suggestions can be of real value.

Your content outline has been developed in the light of an idea, objectives, and audience. You are now ready to make decisions about the single medium or combination of media to carry the objectives and content. Before deciding on the media to use ask yourself such questions as these:

- Will the instructional media be for motivation, information, or instruction?
- Is sound (narration, lip synchronization) necessary, or can a silent medium (film, slides, filmstrip) with titles, captions, and directions be used?
- Is motion important or can still pictures convey the ideas and information?
- Is there to be study by individuals or is the emphasis to be on group use?
- Is color important or will black-and-white be satisfactory?
- Will there be any problems in keeping the materials up to date?
- Will I be able to overcome any technical problems in preparation, or do I know where to get help if necessary?
- Will there be problems of duplication, distribution, or storage of the completed materials?
- Will budget and time permit a good job?
- What problems may be encountered when using the materials (facilities, equipment, size of group, and any others)?

Now consider the various media available to you in terms of characteristics, advantages, disadvantages, and limitations. Then make choices to best serve your objectives.

Review What You Have Learned About Preliminary Planning:

1. This chapter introduces *four* steps in preliminary planning. What are they?
2. What type of thought is probably most important during the creative thinking process?
3. Relate the following elements of preliminary planning on the topic of *Dog Training* to the *four* steps above:
   a. to lead a dog when wearing a collar and leash
   b. stand at "heel" position; give command "sit"; pull on leash and press down on dog's hindquarters; release pressure and give praise
   c. acquire skill for teaching a dog proper behavior
   d. to direct a dog to "sit"
   e. teenagers who have received a young dog
4. What are the *three* broad purposes for which instructional media can be developed?
5. Relate the following items to each of the purposes in number 4:
   a. introducing a new topic to a class of students

   b. engaging a trainee in self-study activity
   c. designing a dramatic, role-playing situation to raise interest
   d. providing background as orientation for new medical treatment
   e. requiring employees to review modified procedures periodically so they can function competently
6. What are *three* roles that personnel should fill during the planning and production of instructional media?
7. What are *three* parts of a properly stated objective?
8. Which verbs might you select when writing instructional objectives?

   ____ a. learn   ____ f. become aware of
   ____ b. assemble   ____ g. name
   ____ c. explain   ____ h. compare
   ____ d. arrange   ____ i. predict
   ____ e. understand   ____ j. master
9. These questions relate to Table 4-1 on page 32.

a. To which category of objectives does each example relate?

b. Are any objectives in the middle column not properly stated and how might they be reworded?

10. What are *four* characteristics of a potential audience for which you might want information when starting your planning?

11. For what *two* reasons can it be useful to examine other materials that relate to a topic you are treating?

12. What suggestion is offered in the chapter as a flexible, visual way to organize the content information relating to the objectives?

13. Select a topic you would like to see developed into instructional media and complete each of the following:

a. Express an *idea* or *purpose* to be served.

b. Write the *objectives* you would want to serve.

c. With what *audience* would you use the materials?

d. What are some of the *audience's characteristics* you feel important to consider?

e. With regard to the three *personnel roles* that must be filled, which role(s) would you fill? What are your qualifications?

f. Would you ask for assistance in any of the three areas?

g. How would you go about finding out whether any instructional media have already been prepared on your topic?

h. Develop a *content outline* for the topic.

i. Relate the *content* to *objectives* on cards of different colors.

REFERENCES

Creativity

Davis, Gary, and Scott, Joseph. *Training Creative Thinking.* Melbourne, FL: Krieger, P.O. Box 9542, 1978.

Fabun, Don. *Three Roads to Awareness.* Beverly Hills, CA: Glencoe, 1970.

Feldhusen, John F., et al. *Teaching Creative Thinking and Problem Solving.* Dubuque, IA: Kendall-Hunt, 1977.

Gardner, Howard. *Art, Mind, and Brain: A Cognitive Approach to Creativity.* New York: Basic Books, 1982.

Van Oech, Roger. *A Whack on the Side of the Head.* Creative Thinking, Menlo Park, CA: P.O. Box 7354, 1982.

Objectives

Bloom, Benjamin S., et al. *A Taxonomy of Educational Objectives: Handbook I, the Cognitive Domain.* New York: Longman, 1977.

Davies, Ivor K. *Objectives in Curriculum Design.* New York: McGraw-Hill, 1976.

Harrow, Anita J. *A Taxonomy of the Psychomotor Domain.* New York: Longman, 1972.

Kemp, Jerrold E. *The Instructional Design Process.* New York: Harper & Row, 1985.

Krathwohl, David R., et al. *A Taxonomy of Educational Objectives: Handbook II, the Affective Domain.* New York: Longman, 1969.

Kryspin, William J., and Feldhusen, John F. *Writing Behavioral Objectives.* Minneapolis: Burgess Publishing Co., 1974.

Mager, Robert F. *Preparing Instructional Objectives.* Belmont, CA: Fearon-Pitman, 1975.

———. *Goal Analysis.* Belmont, CA: Fearon-Pitman, 1971.

Chapter 5

THE KINDS OF MEDIA

- Printed Media
- Display Media
- Overhead Transparencies
- Audiotape Recordings
- Slide Series and Filmstrips
- Multi-image Presentations
- Video Recordings and Motion-Picture Films
- Computer-Based Instruction
- Selecting Media for Specific Needs

Notice the sequence that is being developed. First, establish **objectives** and consider your **audience;** then **organize the content** to fit your objectives. Now **select the specific instructional media** and other experiences to carry through your purposes.

Why this sequence? Because instructional media are channels through which content stimuli are presented to the learner—stimuli to motivate, direct attention, inform, evoke a response, guide thinking, instruct, or whatever. Therefore, only after establishing **what it is that you wish to communicate** are you properly able to select the channel or medium through which the content will most likely elicit the proper response that serves the objective. Review the findings in Chapter 2, relating the effects of media on learning.

If motion is inherent in the subject, consider a motion picture or videotape recording; but if motion is not important, consider materials that demand simpler skills, less time, or less money, yet do the job equally well. To think further: A series of photographic prints, which can easily be studied in detail at a workstation, may be preferable to a filmstrip and less difficult to make. Also, consider using combinations of media to serve your purposes: A series of overhead transparencies that outline a process can be supplemented with a set of slides and the two used concurrently for effective instruction; or for motivational purposes, a dynamic three-screen multi-image presentation may be effective.

On the other hand, perhaps for practice or perhaps because you have certain equipment available, you may wish to prepare a specific material, possibly a series of synchronized slides with tape, a video recording, or a computer-based interactive video program. If this is your starting point, select a subject and establish purposes that will use the medium to its best advantage.

Any one or more of a number of instructional media may be applicable to serve an objective and its content. The decision for selection may be based on your skills, equipment requirements, convenience, or cost. But each of the several types of instructional media makes certain unique contributions to improving communications and subsequent learning. All require careful planning before preparation—some more than others. When selecting the ones to serve your purposes, examine all of them and become aware of their special characteristics and specific contributions to communication and learning.

PRINTED MEDIA

A number of materials, prepared on paper, may serve instructional or informational purposes. They are classified as printed media and consist of three groupings: (1) **learning aids,** (2) **training materials,** and (3) **informational materials.**

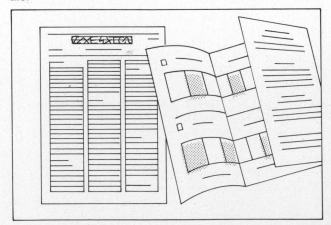

Learning aids comprise resources designed for use by individual students or trainees, as a person follows precise directions for performing a task. A **guide sheet** or **job aid** may be a checklist of steps or procedures to follow when assembling, operating, or maintaining equipment. A more complete learning aid includes line drawings or photographs along with words for better explanation. These materials are often used at job sites or when the need for a handy reference arises after class instruction.

Training materials also relate directly to instruction. **Handout sheets** may be similar to guide sheets. They are usually more informational than procedural. A **study guide** is a set of pages which prepares and directs the student how to proceed with a unit or course of study. An **instructor's manual** provides guidance and assistance to the instructor when preparing for and delivering the instruction. It includes directions and information relating to each topic or unit to be taught.

Group three consists of items that serve informational and motivational purposes. **Brochures** are announcements about a program or service. **Newsletters** report on activities of an organization. They may range from a simple one-page typewritten sheet to a highly professional multipage, full color publication. An **annual report** can also be unpretentious or elaborate with much attention being given to pictures, charts, and diagrams.

DISPLAY MEDIA

Most display media are designed for use by an instructor as information is presented in front of a small class or audience. This category includes the chalkboard, flip chart, cloth boards, magnetic chalkboard, and also bulletin boards and exhibits.

The simplest way of presenting information to a group is to write on a **chalkboard.** With careful planning, by using colored chalk, and disclosing information at the moment the audience should see it, the chalkboard can be an effective communication device.

Visual information, such as outlines, charts, or graphs, can be prepared in advance of a meeting on large sheets of paper which are clipped together and set on an easel. During the presentation with a **flip chart,** the speaker can turn quickly and smoothly from one sheet to another.

On a **cloth board,** outlines, concepts, or processes can be developed by the progressive addition of prepared materials to the surface of the board. Felt, flannel, or sandpaper-backed material will adhere to a **felt or flannel surface.** A preferred cloth board is the **hook-and-loop board.** It consists of a cloth surface containing countless nylon loops. Materials for display are backed with nylon-hook tape. The two nylon surfaces interlock securely, permitting the display of flat or three-dimensional heavy objects which can have a striking visual impact on an audience.

The most versatile display board is the **magnetic chalkboard.** It consists of a thin steel surface coated with a dull-finish paint. Objects and information on cardboard which have small magnets attached to their backs can be displayed easily and also moved about. At the same time, writing or drawing can take place on the chalkboard surface.

A display may also be prepared on a **bulletin board,** which is a two-dimensional surface, or as an **exhibit** that may include three-dimensional elements. The bulletin board and exhibit differ from the other display media in that they must capture the attention of passing individuals so they will stop and read the displayed message.

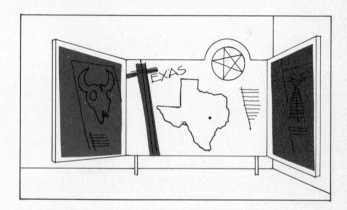

OVERHEAD TRANSPARENCIES

Transparencies are a popular form of instructional media. The use of large transparencies is supported by the development of small, lightweight, efficient **overhead projectors** combined with simple techniques for preparing transparencies and by the dramatic effectiveness of the medium.

The projector is used from near the front of the room, with the instructor standing or seated beside it, facing the group. The projection screen is at the front wall, and room light is at a moderate level. Transparencies are placed on the large stage of the projector, and the instructor may point to features and make marks on the film. The results appear immediately on the screen. **Progressively disclosing** areas of a transparency and adding **overlay** films to a base transparency are special features that make the use of this visual medium effective in many subject areas.

Overhead projectors are especially useful for instructing large groups on all levels. Investigate the range of techniques for preparing transparencies, and select the most appropriate ones for use. Some methods require no special

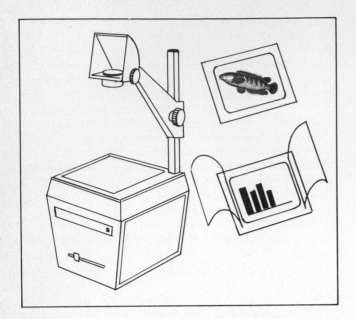

equipment or training, while for others experience in photography and the graphic arts is necessary.

AUDIOTAPE RECORDINGS

Audio materials are an economical way to provide certain types of informational or instructional content. Recordings may be prepared for group or, more commonly now, for individual listening. Small, compact cassette recorders have made the use of audio materials easy and convenient. Increasing attention is being given to recordings, either by themselves or in combination with printed materials.

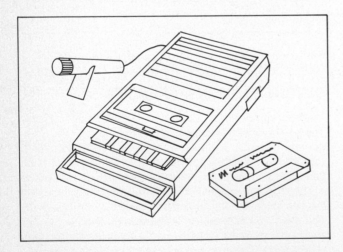

Recordings can be used for many purposes. These include documenting or summarizing a speech, serving as a verbal record of interviews and meetings, recording role-playing situations for follow-up analysis, explaining procedures, and providing drill in various subjects.

For instructional purposes, recordings often are correlated with readings and worksheets in an **audio-notebook** format. When a recording is the central element of a complete instructional program that includes a variety

of learning experiences, an **audio-tutorial** approach is being employed.

Care must be taken in preparing recordings so that they are of high quality and presented in an informal conversational manner. Lengthy verbal lectures should be avoided. Equipment is available to speed up ordinary speech while maintaining its intelligibility. Tape duplicating equipment is necessary when multiple copies, often as cassettes, are required.

SLIDE SERIES AND FILMSTRIPS

Slides are a form of projected media which are easy to prepare; hence they frequently serve as the starting efforts in a media production program. Pictures are generally taken on reversal color film which you can process and mount yourself or send to a film laboratory.

For many uses, any 35mm camera will make satisfactory slides. But for filming some subjects, for close-up and copy work, a 35mm single-lens reflex camera with an appropriate lens is recommended.

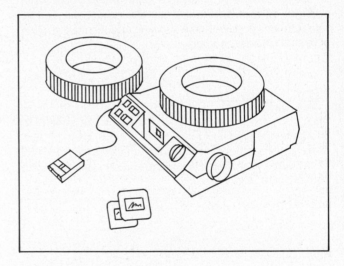

The standard slide dimensions are 2×2 inches. Since the slides are small, they are easily handled and stored. Their sequence can be changed, and slides can be selected from a series for special uses. But this flexibility entails some disadvantages. Slides can become out of order, be misplaced, and sometimes be accidentally projected upside down or backwards. Most of these disadvantages can be overcome by the use of trays and magazines, which store the slides and hold them during use. Also, remotely controlled projectors permit a person making a presentation to change slides by pressing the control button. Tape recordings can be prepared to accompany slides and the latter can be shown automatically as the taped narration is heard. The availability of slide viewers also provides many opportunities for using slides for self-paced learning purposes.

Filmstrips in 35mm are closely related to slides, but instead of being mounted as separate pictures, the film, after processing, remains uncut as a continuous strip. There are two other principal differences. While the same 35mm film is used to prepare each one, the size of the separate pictures on a filmstrip is close to one-half the image area of a 35mm slide. The photography is done with a 35mm **half-frame** camera. Also, the image on each frame is oriented horizontally when the filmstrip is held in position for projection. Slides can be shown either horizontally or vertically.

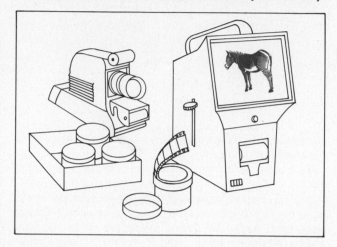

Filmstrips have the advantages of compactness, ease of handling for projection, and low cost for duplication when additional copies are needed. Since pictures are always in order, no wrong positioning can occur, as with slides. On the other hand, filmstrips are not flexible since rearrangement of pictures is not possible.

Filmstrips are more difficult to prepare than are slides. Start by making a slide series. If you have a half-frame 35mm single-lens reflex camera you can copy the slides to filmstrip form yourself. Many commercial film laboratories can convert the slides into a filmstrip. Then a taperecorded narration can be synchronized with the filmstrip. Projectors and viewers are widely available for showing filmstrips to groups and individuals.

MULTI-IMAGE PRESENTATIONS

Combinations of visual materials can be effective when used concurrently for specific purposes. If two or more pictures are projected **simultaneously** on one or more

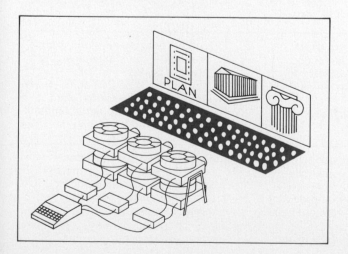

screens for group viewing, the term **multi-image** is used. (When two or more different types of media are used **sequentially** in a presentation or are available as resources in a learning package for self-paced study, the term **multimedia** is used.) The simplest form is to aim two slide projectors at the same screen and *dissolve* from the slide in one machine to one in the other projector, and so on.

Then there may be coordinated slide images on two or three screens with synchronized narration, music, and even sound effects. Brief motion-picture sequences or overhead transparencies may be shown along with slides for special impact or to carry appropriate information. Side-by-side images permit comparisons, relationships, perspective views, or multiple examples. Such multi-image presentations are often motivational by creating a high level of interest in a subject as well as by effectively communicating large amounts of information in a short time.

VIDEO RECORDINGS AND MOTION-PICTURE FILMS

Video and film are both "media of motion." They should be considered for use whenever motion is inherent in a subject, or when it is necessary to communicate an understanding of a subject. Video or film can be more effective than other instructional media for relating one idea to another, for building a continuity of thought, and for creating a dramatic impact. Of the two, the more common production format is video, for reasons that will be given below.

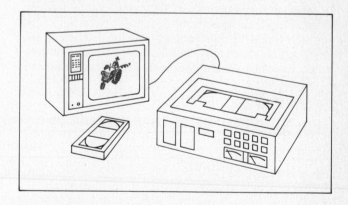

Video is a medium in which images are recorded electronically on magnetic tape along with sound. The essential equipment includes a video camera with microphone, and a videocassette recorder. The immediacy of being able to see what has just been recorded is a key feature that differentiates video from the motion picture. Once recording is completed, the original tape can be used or edited electronically into final form.

Video cameras are compact and portable. Videotape is less expensive than is 16mm film and it can be erased and reused. With television monitors and large-screen video projectors, recordings can be shown to groups or used by individuals for self-pacing their own learning.

The motion picture is still preferred for showings to large groups because of its high resolution and color fidelity. Also, at present, a motion picture camera may give the best results for filming animation sequences or documenting events that take place very slowly (time-lapse photography) or very rapidly. However, video equipment is in development to serve these latter needs.

Two recent developments are video based. First, the combination of both still and moving images may be prepared (from videotape or film original) on a videodisc and viewed through a video system. Second, by integrating computer programming with a videotape or videodisc player, the student can control the pace and order of any instructional sequence. By including participation activities for the student, along with the video material, **interactive video** can be applied on various levels of complexity.

COMPUTER-BASED INSTRUCTION

Computer-based instruction (CBI) refers to any application of computer technology to the instructional process. It includes using a computer to present information, to tutor a learner, to provide practice for developing a skill, to simulate a process which is being studied, and to manipulate data to solve problems. Among instructional media, CBI offers the unique ability to ask a learner a question, record and judge the learner's response, and then use that information to control the sequence of instruction that follows.

Computer-based instructional materials are planned in much the same way as are other media. However, since CBI permits alternative instructional sequences, more sophisticated planning is required. The availability of powerful, low-cost microcomputers for CBI uses requires a basic understanding of computers and computer programming. The required materials, called **software,** are developed by writing programs which are a series of instructions to the computer. Special software packages, called **authoring lan-**

guages or **authoring systems,** are available to simplify the process for the beginner.

Most computer systems will display both text and graphic images and some are capable of simple animation. Computers can also be used to control other forms of instructional hardware, such as slide projectors or video playback equipment. As noted above, interactive video systems which combine the advantages of video and CBI are capable of providing very sophisticated instructional activities.

The nine categories and types of instructional media—printed media, display media, overhead transparencies, audiotape recordings, slide series, filmstrips, multi-image presentations, video or film, and computer-based instruction—have various planning requirements and different degrees of complexity in production and use. Some require close adherence to all steps in the planning process, while others can be developed with a less formal procedure. Refer to the appropriate sections in Part IV of this book for guidance as you study the planning steps that follow in succeeding chapters.

Table 5-1 **SUMMARY OF CHARACTERISTICS OF INSTRUCTIONAL MEDIA**

MATERIAL	ADVANTAGES	LIMITATIONS
PRINTED MEDIA	1. Include common types of materials 2. Have wide variety of applications 3. Simple types quick to prepare	1. Sophisticated types are costly to prepare 2. Requires suitable reading ability
DISPLAY MEDIA	1. Useful in any kind of room without special adaptations 2. Allows user to be flexible for making changes as presentation proceeds 3. Easy to prepare and use materials	1. Limited to use with small groups 2. Requires some showmanship on part of speaker 3. May not be accepted as important media forms when compared with projected types
OVERHEAD TRANSPARENCIES	1. Can present information in systematic, developmental sequences 2. Use simple-to-operate projector with presentation rate controlled by instructor 3. Require only limited planning	1. Require special equipment, facilities, and skills for more advanced preparation methods 2. Are large compared with other projectors

Table 5-1 **(continued)**

MATERIAL	ADVANTAGES	LIMITATIONS
OVERHEAD TRANSPARENCIES (continued)	4. Can be prepared by variety of simple, inexpensive methods 5. Particularly useful with large groups	
AUDIOTAPE RECORDINGS	1. Easy to prepare with regular tape recorders 2. Can provide applications in most subject areas 3. Equipment for use is compact, portable, easy to operate 4. Flexible and adaptable as either individual elements of instruction or in correlation with programmed materials 5. Duplication easy and economical	1. Have a tendency for overuse, as lecture or oral textbook reading 2. Fixed rate of information flow 3. Low fidelity of small portable recorders
SLIDE SERIES	1. Require only filming, with processing and mounting by film laboratory 2. Result in colorful, realistic reproductions of original subjects 3. Prepared with any 35mm camera for most uses 4. Easily revised and updated 5. Easily handled, stored, and rearranged for various uses 6. Increased usefulness with tray storage and remote control by presentor 7. Can be combined with taped narration for greater effectiveness 8. May be adapted to group or to individual use	1. Require some skill in photography 2. Require special equipment for closeup photography and copying 3. Can get out of sequence and be projected incorrectly if slides are handled individually
FILMSTRIPS	1. Are compact, easily handled, and always in proper sequence 2. Can be supplemented with recordings 3. Are inexpensive when quantity reproduction is required 4. Are useful for group or individual study at projection rate controlled by instructor or user 5. Are projected with simple lightweight equipment	1. Are relatively difficult to prepare locally 2. Require film laboratory service to convert slides to filmstrip form unless special camera available 3. Are in permanent sequence and cannot be rearranged or revised
MULTI-IMAGE PRESENTATIONS	1. Can demand attention and create strong emotional impact on viewers 2. Can compress large amounts of information in short presentation time 3. Provide for more effective communications in certain situations than when only a single medium is used	1. Require additional equipment, complex setup, and careful coordination during planning, preparation, and use 2. Equipment and production costs high for complex programs

Table 5-1 (**continued**)

MATERIAL	ADVANTAGES	LIMITATIONS
VIDEO AND FILM	1. Particularly useful in describing motion, showing relationships, and giving impact to topic 2. Allow instant replay of video recording 3. Videotape reusable 4. Easy to record lip sync on videotape 5. May include special filming techniques (animation, time-lapse) 6. Combine still and motion on videodisc 7. Standardized film projector available everywhere	1. High cost for studio producton equipment 2. Resolution limited with video for fine detail close-ups 3. Incompatibility of video format types 4. Value of investment in motion picture equipment reduced as video replaces film
COMPUTER-BASED INSTRUCTION (CBI)	1. Presents text information and graphic images 2. Interacts with learners on individual basis through asking questions and judging responses 3. Maintains record of responses 4. Adapts instruction to needs of learner 5. Controls other media hardware 6. Can interface computer and video for learner-controlled programs	1. Requires computer and programming knowledge 2. Requires essential hardware and software for development and use 3. Resolution of graphic images limited on microcomputer systems 4. Effective when used by only one or few individuals at a time 5. Incompatibility of hardware and software among various systems

SELECTING MEDIA FOR SPECIFIC NEEDS

In this chapter we have examined nine categories of instructional media that can be produced locally. Table 5-1 compares the characteristics, advantages, and limitations of each type. These summaries may be enough to assist you in making decisions about appropriate media to serve your objectives and content.

But many of us select media for use on the basis of what we are most comfortable with or what is conveniently available. The choice is a subjective one, often with little consideration to objective criteria for selection. Can more specific guidelines be established which could offer somewhat closer relationships between the various media and instructional requirements?

This question has been examined by various educational researchers. Their general conclusion is that most media can perform most informational and instructional functions, while no single medium is likely to have properties that make it best for all purposes.

We should not leave the matter of media selection to those who make casual choices, nor can we wait patiently for the results of future research. We need some basis for making logical, educated guesses that will lead to practical media decisions. Fortunately, there are some efforts in this direction.

Media Attributes

The most useful results are derived from a consideration of what Levie and Dickie term **media attributes.** These are the capabilities of a medium to exhibit such characteristics as motion and color, and include sound. The important media attributes are the following:

- pictorial representation—photographic or graphic
- factor of size—nonprojected or projected
- factor of color—black-and-white or full color
- factor of movement—still or motion
- factor of language—printed words or oral sound
- sound/picture relationship—silent picture or picture with sound

The question of what media attributes are necessary for a given learning situation becomes the basis for media selection. After the appropriate media attributes have been specified, the medium, or group of media, which best incorporates these attributes can be identified.

Application of this approach to media selection has been attempted by some developers. Tosti and Ball identified six "dimensions of presentation" and broadly related media decisions to them; Reiser and Gagné use objectives, domains of learning, instructional setting, and reading ability of learners as the bases for selecting media. Flow diagrams

MEDIA SELECTION DIAGRAMS
Based on learning objectives and subject content, what attributes are required in the resources?

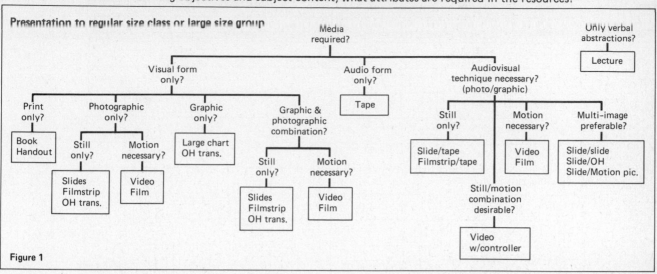

Figure 1 — Presentation to regular size class or large size group

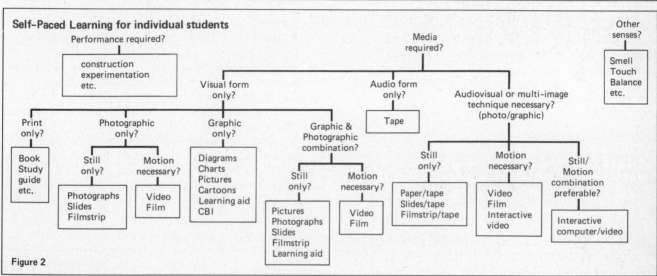

Figure 2 — Self-Paced Learning for individual students

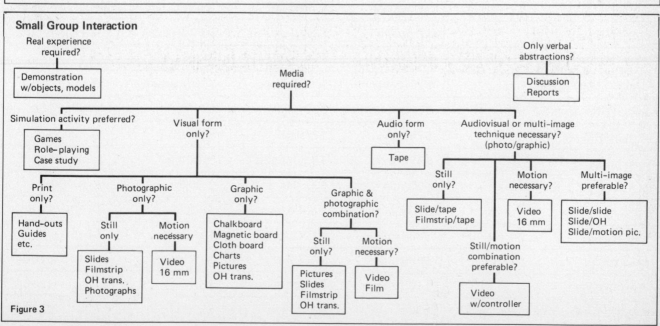

Figure 3 — Small Group Interaction

Table 5-2 **FINAL MEDIA CHOICE WITHIN A CATEGORY**

	ALTERNATE MATERIALS		
CRITERIA	PHOTOGRAPHS	SLIDES	FILMSTRIP
COMMERCIALLY AVAILABLE			
PREPARATION COSTS			
REPRODUCTION COSTS			
TIME TO PREPARE			
SKILLS, SERVICES REQUIRED			
VIEWING, HANDLING			
MAINTENANCE, STORAGE			
STUDENT PREFERENCE			
INSTRUCTOR PREFERENCE			

with questions lead to media decisions. Bretz employs a similar technique as questions point the way through levels, terminating in a particular medium or group of related media. A more expansive approach of this method has been designed by Anderson.

Selection Procedure

A similar practical approach for media selection can start with answers to three general questions:

1. Which teaching/learning pattern (page 5)—presentation, self-paced learning, or small group interaction—is selected or is most appropriate for the objective and the nature of the student group?
2. Which category of learning experiences—realistic experiences, verbal or printed word abstractions, or sensory media experience—is most suitable for the objective and instructional activity in terms of the selected teaching/learning pattern?
3. If sensory experience is indicated or selected, which attributes of media are necessary or desirable?

Upon answering the three questions, refer to the appropriate Media Selection diagram (Figures 1, 2, and 3) on

Table 5-3 **PLANNING, PRODUCTION, AND DUPLICATION COSTS (ON A RELATIVE BASIS)**

	PLANNING	ORIGINAL	DUPLICATE	EQUIPMENT FOR USE GROUP/INDIVIDUAL
PRINTED MEDIA	$25/page	$25/page	$0.05/page	— —
DISPLAY MEDIA	$25/display	$10/display	—	$50+ —
OVERHEAD TRANSPARENCY	$10/transp	$25/transp	$10/transp	$350 —
AUDIOTAPE RECORDING	$20/min	$30/min	$4/tape	$350 /$50
COLOR SLIDE/TAPE	$25/slide	$3/slide	$.60/slide	$1200 /$600
COLOR FILMSTRIP/TAPE	$25/frame	$3.50/frame	$.30/foot	$600 /$250
MULTI-IMAGE (3 SCREENS)	$50/min	$3/slide	$.60/slide	$10,000+
VIDEOTAPE	$200/min	$1000/min	$3/min	$1000/$1000
16MM SOUND	$200/min	$1200/min	$10/min	$1200 —
VIDEODISC	$500/min	$3000/disc	$15/disc	$1000/$1000
INTERACTIVE VIDEO	$1000/min	$2000/min	$5/min	— $2000
COMPUTER-BASED INSTRUCTION (CBI)	$300/min	$300/min	$5/disc	— $1200

page 43. Each diagram is a sequence chart for a teaching/learning pattern. Questions which match media attributes to learning objective needs are answered at various levels and lead to media choices.

Often the decision reaches a group of related media, such as still pictures. Each still-picture form—for example, photographs, slides, or a filmstrip—would provide equally effective instruction for an individual. The choice, then, is based on the most practical form to use, considering the relative merits of a number of empirical factors, such as those shown in Table 5-2 on the previous page.

Costs are always an important consideration in media decisions. Actual costs are difficult to specify, because each situation is different and prices are changing continually. Table 5-3 on the previous page, offers some *relative* commercial costs that should be used *only* for comparison purposes.

Finally, media decisions should be made not for a gross entity of learning as large as a *topic,* but rather for groups of objectives that collectively make up the topic. Within a given topic, carefully designed combinations of media, where each performs a particular function based on its attributes and reinforces the learning effects of the others, may be required to achieve the kind of communication or instruction for a group or individual that is most effective.

As an example for media selection, let us consider the unit on solar energy. The objectives on page 29 relate to collecting solar energy. The content in terms of the objectives is listed on page 33. Applying the approach to media selection recommended above, we reach a decision as follows:

1. Teaching/learning pattern–self-paced learning by students (page 6)
2. Learning experiences required—media
3. Attributes of media—graphic and photographic still pictures in color with sound
4. Media category—from Media Selection diagram Figure 2; choice from box containing tape/print, tape/pictures, tape/slides, and so on
5. Final media choice—comparing costs and other factors for alternatives in Table 5-2; final choice is tape/slides

---

Review What You Have Learned About The Kinds of Instructional Media:

1. Of the nine types of instructional media described in this chapter, which ones best apply to each category?
   a. primarily large group use
   b. primarily small group use
   c. primarily for individual use
   d. may be for any of the above
2. Which of the following "attributes" of media are helpful in the selection process?
   _____a. black-and-white or color
   _____b. still or motion
   _____c. instructor or student preference
   _____d. photographic or graphic
   _____e. self-prepare or purchase
   _____f. silent or sound
   _____g. symbolic/verbal or real
3. Refer to the Media Selection diagrams (Figures 1, 2, and 3) on page 43. For each situation indicate (1) the questions you would ask in sequence, leading to a decision, and (2) your media choice.
   a. presentation to a group—visual diagram of an industrial-flow process with segments shown sequentially
   b. self-paced learning—still color pictures with narration of architectural styles for 2000 students

   c. small-group interaction—a report on community recreational facilities, showing activities and interviewing individuals
   d. presentation to a group—the services of a major charity organization for motivating interest at large fund-raising meetings
   e. self-paced learning—the procedures for handling new-type savings accounts for bank employees (to be studied during free time or at home)
4. Which factors are useful for making a final media decision?
   _____a. purchase cost
   _____b. preparation time
   _____c. instructor preference
   _____d. length of material
   _____e. equipment maintenance requirements
   _____f. handling materials by students
   _____g. proven to be a better way of learning
5. In question 13 following Chapter 4 on page 35, you were asked to select a topic and start to plan the preparation of instructional media. Now, what media form would you choose? What are the objective reasons for this choice?

REFERENCES

Instructional Media

Brown, James W., et al. *AV Instruction: Technology, Media and Methods.* New York: McGraw-Hill, 1983.

Heinich, Robert, et al. *Instructional Media: The New Technologies of Instruction.* New York: Wiley, 1983.

Wittich, Walter A., and Schuller, Charles F. *Instructional Technology: Its Nature and Use.* New York: Harper & Row, 1979.

Selecting Media

Anderson, Ronald H. *Selecting and Developing Media for Instruction.* Cincinnati: Van Nostrand Reinhold, 1983.

Bretz, Rudy. *A Taxonomy of Communications Media.* Englewood Cliffs, NJ: Educational Technology Publications, 1971.

Levie, W. Howard, and Dickie, Kenneth E. "The Analysis and Application of Media." In *Second Handbook of Research on Teaching.* Robert M. W. Travers, ed. Chicago: Rand McNally, 1973, pp. 858–882.

Reiser, Robert A., and Gagné, Robert M. *Selecting Media for Instruction.* Englewood Cliffs, NJ: Educational Technology Publications, 1983.

Tosti, Donald T., and Ball, John R. "A Behavioral Approach to Instructional Design and Media Selection." *AV Communications Review* 17 (Spring 1969):5–25.

Chapter 6

DESIGNING THE MEDIA

- Plan for Participation
- Write the Treatment
- Make a Storyboard
- Develop the Script
- Consider the Length
- Prepare the Specifications

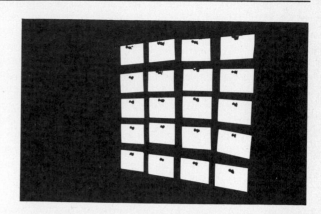

You have your content organized in terms of objectives and the audience. You are aware of the kinds of instructional media you may consider for preparation, their characteristics and particular contributions, advantages, and limitations. And you have some empirical procedures for media selection. In putting all these together you must decide which materials can best communicate the content of specific objectives. Your plans may require that a single medium (video, slides, transparencies, or such) carry your message. Or, in the design approach, a number of media may be integrated, each serving one or more specific objectives and content.

Examine the cards you prepared listing objectives and content. Then decide on the medium or media to use.

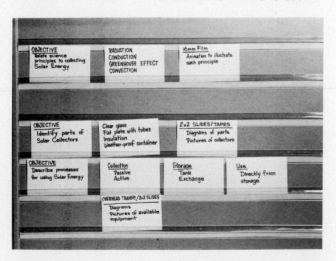

If more than a single medium is to be employed, make cards for each one and organize the objective and content cards with the appropriate medium cards. This plan will give you a visual reference to the flow and relationship of elements within the total topic. Now, start planning for production.

There is no single best manner in which the details of the content outline can be transformed into meaningful and related pictures and words. Two approaches have been established through experience, but they are by no means the only sound ones. First, many successful materials carry an audience from the known to the unknown; they start with things familiar to the audience (perhaps by reviewing the present level of understanding) and then lead to as many new facts and new relationships as the material is meant to achieve. Second, many materials are successfully built around three divisions—the introduction, which captures the attention of the audience; the developmental stage, which contains most of the content and in effect tells the story (or involves the viewers in active participation); and, finally, the ending, which may apply, summarize, or review the ideas presented and suggest further activity.

Recall from Chapter 3 that there is evidence from research to show that detailed introductions and summaries in films, and probably in other instructional media, do not add much to effectiveness. This finding may be particularly true of a series of integrated materials, each of which serves a specific objective or a related series of objectives. The brief film or video recording illustrating a process, and the short audiotape recording containing explanatory information for a topic are examples of the latter kinds of materials. A printed study guide or an instructional module can introduce, relate, summarize, and direct student participation. The material itself contains just the essential facts, explanations, demonstrations, or whatever, without any embellishments.

PLAN FOR PARTICIPATION

A number of investigations of the effectiveness of instructional media designed for instruction have shown the value of having the learner participate in some way during, or immediately after, studying the material. These experiments, without reservation, have proven that *active participation definitely helps learning.* But most producers of commercial instructional media continue to ignore this principle. Video recordings, films, filmstrips, and sets of transparencies are designed primarily to present information and, within themselves, provide no opportunity or directions for other than passive activity—mentally or mechanically following the presentation. Hopefully, the treatment of the subject is so motivational that interest is maintained throughout. But often it is not. It is then left up to the instructor or presenter to plan for pre- and post-use activities.

The way to create participation is to make involvement an inherent part of the material itself. The expression **interactive learning** is widely used. It means that while working with the material, a student is frequently directed to make a choice, answer a question, or otherwise engage in an activity. The purpose is to ascertain understanding and competency to use or apply the information or skill being presented. While interactive learning is an integral part of computer-based instruction and the use of computer/video programs, it can be applied to the design of all types of instructional media.

Here are some suggestions for developing participation in instructional media:

- Include questions, requiring an immediate written or oral response.
- Direct other written activity (explain, summarize, give other examples, and so forth).
- Require selection, judgment, or other decisions to be made from among things shown or heard.
- Require performance related to the activity or skill shown or heard.

These participation techniques often require a break in the presentation—having the student stop the projector or recorder to do something, or promoting immediate activity after studying a section of the material.

Also, be sure to plan for evaluation of the participation results and provide *feedback* to the student indicating the correct reply or a comparison of measurement for the level of accomplishment.

If your materials are designed for motivation or information rather than for instruction (as described on page 28), you should still want a response. Possibly it is acceptance as expressed by applause, or an action to follow the viewing. Whatever the purpose, plan for the desired outcome at this time.

WRITE THE TREATMENT

Carefully examine the example of a content outline (page 33) and form an idea of how you might develop the generalizations visually. This is called a **treatment.** It is a descriptive synopsis of how the content of the proposed instructional media can be organized and presented. A treatment is similar to a verbal summary in which you describe to someone the story line of a book you have read.

Two or more treatments that handle the subject in different ways (expository, personal involvement, or dramatic)

can be written so as to explore different approaches to the topic. See example on the following page.

Commercial television has a great impact on the "visual literacy" of most people. We have learned to accept fast pacing of commercials and dramatic programs. The sophistication that results from these experiences requires that many instructional media be planned to move briskly both in content treatment and visual techniques.

Writing the treatment is an important step since it causes you to think through your presentation, in terms of the project objectives, putting it in a sequential, organized form that you and others can follow easily. The treatment provides the framework for the planning elements that follow.

While deciding how to treat the subject (or later, when the script is being written) the idea of including some already available sequences may come to mind. Historical and political subjects, wildlife scenes, space exploration, and underwater footage are some of the areas found in stock slide, film, or videotape collections. Stock visual materials can be purchased from many sources. See the list on page 54.

MAKE A STORYBOARD

As you develop your story, *try to visualize the situations you are describing.* Remember, you are preparing an audiovisual material—with the emphasis on the word *visual.* Most people normally think in words, but now you may have to reorient yourself to learn to think in pictures—not in vague general pictures but in specific visual representations of real situations. Visualization can be aided by making simple sketches or by taking pictures (instant pictures on self-processing film are ideal for this purpose) that show the treatment of each element or sequence. These sketches or pictures, along with narration notes, become the **storyboard.**

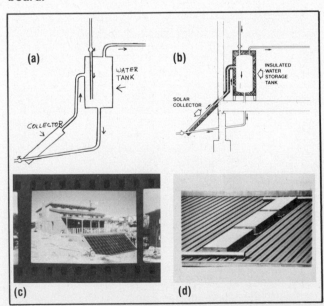

(a) Simple sketch, (b) Detailed sketch, (c) 35mm contact print (or 2 × 2 slide), (d) Self-processing film print

Examples of
Treatment

EXPOSITORY TREATMENT

About half of the heat and light generated by the sun reaches the earth and can be used if collected. A solar collector consists of a box covered with glass or plastic. Inside the box are tubes filled with fluid and attached to a dark-colored flat plate. Insulation prevents the loss of heat. Collectors can be placed on the roof of a new or older house, on a wall, or even on the ground. Solar collectors can be used efficiently for collecting heat in northern latitudes and at high elevations. Their size should be about 100 square feet for heating water and from $\frac{1}{3}$ to $\frac{1}{2}$ the floor area of the house for space heating. The collectors are most beneficial when facing south and tilted to an angle equal to the latitude of the location plus 15 degrees. In addition to house and indoor water heating, solar collectors are used to heat swimming pool water.

PERSONAL INVOLVEMENT TREATMENT

How much energy generated by the sun is available for use to meet the growing energy shortage? How would you go about capturing this energy? Two students decided to investigate this problem. They visited a local company that builds and installs solar collectors. They were shown what a collector looks like and its essential parts. On a tour to see how collectors were being used, they saw them on the roofs of new and older houses, on an apartment building wall, on the ground adjacent to a house; one was being used to heat water for a swimming pool. They learned these facts about installing collectors: (1) the collector should be placed facing south for gathering the most sunlight; (2) the collector should be tilted according to the degrees of latitude for the location plus 15 degrees; and (3) collectors should cover $\frac{1}{3}$ to $\frac{1}{2}$ of the floor area of the house for space heating and should be about 100 square feet in size for heating water. From the literature they received they were able to design a solar collector of their own using the four main parts. The students next planned to build a collector to heat water.

DRAMATIC TREATMENT

What's going to happen when the world's oil supply runs out or when natural gas is no longer available? One source for energy that must get more attention is the sun. Forty-seven percent of the sun's energy reaches the earth and can be collected and used. Anyone can make use of it *now.* Solar collectors are placed on the roofs of new or even older buildings, on the ground near a house, or on the side wall of a building. Collectors covering $\frac{1}{3}$ to $\frac{1}{2}$ of the floor area of a house are necessary for space heating and about 100 square feet in size for heating water. The greatest amount of the sun's energy is gathered when the collector faces south and is tilted the number of degrees of latitude for the location plus 15 degrees. By applying this information, collectors can be efficient devices to gather heat in northern latitudes and at high elevations. When the four main parts of a solar collector are properly assembled and installed in sufficient numbers, each household has the potential, in part, for serving their own energy needs, which in turn helps to reduce the energy problem for the country.

Put the storyboard sketches or pictures on cards, as was suggested for the objectives and content on page 32. Use 3×5 or 4×6 inch cards, or 8½×11 inch paper. These proportions approximate the format of most visual materials, the pictures being wider than they are high. Use a card with an area blocked off for the visual and having space for narration notes or production comments. (See sample for storyboard cards inside the back cover of this book. Duplicate them for use.)

Every sequence should be represented by one or more cards. Include separate cards for possible titles, questions, and special directions (such as indications for student participation). It may not be necessary to make a card for all anticipated scenes. The details, like an overall picture of a

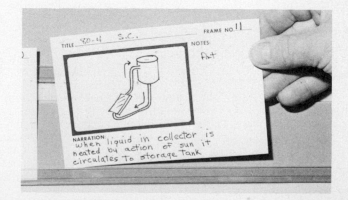

subject followed by a close-up of detail within the subject, will be handled in the script that follows.

Here is an example of a partial storyboard for *Collecting Solar Energy.*

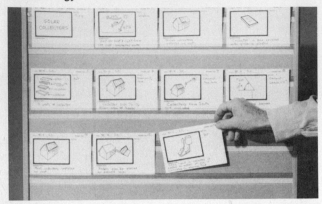

The storyboard is another important **checkpoint** stage in the design of your instructional media. The first was when the content outline (page 32) was completed. Display the storyboard; make it easy to examine. (For suggestions on preparing and using a planning board for displaying storyboard cards, see Chapter 13 in Kemp, *The Instructional Design Process.*)

Reactions and suggestions from those involved in the project or from other interested and qualified persons are valuable at this point. These people may offer assistance by evaluating the way you visualize your ideas and the continuity of your treatment. Often people studying the displayed storyboard point out things that have been missed or sequences that need reorganization. Rearranging pictures and adding new ones are easy tasks when the storyboard is prepared on cards.

DEVELOP THE SCRIPT

Once the treatment and the storyboard continuity are satisfactorily organized, you are ready to write your detailed blueprint, the **script.** This script becomes the map that gives definite directions for your picture-taking, art work, audio recording, or video recording and filming. The script is a picture-by-picture description with accompanying narration and audio effects. As was indicated for storyboarding, first plan what will be seen, then what will be said or otherwise heard.

One procedure is to prepare the script in **block format.** The visual description of a scene is written from margin to margin across the page, often typed in capital letters. The narration and indications for music and sound effects are shown in narrow columns centered below the visual. Some people prefer this format as the eye moves down the page, easily noting the continuous flow of action.

An Example of
a Block Format Script

Music Up

FADE-IN
1. Title: RECONSTITUTING A DRY DRUG
DISSOLVE
2. Produced by Instructional Media Productions
DISSOLVE
3. Still Picture: SOLVENT BEING ADDED TO A VIAL
    Overprint: While viewing this tape have module #12 at hand

Music out

4. MS. NURSE REFERS TO ROUTE SHEET AND PICKS UP VIAL

Narration

    Always compare route ordered in medication order form
    with route for which medication may be used on label of
    vial.
5. CU. LABEL OF VIAL
    While reading a label, check expiration date on the vial.
6. ECU. LABEL WITH INFORMATION ON RECONSTITUTING
    Amount and type of solvent to be used may be indicated on
    the label. If it is not . . .
7. MS. NURSE TAKES REFERENCE BOOK FROM SHELF
    Refer to the pharmaceutical literature for reconstitut-
    ing information
8. CU SECTION OF PAGE WITH INFORMATION
9. MS. NURSE SELECTS BOTTLE OF SOLVENT FROM SHELF
    Locate the correct solvent (and so on . . .)

A more widely used format is the **two-column script** design. Camera positions and picture descriptions are placed on the left half of the page and narration on the right, opposite the appropriate scene description. (See the example on the following page.)

The placement of the camera for each picture, with respect to the subject, should be indicated. If the subject is to be at a distance from the camera the picture is a **long shot** (LS); if the camera covers the subject and nothing more, a **medium shot** (MS); while a **close-up** (CU) brings the camera in to concentrate on a feature of the subject. Whether the scene is to be photographed from a **high angle** or a **low angle,** or **subjective** position (from the subject's viewpoint, as over the shoulder) can also be specified. (For further information on camera positions, see page 91).

At this stage the narration need not be stated in final, detailed form. It is sufficient to write **narration ideas,** or brief statements, that can be refined later. Only when an explanation requires that a motion-picture or videotaped scene be of a specific length must the narration be written carefully and in final form at this point.

Because a script must frequently be revised or completely restructured, a word processor can be a real time saver. The script is typed on the keyboard and appears on the display screen. At the same time, the script is held in the computer memory. A copy can be put on paper through an attached printer. The script can also be stored on a disk and, when necessary, recalled to the screen for rewriting, making deletions, and repositioning sections. After these changes are made, a new copy can be printed. Much time spent on typing can be saved.

As you write the treatment and the script, keep in mind these suggestions:

- Open by capturing the attention of the audience with a dramatic sequence, by illustrating the importance of the topic to the viewer, by carefully using music and other sounds.
- Plan for variety in visuals (long shots, close-ups, low angle shots) and a frequent change of pace as the narration proceeds.
- Realize that few words, or no narration or other sound, can satisfactorily carry meaning in some scenes.
- Use easily understood words and phrases.
- Introduce a technical term by showing it as an "overprint" on the screen, in context with a scene it represents.
- Do not belabor the message being presented; illustrate and explain a concept, then move on.
- In visualizing a procedure, show common errors in action, but *always* end with the proper performance.
- Humor and unusual situations will create and maintain interest.
- Consider the value of music as a complement to narration (see page 150 for reasons to include music in a script).
- Explain the importance, to the viewer, of the information being presented.
- Recognize that there may be a limitation to the amount of detailed information that can be communicated

through the media; consider accompanying materials to carry details.

Narration is important not only for the part it plays in explaining details as the *audio* of audiovisual; it also may call attention to relationships and indicate emphasis that should be given in some pictures (center of attention or camera position) when filming. There should be a unity of words and pictures to communicate information effectively.

Be alert to problems that may arise with narration. If it is not related closely to the visual so as to reinforce the visual, the narration may interfere or inhibit learning. Review research evidence in Chapter 3 about the relation between visual and audio channels in instructional media.

When developing computer-based materials or an interactive video program, either storyboarding or scripting procedures may require the applications of **flowcharting** methods (including linear and branching techniques). These may precede the writing of scene or "frame" descriptions. See page 249 for further discussion of flowcharting methods.

If you must obtain approval of the script before picture-taking starts (another possible **checkpoint**), either read the narration or put it on audiotape for the client to hear while viewing the storyboard pictures. A script is to be heard. It's effect is much different than when being read. On the facing page is an example of a script for *Solar Collectors,* developed chiefly from the expository treatment, with some details adopted from the other treatments.

CONSIDER THE LENGTH

The content to be treated in instructional media affects the time needed to present it. An instructional presentation may require extensive explanation or sufficient time for the user to study details. On the other hand, a motivational or informational presentation must have a dynamic movement and include a variety of scenes in order to maintain a high interest level.

Some people do not realize that carefully planned and produced instructional media can effectively present a large amount of information in a short time. For example, a 10–12 minute presentation is a "long time" in media terms and can be a sufficient period of time to fully communicate the content for a topic's objectives.

There is only one rule for length. A scene or a group of scenes, forming a sequence, should be on the screen long enough to present the required information for satisfactory viewer understanding. This should not be so long that the audience will lose interest and the picture lose its effectiveness. Or, if images change too rapidly, their communications value may be lost and the viewer can become confused.

The length will be determined by the time needed to develop the subject visually in combination with the sound. You will learn from experience that, for example, a projected slide may remain on the screen for only a few seconds or that a complex image may hold attention for about

An Example
of a Two-
column
Script

COLLECTING SOLAR ENERGY

Visuals	Narration ideas
1. Title: <u>Solar</u> <u>Collectors</u>	Refer to objectives in workbook.
2. Art: Energy from sun; arrow with 34%	Light and heat energy from sun; 34% reflected back into space.
3. Art: Add arrow with 19%	19% remains in atmosphere.
4. Art: Add arrow with 47%	47% reaches the earth's surface and may be captured for use.
5. MS: Rooftop with solar collectors	Device used is a solar collector; can be placed on rooftop of new building
6. MS: Older building with collectors	or installed on older buildings.
7. LS: Collector on ground beside house	Collector panels can be on the ground
8. LS: Apartment building with collectors on wall	or attached to walls of a building.
9. LS: House with snow on rooftop collectors	Solar energy can be collected in northern latitudes and at high elevation.
10. CU: Worksheet, Part A Overprint: <u>Stop</u> <u>the</u> <u>tape</u>	Complete worksheet, Part A.
11. CU: Section of collector	A collector is like a box.
12. Art: Exploded view showing parts of collector with labels	It has four parts: clear glass or plastic top, tubes filled with fluid attached to a dark, coated plate, insulation, and a waterproof container.
13. LS: House with collectors on roof. Overprint: <u>Face</u> <u>south</u>	Collectors should face south and be unshaded.
14. MS: Another house with collectors. Overprint: $\frac{1}{3}-\frac{1}{2}$ <u>floor</u> <u>area</u>; <u>100</u> <u>square</u> <u>feet</u>	For space heating, size of collector should be one-third to one-half of the floor area of house. About 100 square feet are needed for heating water.
15. MS: Another house with collectors. Overprint: <u>Degrees</u> <u>latitude</u> <u>plus</u> <u>15</u>	For efficiency, collector tilted up according to degrees of latitude at location plus 15.
16. CU: Collector on roof for swimming pool.	A major use of inexpensive collectors is
17. MS: House and swimming pool	to heat swimming pools.
18. CU: Worksheet, Part B Overprint: <u>End</u> <u>of</u> <u>program</u>	Complete worksheet, Part B.

30 seconds. The average film or video scene runs for about seven seconds.

When you (and your team) appraise your script against the amount of time desirable or available, you may need to review it to see whether the needed time can be shortened, or the content divided into two or more outlines for a series of presentations.

PREPARE THE SPECIFICATIONS

You have prepared a map, the script. Now you face the questions: What specific things are to be done now, next, and thereafter until the instructional media are ready to be used? What is to be made or purchased? The answers to these are the specifications.

The more complex the projected media, the more numerous the specifications. Naturally they need to be organized and classified. Some classes of specifications have no bearing on some kinds of materials, while on others (slide series, filmstrips, motion pictures, video recordings, multi-images) all may be needed. Here are some examples of specifications, with detailed and specific points that must be considered and choices that must be made:

- **Type of instructional media:** learning aid, training material, informational material, display media, slide series, filmstrip, audiotape recording, overhead transparencies, video recording, multi-image presentation, computer-based instruction, or interactive video unit.
- **Material and size:** 35mm Ektachrome 64 film, thermal transparency film, $\frac{1}{2}$ inch VHS videotape, and so on.

- **Sound:** tape-recorded narration, synchronous sound, silent reading matter, titles, captions, and so on.
- **Length:** approximate number of pages, photographs, slides, filmstrip frames, transparencies; running time for audiotape or video recording.
- **Facilities and equipment:** locations for filming and recording, camera equipment, recorders, and accessories, graphic and photographic studio equipment, computer facility.
- **Special assistance required:** for acting, filming, lighting, graphics, sound recording, film processing, editing, programming, printing, duplicating, secretarial.
- **Completion date:** planned for or *must* be completed by when?
- **Budget estimate:** including film and other materials, equipment purchase or rental, film laboratory and other services, salaries (if applicable), and overhead charges.

An Example of Simple Specifications

```
                    SPECIFICATIONS

30 2×2 color slides

Five or six titles and overprint captions and four or five drawings for
   close-up photographic copy work

Synchronized tape-recorded narration of 5-6 minutes duration

Materials to be prepared during May for use in September

All facilities and equipment available at no cost; most picture tak-
   ing in the community

Budget:
   Two rolls 35mm 36-exposure Ektachrome 200 film,
      with processing . . . . . . . . . . . @ $12    $24.00
   Two sets slide duplicates . . . . . . . . @ $15     30.00
   Art supplies . . . . . . . . . . . .                10.00
   One roll 300 feet × ¼-inch magnetic recording tape   7.50
   Two C30 audiocassettes . . . . . . . . . @$2.50      5.00
                                                     _____
                                                      $76.50
```

Review What You Have Learned About Designing Media:

1. What *three* elements of media planning are introduced in this chapter? What is the meaning and value for giving attention to each one?
2. *Two* general methods for handling the content for instructional media are described near the beginning of this chapter. What are they?
3. What do we mean when we say "participation" relative to using instructional media?
4. What is the value of having students participate while using or immediately after using the media materials?
5. What are *three* forms in which a treatment may be written?
6. In what *four* visual ways can a storyboard be prepared?
7. What *four* kinds of information may be put on a storyboard card?
8. Mark each statement as *True* or *False*.
   a. A script is always prepared in two columns.
   b. A two-column script has visual descriptions on one side and sound elements on the other side.
   c. The visual side of a script includes camera placement and scene description information.
   d. Carefully write narration in final, complete form whenever a script is prepared.
   e. Every scene requires some narration.
   f. It is preferable to record the narration and have other persons listen to it, for evaluation, than to have them read the script.
   g. A script can easily handle detailed explanations for any information on a topic.
   h. Two key concepts for a good script are *variety* and *change of pace.*
9. What is a good rule to follow in determining the proper length of a scene?
10. What is the purpose for setting *specifications* for instructional media?
11. What *eight* categories for specifications are listed?

12. Show that there is more than one way of handling your topic's content. Prepare two brief treatments of the topic you selected on page 49, each having a different approach and giving a different emphasis to the topic. Keep in mind your audience and your purposes.
13. Consider the type of instructional media you would like to prepare. Does it fit the presentation of the content?
14. Sketch a few sequences of your storyboard on cards.
15. Prepare a script from the treatment and storyboard. Describe the scenes carefully, using letter abbreviations for camera positions. Write narration ideas if appropriate.
16. List the specifications necessary for your materials.

REFERENCES

Designing the Media

Effective Visual Presentations. Slide/tape program V10-10, Rochester, NY: Eastman Kodak Co., 1980.

The Impact of Visuals on the Speechmaking Process. Slide/tape program V10-71. Rochester, NY: Eastman Kodak Co., 1982.

Speechmaking . . . More Than Words Alone. Pamphlet S-25. Rochester, NY: Eastman Kodak Co., 1979.

Witt, Gary A. "How to Design a Production that Viewers Will Remember." *Instructional Innovator* 26 (October 1981): 37–43.

———. "How to Present Information That Viewers Will Remember." *Instructional Innovator* 26 (November 1981):37–43.

Storyboard

Kemp, Jerrold E. *The Instructional Design Process.* New York: Harper & Row, 1985.

Ruark, Henry C. "Paper Storyboards." *Functional Photography* 16 (January 1981):38–41.

Scripting

Edmonds, Robert. *Scriptwriting for the Audio-Visual Media.* Totowa, NJ: Teacher's College Press, 1978.

Lee, Robert, and Misiorowski, Robert. *Script Models: A Handbook for the Media Writer.* New York: Hastings House, 1978.

Matrazzo, Donna. *The Corporate Scriptwriting Book.* Philadelphia: Media Concepts, 1980.

Miller, William. *Scriptwriting for Narrative Film and Television.* New York: Hastings House, 1980.

O'Bryan, Kenneth G. *Writing for Instructional Television.* Washington, DC: Corporation for Public Broadcasting, 1981.

Scriptwriting Techniques. Slide/tape program V10-31. Rochester, NY: Eastman Kodak Co., 1983.

Swain, Dwight V. *Scripting for Video and Audiovisual Media.* Boston: Focal, 1981.

"Visualizing Your Way to a Script." *Audiovisual Notes from Kodak,* publication 91-1-1, 1981.

Wolfe, Glenn M. *AV Scriptwriting Kit.* Duncan, OK: Haas-Haus Productions, 1981.

Media Specifications

Chamness, Danford. *The Hollywood Guide to Film Budgeting and Script Breakdown.* Los Angeles: Stanley J. Brooks Co., 1460 Westwood Blvd., Suite 303, 1977.

Hutcheson, James W. "Film Budgeting: How Much Will It Cost?" *Photomethods* 27 (January 1984):33–36.

Sources for Stock Footage

Film Search, 21 West 46th Street, New York, NY 10036

Media Research Association, Inc., 1629 K Street, N.W., Washington DC 20036

National Archives and Records Service, Audiovisual Archives Division, Stock Film Library Branch, 1411 South Fern Street, Arlington, VA 22202

National Geographic Society, Stock Shot Film Library, 17th and M Streets, N.W., Washington, DC 20036

Chapter 7
PRODUCING THE MEDIA

- Schedule the Picture Taking
- Take the Pictures
- Work with Actors
- Keep a Record
- Obtain Permission for Pictures
- Edit the Pictures
- Edit the Narration and Captions
- Prepare Titles and Captions
- Conduct Formative Evaluation
- Record the Narration
- Mix Sounds
- Complete Photographic Work

Production is the point at which some people start their work on instructional media. Unfortunately, this is a mistake, but it is difficult to convince them that they need to do some thoughtful planning before making pictures. They may feel they do not have the time to plan, or that they are too knowledgeable about their subject to have to plan. Or they may be so taken with the mechanical phases of production (camera use, audio or video recording, synchronizing slides with sound, and so forth) that they have an insatiable urge to do something and see results *right now!* If you deal with such persons, you probably will not be able to influence them differently. Let them go directly to production. Most often the results will show the fallacy of this approach and the added expense of unplanned production. Experience is sometimes the best teacher!

There is one situation in which unplanned picture taking may be necessary. It is when action over which you have no control is to be recorded. This is similar to the way an event is documented for the evening television news. See the discussion of *documentary* production on the next page.

You will generally find, in contrast to what results from precipitous picture taking, that preliminary planning, storyboarding, and writing a script will enable you to visualize your ideas more clearly and that the final result will serve your purpose better because of its coherence and completeness. In the long run you will probably also save time and money by eliminating errors, reducing the need for retakes, and by not forgetting scenes when picture taking.

Detailed production techniques and the preparation of specific instructional media are presented in Parts III and IV of this book. The general procedures that relate to the production of most types of media, with the exception of printed media, display media, overhead transparencies, audiotape recordings, and CBI are considered here.

SCHEDULE THE PICTURE TAKING

From the script, make a list, grouping together all scenes to be made at the same location, those having related camera positions at a single location, or those with other similarities. Then schedule each group for filming at the same time. Preparing and using this list will save you time. You will achieve further economies if you visit locations, check facilities, and gather items (props) for use prior to the time of filming. But remember—if you take pictures out of script order you must be especially careful to edit them back into their correct position and relationship with other scenes.

If you visited locations during preliminary planning you will be prepared to overcome obstacles that might delay picture taking. Also, by meeting persons in charge at these locations, you can better ensure their cooperation while filming.

TAKE THE PICTURES

In general, the script outline of scene content should be followed. Sometimes, however, as you prepare to take a picture, it becomes evident that a script change is needed. If a single picture as planned does not convey the intent of the script, two pictures may be necessary. Don't hesitate to take them. Or, if you are uncertain about exposure, make an extra take of the same scene with different camera settings. Also, if desirable, repeat a scene, shooting it from a second position or with a change in action. In such cases, the entry on the record sheet (see page 57) should indicate **take 1,** then **take 2,** and so forth, *all of the same scene.* Be flexible in your picture taking. You spend less time and energy making adjustments at this time than you would spend retaking pictures later if they are found to be unsatisfactory—and you have extra pictures from which to make choices during editing.

Example of a
Filming Schedule

FILMING SCHEDULE—Solar Collectors

10/12 (in driving order)
    Scene 16—Williams house
         17—Williams house
          8—Village apartments
          5—Quimby Road

10/15
    Scene   6—James house
         11—Community Center

To arrange
    Scene   7—at beach house
            9—at mountain house

Artwork   Titles
    Scene   2   Scene   1
            3   10
            4   13
           12   14
                                15
                                18

When recording on videotape or filming, rehearse each scene before shooting. Rehearsals permit a check on the action so that the cameraperson can "set" the shot and the actor can "feel" the action being performed. For still pictures, carefully set the scene and check the appearance through the camera viewer before clicking the shutter.

There may be times when a **documentary** approach to all or portions of instructional media is necessary. In such cases, pictures are taken of events as they happen, without a detailed script. It is recommended that this approach be followed only for special cases (athletic events, meetings, ceremonies, production line work, and other uncontrolled situations). Sometimes two or more cameras are used and coordinated so as to capture all important action.

The documentary technique puts an extra burden on the photographers, and the editing stage becomes extremely important for making decisions that may affect the final treatment of the subject (see page 58 for suggestions concerning preparing a script *after* documenting an activity on tape or film). The producer of documentary media needs greater experience and must give more attention to technical details than does the producer of a planned and scripted material, since in a documentary production there can seldom be any retakes. The cameraperson must think in sequences and not in individual shots. It is necessary to recall what went before and anticipate what will be happening and be ready to record it. Contrast to this the making of pre-planned material in which all scenes are thought out in advance and action is controlled.

WORK WITH ACTORS

The individual, whether a professional actor or amateur, who appears in front of a camera must portray normal behavior for the situation being recorded. Here are some suggestions for accomplishing this:

● Treat the person with respect.
● If possible, allow the person to observe the filming of other scenes which can help to put him or her at ease.
● Help the person to relax by maintaining a calm atmosphere at the location.
● Inform the actor of the purpose of the scene or sequence to be filmed.
● Invite suggestions relative to positions and how to carry out the action.
● Use prompting cards if necessary, but keep wording to a minimum, allowing the actor to speak naturally.
● Rehearse the scene until the actor feels comfortable.
● Allow for relaxation breaks if tension seems to build.

KEEP A RECORD

Keep a careful record, a **log sheet,** of all pictures taken. It indicates the order in which scenes were photographed, and it includes data on exposure settings (unless a camera with automatic settings is used) and remarks about the suitability of the action. This record will be useful to you while evaluating picture quality and content when selecting scenes. These items are important:

● Scene number (according to the script)
● Take number (change each time the same scene is filmed)
● Light intensity (reading from light meter, if used)
● Camera settings (lens, shutter speed, and distance)
● Remarks (notes about the action, scene composition, and reminders for editing)

Example of a
Log Sheet

```
                         LOG SHEET
          Title: Solar Collectors     Date: 10/12
          Film: Ektachrome 200        Location: Community
          Camera: Olympus OM-1

          Scene   Take    f/stop   Shutter   Remarks
           16      1       16       1/125    normal
                   2        8       1/125    increase exposure
                   3       22       1/125    less exposure
           17      1       16       1/125    ok
                   2       16       1/125    change position
            8      1       11       1/125    possible reflections
                   2       11       1/125    change angle
                   3        8       1/125    change angle
                                             increase exposure
```

OBTAIN PERMISSION FOR PICTURES

Almost everyone has the right to control the use of pictures of himself or herself, or of one's property. If you are making instructional media you must respect this right. If you fail to do so, you may expose yourself to personal, professional, or financial embarrassment. Specifically, a person may either permit you or forbid you to show pictures of himself or herself, one's children, or one's property. It does not matter whether you show them free or for a compensation, to a large or a small audience, or whether you show them yourself or turn them over to someone else to be shown.

Most people readily agree to being filmed and to having the pictures used, but you should protect yourself and your associates by having them sign a **release form.** (A suitable release form is shown below.) The release authorizing the use of pictures of a minor child must be signed by the parent.

A special kind of property that must be covered by a signed release is the property in copyrighted materials—commonly books, magazines, other printed matter, musical records, and footage from commercial products including films, video recordings, slides, and filmstrip frames. In this case you must get the clearance from the owner of the copyright, *not* the owner of the object. (You may own this book; but you do not own the copyright in it—see the back of the title page.) You will be wise if you assume that books and the like are copyrighted, and seek clearance before you use pictures of them or parts of them in your instructional media.

The provisions of the present copyright law are complex, and guidelines relating specifically to instructional media were not included, other than for off-air reproduction of television programs and copying phonorecordings. The most common interpretation for using copyrighted items is that an instructor can make, or have made, a *single* copy of *one* visual (photograph, chart, diagram, cartoon) from *one* book, periodical, or newspaper for purely noncommercial teaching use as long as the materials are for use *in the classroom.* This is the doctrine of "fair use" and applies to any instructional situation, whether in education or training.

The law does not permit the extension of this right of free reproduction to materials used by other than a teacher in *direct* classroom instruction. This means that copyrighted materials for use in motion pictures, slide series, filmstrips, audiotape recordings, or video recordings cannot be reproduced without permission. This applies to materials designed for individualized learning by students, for use in audio listening centers, to be used by an instructor *other than the one preparing the material.*

A Sample
Release Form

```
                                        Date:

          I hereby give the unqualified right to (insert name of indi-
          vidual, group, or institution making the instructional media
          to make pictures of me, of my minor child (insert name of
          child), or of materials owned by me and to put the finished
          pictures to any legitimate use without limitation or reser-
          vation.

                    Signature:
                    Name printed:
                    Address:
                    City:                        State/Zip:

          Project:
          Director:
```

To be on the safe side, if you wish to use copyrighted materials, write to the copyright owner and request permission to use the materials. Describe the use you will make of the material. In many instances you will receive permission to use the materials as long as you include an acknowledgment of the copyright. In other instances, a charge may be assessed. If your product will have widespread use, and especially if it will be marketed, you may be asked to pay a license fee for the inclusion of the copyrighted material.

There are sources of noncopyrighted materials or those for which permission to use is easily acquired. Clip-book art (page 112) can be reproduced. A directory of pictures for public use has been compiled. (See page 31.) Certain collections of slides and stock motion picture footage are available for purchase and free use. (See page 54.)

A particularly important area is the use of copyrighted music on sound tracks for instructional materials. Popular recordings you may have in your personal library fall into this grouping, and their use in media productions is often a violation of the copyright law. There are music libraries, which consist of many records containing all types of short selections, including sound effects. These can be purchased for unlimited use. See sources on page 155. Media production companies and film laboratories provide the use of music from recordings for a "needle-drop" fee. A needle-drop refers to each use made of part or all of a musical selection in a media presentation.

The other side of copyright—how to protect your own product—is considered on page 67.

EDIT THE PICTURES

Once the pictures are prepared you are ready for the next major step in the production process. During filming the pictures may have been taken out of script sequence, some scenes not indicated in the script may have been filmed, and more than one take may have been made of some subjects. During preparation of the script, the narration or captions were written in rough form or noted only as ideas. These changes and unfinished work give rise to the need for examination, careful appraisal, selection, then organization of all pictures, and the refinement of narration and captions. These activities become the all-important **editing** stage.

Using the script and the log sheets completed while filming, put all pictures in proper order. Arrange slides on a viewbox (page 203). Prepare contact prints or proof copies of still pictures (page 102). Review video and motion picture·footage.

Now choices must be made from among multiple takes of a scene. Examine all your work critically. You must be impersonal in your judgment and eliminate pictures that fail to make a suitable contribution to your specific purposes, and you must be firm in rejecting pictures that do not meet your standards.

If you have made changes in the original script (page 50) by adding scenes or changing the sequence, rewrite the picture side to fit the edited picture version. Then complete the narration or caption side of the script as described below.

EDIT THE NARRATION AND CAPTIONS

Since your original script may have included only rough drafts or mere ideas for narration, there is need for further developing and rewriting the narration or captions in order to correlate words with the edited pictures.

As was indicated in explaining the storyboard on page 48, most of us think in words, and therefore we have a tendency to attempt to communicate with words more readily than with pictures. Even so, we comprehend things more effectively and retain information much longer when major ideas are presented visually in the form of pictures while being supplemented with written or spoken words. Words thus have an important part to play in your materials, but generally they should be secondary to the pictures. They can direct attention, explain details, raise questions, serve as transitions from one picture or idea to the next, and aid in preserving the continuity of the materials. If you find you have to use many words to explain what a picture shows, or to describe things not shown in a picture, perhaps the picture deserves another critical evaluation and possible replacement, or you may need to add a supplementary picture. **Let the pictures tell the greater part of your story.** If they do not, you will have a lecture with illustrations and you will fail to use the medium to its best.

Keep in mind your anticipated audience and its background as you refine the narration. The audience will have bearing on the vocabulary you use and on the complexity and pacing of the commentary. Lengthy narration and long captions are detrimental to the effects of visual materials. Distill concepts down to their essentials. As previously explained, the average narration should require viewing a slide for no more than 30 seconds; a caption or title frame should be no longer than 12 to 15 words. The average length of a video or motion-picture scene is 7 seconds; individual scenes, depending upon action and narration required, range from 2 seconds to possibly 30 seconds running time.

Here are some suggestions for refining narration:

- Write for the ear rather than the eye. Be conversational by writing the narration in simple, easy-to-understand phrases, and straightforward English—the way people talk.
- Avoid such expressions as "here we see," or "in the next slide. . . ." There is no need to tell the viewer what is being seen when it is obvious.
- Identify the picture subject being shown (especially if unusual) as quickly as possible with cue words or phrases. Identification that comes late in a written or spoken line may find the viewer lost in the attempt to understand what is being shown. (Example: "The rapid encoding of information on microfilm is being accomplished by *lasers*." change to: "*Lasers* are accomplishing the rapid encoding of information on microfilm.")
- Carefully select the best word or phrase for quick understanding because hearing a narration is not like reading a book or newspaper. There is often little opportunity for the viewer to either go back and hear again what was said or have the opportunity to reflect on what was said.
- Be alert to where verbal transitions, sound effects, or

- musical bridges are needed to lead the user from one section of the presentation to the next.
- Although proper grammar and expressions are preferable, there may be times when sticking strictly to an accepted rule may make the sentence sound awkward or unnatural, and not appropriate to the particular audience.
- Keep sentences short (10–15 words) and avoid multiple clauses. Place the subject and verb close together.
- Write enough to carry the picture as necessary—then stop writing. Do not assault the audience with a continuous string of facts.

- Have some pauses in narration, otherwise the audience will stop listening.
- Realize that one bit of narration can cover a number of pictures, and that narration can carry over from one scene to the next.
- Read the narration aloud to test expressions, the pacing, the emphasis, and to make sure there are no tongue twisters.
- Periodically plan to summarize and indicate relationships among the concepts and information presented.
- If music will be included on the sound track, refer to the suggestions on page 150.

Part of a Revised Script, with Pictures for Reference

COLLECTING SOLAR ENERGY

VISUAL	SLIDE	NARRATION
Title: <u>Solar Collectors</u>		Before starting to view the slides, turn to page 4 in the workbook and read over the objectives for this presentation. When you are ready, restart the tape and change slides as the numbers are called. If you wish, you can go back to listen to any part or view any slide again.
2. Art: "Energy from sun"; add arrow with "34%"	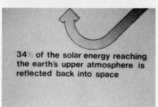	Slide 2 Much of the energy from the sun is generated as light and heat. Thirty-four percent of this energy is reflected back into space.
3. Art: Add arrow with "19%"	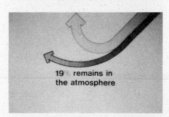	Slide 3 Nineteen percent is absorbed by the atmosphere surrounding the earth.
4. Art: Add arrow with "47%"		Slide 4 And 47 percent of the sun's energy reaches the earth's surface. It may be captured for use.
5. MS. Rooftop of new house with solar collectors		Slide 5 A solar collector, can be placed on a roof to absorb the energy from the sun and convert it into heat.

6. MS. Older building with collectors on flat roof

Slide 6
Collectors can be installed on older buildings as well as new ones, and a flat roof often facilitates installation.

7. LS. Collectors on ground level beside house

Slide 7
Sometimes panels of collectors are constructed on ground level,

8. LS. Apartment building with collectors on side wall

Slide 8
or attached to walls of apartment buildings.

9. LS. House with snow on rooftop collectors

Slide 9
Even in northern latitudes and at high elevations, solar energy can be collected effectively.

10. CU. Worksheet, Part A. Overprint: <u>Stop</u> <u>the</u> <u>tape</u>

Slide 10
Now turn to the worksheet on page 4 and complete Part A. Stop the tape. After you have checked your answers, start again.

11. CU. Section of a collector

Slide 11
Solar collectors look like glass-covered boxes.

12. Art: Exploded view of collector showing parts with labels

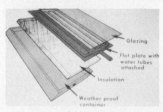

Slide 12
Each section consists of four basic parts—at the top, clear glass or plastic to trap the heat; next, tubes attached to a metal plate that conduct water or other liquid to be heated; behind the heat absorber, insulation to reduce the loss of heat, and around the whole device, a weatherproof container.

13. LS. House with
collectors on roof.
Overprint: <u>Face</u>
south

Slide 13
Solar collectors are most
effective when they are
oriented south and are
unshaded by trees or
buildings.

14. <u>MS.</u> Another house with
collectors.
Overprint: <u>1/3-1/2</u>
<u>floor</u> <u>area</u> <u>for</u> <u>space</u>
<u>heating</u>

Slide 14
To capture sufficient heat,
the area of a collector should
equal from one-third to
one-half of the inside floor
area of a house.

15. CU. Collectors on
roof.

Slide 15
About 100 square feet of
collector are required to heat
water for the average family.

16. MS. Another house with
collectors.
Overprint: °latitude
plus 15°

Slide 16
By tilting a collector at an
angle that is equal to the
degrees of latitude at the
location plus 15 degrees,
maximum use will be made of the
sun's rays.

[and so on] . . .

PREPARE TITLES AND CAPTIONS

Titles serve to introduce the viewer to the subject. The **main title** presents the subject. **Credit titles** acknowledge contributions of those who participated in or cooperated with the project. **Special titles** and **subtitles** introduce individual sequences and may serve to emphasize or to clarify particular pictures. An **end title** gives your instructional media a completed appearance.

A little thought and some care in presentation and in filming will result in neat, professional-looking titles, captions, and labels. They should be simple, brief, easily understood, and large enough to be read when projected. Complex, vague, illegible titles confuse the audience rather than arouse its interest in an otherwise good production.

Other sections of this book will assist you in the preparation and the filming of titles and graphic materials:

- Making titles (page 124)
- Preparing artwork (page 112)
- Legibility standards for lettered materials (page 121)
- Photographic close-up and copy work (page 96)

Some materials, such as those to be less formally presented by integrated use within an instructional design, should not require extensive, formal titles. Keep these materials as simple as possible with an identifying number or brief title and incorporating captions, labels, questions, and directions as necessary. A formal end title may not be needed.

CONDUCT FORMATIVE EVALUATION

As the production nears completion, judgments should be made as to how satisfactorily the materials do their intended job. Consider such questions as:

- Does the material satisfactorily serve the original objectives?
- Is the content treated accurately?
- Is there a smooth flow from one picture or idea to the next one?
- Does the narration aid in continuity and support the visuals?
- Is the material of suitable length, or is it too long overall, requiring deletions?
- Have important points, not apparent before, been left out?
- Should some of the pictures be replaced or are additional ones needed for satisfactory communication?
- Is the material technically acceptable?

An Example of a Preview-Appraisal Questionnaire

COLLECTING SOLAR ENERGY

This set of slides is designed to allow students to study by themselves.

Objectives
1. To identify the four parts of a solar collector and their functions.
2. To describe the placement and size of collectors on various type buildings.

How well do you feel these objectives are accomplished?
_____

Audience
1. These slides and the narration are designed for adults. Is the content appropriate?
_____
Are the pacing of the narration and the vocabulary appropriate?
_____

Content and technical quality
1. Has any important information been left out?
_____
2. Are there any errors or inconsistencies in the presentation?
_____
3. Would you suggest any reorganization? If so, specify.
_____
4. Is the material technically acceptable? Would you replace any slides?
_____
5. What is your rating of the slides and the narration in terms of the objectives to serve the indicated audience?

Excellent   Good   Fair   Poor

You may do better than quiz yourself. Here is another good checkpoint for an evaluation of your materials, in terms of your objectives, by other subject and media specialists, or even by a potential audience group. Show the edited visuals and read the narration or captions. A brief questionnaire for reactions and suggestions may be helpful. See the sample questionnaire on the following page.

This **checkpoint** is often called the **formative evaluation** step. Feedback and reactions may reveal misconceptions that are conveyed, shortcomings, or other needs for improvement. At this time you can also observe how convenient the equipment is for use. You can still make changes before your materials are in final form and reproduced for actual use. The final product will be a better one for your having done a careful formative evaluation.

RECORD THE NARRATION

Once the narration has been refined and, if possible, tested along with the visuals on a typical audience group, you are ready to record it on tape. Prepare the narration to be read by the narrator. Here are some suggestions:

● Type the narration on good quality paper. Use one side of the paper. Doublespace lines.
● Do not carry a sentence from the bottom of one page to the top of the next.
● Spell phonetically any words you think will be unfamiliar to the narrator.
● Mark places that will require cueing, pauses, and special emphasis (see the example).

Then make the recording. See the suggestions in Chapter 12 for working with a narrator and for preparing the recording.

MIX SOUNDS

Along with the narration, the script may indicate music as an introduction and as background under the narration. The

NARRATION—COLLECTING SOLAR ENERGY

// Point at which cue will be given to start reading.

   —Underlined words are to be emphasized.

   ◯◯◯ Brief pauses between phrases.

// Before starting to view the slides, turn to page 4 in the workbook and read over the objectives for this presentation. When you are ready, restart the tape and change slides as the numbers are called. If you wish, you can go back to listen to any part or view any slide again.

// Slide 2 ◯◯◯ Much of the energy from the sun is generated as light and heat. <u>Thirty-four</u> percent of this energy is reflected back into space.

// Slide 3 ◯◯◯ <u>Nineteen</u> percent is absorbed by the atmosphere surrounding the earth.

// Slide 4 ◯◯◯ And <u>47</u> percent of the sun's energy reaches the earth's surface. It may be captured for use.

// Slide 5 ◯◯◯ A solar collector can be placed on a roof to absorb the energy from the sun and convert it into heat.

// Slide 6 ◯◯◯ Collectors can be installed on older buildings as well as new ones, and a flat roof often facilitates installation.

// Slide 7 ◯◯◯ Sometimes panels of collectors are constructed on ground level ◯◯◯

// Slide 8 ◯◯◯ or attached to walls of apartment buildings.

// Slide 9 ◯◯◯ <u>Even in northern latitudes</u> and at high elevations, solar energy can be collected effectively.

// Slide 10 ◯◯◯ Now turn to the worksheet on page 4 and complete Part A. <u>Stop the tape</u>. After you have checked your answers, start again.

// Slide 11 ◯◯◯ Solar collectors look like glass covered boxes.

// Slide 12 ◯◯◯ Each section consists of four basic parts—at the top, clear glass or plastic to trap the heat; next, tubes attached to a metal plate that conduct water or other liquid to be heated; behind the heat absorber, insulation to reduce the loss of heat; and around the whole device, a weatherproof container.

// Slide 13 ◯◯◯ Solar collectors are most effective when they are oriented south and are unshaded by trees or buildings.

63

//Slide 14 ⊙⊙⊙ To capture sufficient heat, the area of a collector should equal from <u>one-third to one-half</u> of the inside floor area of a house.

//Slide 15 ⊙⊙⊙ About 100 square feet of collector are required to heat water for the average family.

[and so on] ⊙⊙⊙

purpose of a musical introduction is to settle the audience down and prepare its members to be receptive to the ideas that will be presented. The musical introduction also allows the user or projectionist to set the volume level when the presentation starts. Also, various sound effects (crowd noise, equipment sounds, animal sounds) may be required at certain points.

It is now necessary to assemble all these separate recordings onto a single tape as the final "master mix." Equipment for mixing sounds and the procedure to follow are explained on page 151. The final, edited version of the script becomes the guide sheet as you blend various sounds at appropriate times to create necessary moods. This process, as do other phases of media production, requires both creative and technical skills.

COMPLETE PHOTOGRAPHIC WORK

If for your editing you used proof copies of photographs or slides or a workprint of the motion picture, now prepare the final materials or have them prepared in the necessary number of copies. (See the appropriate sections in Part IV for instructions.) Then check the final prints with the recorded sound. You are now ready for your first formal showing.

A *final bit of advice:* You have spent much time and, no doubt, have gone to some expense to prepare your instructional media. Protect the time, money, and work against loss by preparing at least one duplicate set of all materials, both audio and visual, and file all the originals in a safe place for any future need.

Review What You Have Learned About Producing Media:

1. Here is a list of the matters that should receive attention during the production of many types of instructional media. Write a number before each one to indicate the order in which each should be given attention.
   _____ **a.** edit pictures
   _____ **b.** keep a record of scenes filmed
   _____ **c.** try out materials with an audience group
   _____ **d.** prepare a schedule for making pictures
   _____ **e.** do the filming and sound recording
   _____ **f.** mix all sound
   _____ **g.** record narration
   _____ **h.** obtain permission to use copyrighted material
   _____ **i.** prepare and film titles
   _____ **j.** work with persons who will appear in scenes
   _____ **k.** prepare final copies of media for use
   _____ **l.** have those appearing in scenes sign release
   _____ **m.** edit sound portion of script
2. Answer *True* or *False* for each statement as relating to a phase of the production process.
   _____ **a.** It is frequently advisable to repeat scenes, shooting from different positions or for different action.
   _____ **b.** "Fair use," relative to copyright, gives the teacher permission to use single copies of visuals from printed sources before the class.
   _____ **c.** The reason for making a filming schedule is to

be certain all scenes are shot in script order.
   _____ **d.** It is rarely necessary to make duplicate copies of instructional media before use.
   _____ **e.** The expression "formative evaluation" is used when materials are tested or tried out before final completion.
   _____ **f.** Reasons for using titles include calling attention to or clarifying parts of a visual.
   _____ **g.** *All* published materials are copyrighted. You need permission to use any of them.
   _____ **h.** Explain to a person appearing in a scene the purpose of the action before doing the filming.
   _____ **i.** A *documentary* approach means to shoot pictures *precisely* as specified in the script.
   _____ **j.** According to the copyright law, you can duplicate any number of photographs in a book for instructional use.
   _____ **k.** Make your own clear decision about how to film a scene without any interference from an actor or other persons.
   _____ **l.** When using a documentary approach to filming, the photographer should relate a scene being filmed to what was previously shot and try to anticipate what is coming next.
   _____ **m.** The narrator reads the narration from the two-column printed script.

_____**n.** A selection from your own music collection can be put on tape as background in a production without any restriction.

_____**o.** Music is used at the beginning of a media presentation both to settle the audience and allow the sound level to be set properly.

_____**p.** There are legal reasons for having a person sign a release when appearing in a media production.

_____**q.** Editing means _both_ to arrange scenes in order according to the script and to select the "best" _take_ of a scene.

_____**r.** Avoid introducing a scene with such an expression as "now you see . . . ".

3. Check the statements that apply _positively_ to narration:
   _____**a.** be conversational
   _____**b.** identify the picture subject at the beginning of the scene
   _____**c.** avoid pauses in narration
   _____**d.** test the narration by reading it aloud
   _____**e.** limit the use of transitions from one section to another; they confuse the audience
   _____**f.** each scene should have its own complete narration

_____**g.** choose words that communicate the intent of a message quickly to the audience

4. For what _five_ reasons might you want to field test instructional media as it nears completion?
5. As you proceed with planning and production, _checkpoints_ have been suggested:
   a. What purposes do such checks serve?
   b. List places at which checkpoints are recommended.
6. You are now ready for the production phase of the instructional media you have been planning. Refer to the how-to-do-it sections in Parts III and IV as you proceed:
   a. make a filming schedule
   b. select the persons who will appear in scenes
   c. do the filming and sound recording
   d. keep a log sheet
   e. obtain releases and copyright permission
   f. edit the pictures and narration or captions
   g. prepare titles
   h. if possible, conduct a formative evaluation of your materials—develop an evaluation form, conduct a tryout, examine results, and make any changes accordingly
   i. record narration
   j. mix sounds
   k. complete the project

REFERENCES

Working with Actors

Carlberg, Scott. "Directing Nonprofessional Video Talent." _Photomethods_ 24 (May 1981):51–53.

Copyright

Copyright and Educational Media: A Guide to Fair Use and Permissions Procedures. Washington, DC: Association for Educational Communications and Technology, 1977.
Johnston, Donald J. _Copyright Handbook._ New York: R.R. Bowker, 1978.

Juliette, Ronald A. "Copyright: Knowing the Basics." _Performance and Instruction Journal_ 20 (March 1981):32–34.

Formative Evaluation

Dick, Walter. "Formative Evaluation." In _Instructional Design: Principles and Applications._ Leslie J. Briggs, ed. Englewood Cliffs, NJ: Educational Technology Publications, 1977, pp. 311–333.
Seyer, Philip C., and Smith, Carole R. _Developing Opinion, Interest, and Attitude Questionnaires._ San José, CA: San José State University, Faculty and Instructional Development Office, 1981.

Chapter 8

USING AND EVALUATING MEDIA

- Group Use
- Individual Use
- Evaluate Results
- Copyright Your Materials

Now you are ready to use your instructional media with the intended audience. You—and your team—have spent much time in planning and in preparation. In order to insure a successful reception, it is important that you arrange for the mechanics of the presentation, whether your materials are for use with a group or for self-paced learning. Then after use, plan to evaluate the effectiveness of the materials.

GROUP USE

If you are not familiar or experienced with the use of audiovisual projection or video viewing equipment, you probably can obtain technical assistance. But if you must take responsibility for arranging a room and showing the materials yourself, here are some pointers that may make this and subsequent uses of your instructional media successful:

- If you are not familiar with the room in which the materials will be used, try to visit it in advance. Check for electrical outlets, screen placement, seating arrangements, viewing distances, appropriate placement for the projector and other equipment. Also, find out how the room lights are controlled.
- Arrange for necessary equipment—projector, tape recorder, stands, screen, video monitors, speakers, extension cord, adapter plugs, and extra projection lamp.
- If necessary, find out who will assist you with projection and other services, and give instructions accordingly.
- Rehearse your use of the materials (if possible in the setting in which they are to be used).
- Arrange materials for use in proper sequence and in proper position.

- Provide for distribution of handout materials, if appropriate.
- Provide for the proper physical comfort of the group—ventilation, heat control, light control, and other conditions.
- Through preprogram study or introductory remarks, prepare the group for viewing the materials.
- Make your presentation, using good projection techniques (centering of the image on the screen, focus, sound level, and the like).
- After the presentation, discuss the materials and, if possible, provide for related activities.

INDIVIDUAL USE

When instructional media are designed for self-paced learning, their successful use will be better guaranteed if attention is given to the following:

- Make sure the media are packaged for convenient use and correctly labeled. Color coding packages and individual items for quick identification is worthwhile.
- Prepare instruction sheets and guides for media use unless the media are part of a packaged unit of instruction like a study module.
- Provide a convenient method to assign or check out the materials for student use.
- Make available the necessary equipment with which to use the media in a convenient study area.
- Introduce students to use of the equipment.
- Have a qualified person available who can answer questions and give assistance, either while students are studying or on an assigned-time basis.

● When students complete their use of the materials, check each item for its condition and proper placement in the package.

EVALUATE RESULTS

Finally, evaluate the effectiveness of your instructional media. Recall that a suggested checkpoint during the *formative* evaluation stage of development was to gather reactions from colleagues and from a student group for improving the materials at the time (page 61). Now, after use, encourage reactions from those viewing and using the materials. Determine changes in audience and individual student behavior in terms of the objectives originally established. Accomplish this by observing specific actions of members of the audience and by administering performance or written tests to individuals. The results will allow you to answer the question, "How well have the materials done the job for which they were designed?"

Here are suggestions for items to include when evaluating the effectiveness of your instructional media:

● How well do students accomplish the objectives upon which the materials are based?
● Do reactions indicate the materials are appealing to the audience or to individual students?
● If the materials do not meet the criteria of the objective(s), or lack appeal, what revisions can be made?
● Are the arrangements for use of the materials convenient for instructor and students (applicable to individualized learning)?
● Was any difficulty encountered in using equipment?
● Were the facilities satisfactory?
● What were the development costs (professional and staff time, materials, services)?
● What are the operational use costs (staff time, materials, facilities use)?

When seeking answers to these questions, it will be necessary to design appropriate evaluation instruments and gather data from members of the audience. The instruments may take any of a number of forms, depending on the purpose of the question. Selection can be a checklist, a rating scale, or a questionnaire (consisting of either open-ended or alternative-response questions). See the references under Program Evaluation for assistance in developing such instruments.

On the basis of the results of the evaluation, revise the materials as necessary. Repeat the evaluation periodically to maintain a standard of effectiveness. Keep the materials up to date by adding or substituting new content when appropriate and eliminating the obsolete. Only by revision will your instructional media be kept timely and maintained at your standards of quality and effectiveness.

COPYRIGHT YOUR MATERIALS

If you have developed instructional media that you would like to protect, copyright it. It is a simple matter to let other persons know that this property is yours. Put a copyright notice on it. Use these three elements:

1. "Copyright" or the symbol ©
2. Name of the copyright owner or organization
3. Year of production or publication

A copyright notice should read:

● © Your name 1985

For a sound recording, use these elements:

1. The symbol p
2. The year of production or publication
3. The name of the copyright owner

The notice for a recording looks like this:

● p 1985 Your name

If a tape recording is part of a slide/tape, filmstrip/tape or similar program, then the recording is copyrighted as part of the visual materials. The © symbol is used.

This notice should appear on the master material and all copies, generally on the title frame or first picture and also on the container label. If you plan to sell or otherwise distribute copies without this notice, you have no protection from anyone who duplicates the materials. (See the discussion about using copyrighted materials on page 57.)

Although this notice appearing on your materials is generally all that might be necessary to establish your rights, there are legal advantages to registering the copyright formally with the Copyright Office. This can be especially important if you anticipate selling or otherwise widely distributing your materials. You cannot file suit for infringement until your copyright is registered.

An application form for copyright protection is available from the Information and Publications Section, Copyright Office, Library of Congress, Washington, DC 20559. Request the form for the type of material that you wish to register. Two copies of the work and a fee of $10 are submitted. The term of copyright ownership is the life of the author plus 50 years.

Review What You Have Learned About Using and Evaluating Media:

1. Assume you are ready to use your materials for a group showing. List the factors you should consider in preparation for use.

2. For what matters would you like to gather information in evaluating the effectiveness of your materials?
3. What procedure would you follow to copyright instructional media you produce?

REFERENCES

Using Media

Audiovisual Projection. Publication S-3. Rochester, NY: Eastman Kodak Co., 1982.

Brown James W., et al. *AV Instruction: Technology, Media, and Methods.* 6th ed., New York: McGraw-Hill, 1983.

Effective Projection. Slide/tape presentation V10-23. Rochester, NY: Eastman Kodak Co., 1982.

Heinich, Robert, et al. *Instructional Media: The New Technologies of Instruction.* New York: Wiley, 1983.

Program Evaluation

Bloom, Benjamin S., et al. *Handbook on Formative and Summative Evaluation of Student Learning.* New York: McGraw-Hill, 1971.

Guba, Egon C., and Lincoln, Yvonna S. *Effective Evaluation.* San Francisco: Jossey-Bass, 1981.

Henerson, Marlene E., et al. *How to Measure Attitudes.* Beverly Hills, CAL: Sage, 1978.

Seyer, Philip C., and Smith, Carole R. *Developing Opinion, Interest, and Attitude Questionnaires.* San José, CAL: Faculty and Instructional Development Office, San José State University, 1981.

Chapter 9
PLANNING AND PRODUCTION SUMMARY

The successful planning and production of instructional media follow a logical sequence. For some types this is a detailed step-by-step procedure; for others it may be simplified and brief. Also, the degree to which any materials are formally completed (including titles, music, duplicates, laboratory services, and the like) is determined by the specific purposes to be served. For example, a short video recording designed to develop a single concept for direct instruction might be used with limited editing and without extensive titles other than an identifying label.

A checklist and outline of steps is given in tabular arrangement in Table 9-1. It covers all possible steps; but your purposes and the treatment you give the subject may permit you to omit some of these steps. The numbers in the various columns refer to pages containing a discussion of the things to be done at each step. Study these pages as you develop the planning and prepare to carryout the production of specific materials.

Table 9-1 **SUMMARY OF PLANNING AND PRODUCTION STEPS FOR INSTRUCTIONAL MEDIA**

Step	Printed Media	Display Media	OH Transparencies	Audio Recordings	Slide Series/ Filmstrips	Multi-Image	Video-Recordings	CBI
Planning								
1. Express your idea	28	28	28	28	28	28	28	28
2. Develop the objectives	29	29	29	29	29	29	29	29
3. Consider the audience	30	30	30	30	30	30	30	30
4. Get some help	30	30	30	30	30	30	30	30
5. Prepare the content outline	32	32	32	32	32	32	32	32
6. Select the medium	42	42	42	42	42	42	42	42
7. Write the treatment	—	—	—	—	48	48	48	48
8. Make a storyboard	—	48	48	—	48	48	48	48
9. Flowchart preparation	—	—	—	—	—	—	—	249
10. Develop the script	159	—	—	50	50	50	50	50
11. Prepare the specifications	53	53	53	53	53	53	53	53
12. Schedule the picture taking	—	—	—	—	55	55	55	55
Production								
13. Take the pictures	—	—	178	—	55/196	55/215	226	253
14. Process the film	—	—	—	—	99/201	99	—	—
15. Edit the pictures	160	—	—	—	58/203	58	236	255

Table 9-1 **(continued)**

Step	Printed Media	Display Media	OH Trans-par-encies	Audio Record-ings	Slide Series/ Film-Strips	Multi-Image	Video-Re-cordings	CBI
16. Edit narration and captions	160	—	—	58/151	58	58	58	255
17. Prepare artwork, titles, and captions	161	106	175	—	61	61	61/234	255
18. Record narration and sounds	—	—	—	149	62/205	62/217	236	—
19. Mix sounds	—	—	—	151	62	62	62	—
20. Sync programming	—	—	—	—	205/210	217	—	—
21. Prepare final copies	164	—	177	154	203/209	219	243	—
Followup								
22. Prepare for use of the materials	168	172	190	195	207/210	219	244	258
23. Use the materials	—	172	190	195	66	66	66	66
24. Evaluate for future use	67	67	67	67	67	67	67	67
25. Revise as necessary	67	67	67	67	67	67	67	67
26. Copyright materials	67	67	67	67	67	67	67	67

Part Three

FUNDAMENTAL PRODUCTION SKILLS

Chapter 10

PHOTOGRAPHY

- Your Camera
- Camera Lenses
- Camera Settings
- The Film
- Correct Exposure
- Artificial Lighting
- Camera Shots and Picture Composition
- Close-up and Copy Work
- Processing Film
- Making Prints

Advances in photographic equipment allow you to give more attention to the important job of taking pictures, rather than being concerned with routine matters. Your emphasis now can be on composing the picture and catching the decisive instant of action.

While recognition is given in this chapter to many newer developments, an understanding of the fundamental elements of photography will help you to make choices and decisions that can result in high-quality visual materials, prepared with a sense of pleasure and satisfaction.

YOUR CAMERA

Although cameras can be classified into groups according to film size or operating characteristics, all cameras are basically similar and include seven essential parts:

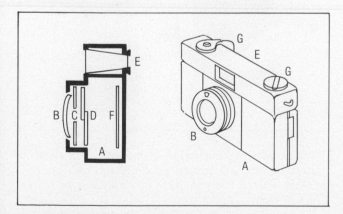

A. Light-tight enclosure (camera body)
B. Lens
C. Lens diaphragm
D. Shutter

E. Viewfinder
F. Film-support channel
G. Film-advance and rewind knobs

In addition, all cameras except the simplest have some means for changing the distance between the lens and the film plane in order to focus the image on the film. The more expensive kinds have features and attachments that offer greater versatility when filming. These include manual or automatic methods of changing lens-diaphragm openings, shutter speeds, and focus. Adjustable diaphragms and shutters are desirable because they can be adapted to changing light conditions and to different kinds of subjects. Lens quality differs over a wide range and has important bearing on the price of a camera.

Small-Format Cameras (35mm)

This is the most widely used type of camera for preparing visual materials, principally as slides for slide series or multi-image presentations, and for conversion to filmstrips. Color and black-and-white prints can also be made from appropriate 35mm film.

These cameras have either **window-type viewfinders** or **a prism and mirror system** that allows the subject to be framed and focused through the camera lens for more accurate viewing. The latter type of camera is called a **single-lens reflex** (SLR) and is especially convenient for close-ups and copy work. This is the type of camera generally used for preparing instructional media. Lenses and shutters of higher-priced cameras can be adjusted over a wide range of settings, which broadens the possibilities for taking pictures under varying conditions.

Some models of single-lens reflex cameras have automated features, controlled by a microchip, that permit exposure and focus settings to be made rapidly and precisely. There are models of window-type viewfinder cameras that are very compact, fitting easily into a coat pocket.

Quick operation is a feature of 35mm cameras, and there is an economy with film in that many exposures can be made on a single roll before reloading. For more details on the characteristics of 35mm cameras, see page 196.

Medium-Format Cameras

These cameras use a larger size film (generally 120) than 35mm cameras, thus providing $2\frac{1}{4} \times 2\frac{1}{4}$ inch (6×6 cm) or $2\frac{1}{4} \times 2\frac{3}{4}$ inch slides, color negatives, or black-and-white negatives. The larger size may be preferable when considerable enlargements and exceptionally sharp images are required. One disadvantage of this slide size is the higher cost and, often, the unavailability of projectors

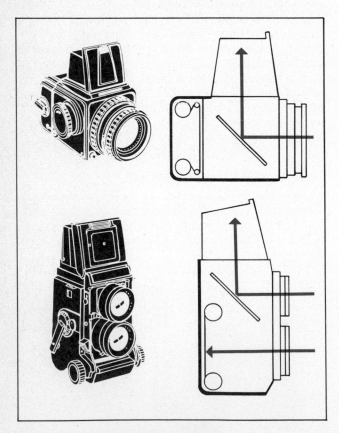

There are two subcategories in this group. Both focus the image, when viewing, on a ground glass surface. The most

useful type has a **single lens,** with a mirror that reflects light to the ground glass. The mirror moves upward to clear the film plane just before the shutter opens as in a 35mm single-lens reflex camera. An advantage of most of these cameras is the ease with which the camera back can be removed and interchanged. This allows the use of different films (black-and-white, color reversal, or instant Polaroid) without unloading the roll presently in the camera.

The other type of medium format is a **twin-lens** camera. Its lower lens (*A*) is for taking pictures, while the upper one (*B*) carries light rays to the ground glass. The parallax problem caused by the separation between viewing and filming positions (page 96) is a limitation of the twin-lens camera when used for close-ups.

On the other hand, with the twin lens the subject can be viewed as the shutter is released during exposure. Because of mirror movement in the single lens camera, at the instant of exposure, viewing is eliminated momentarily.

The method of viewing the subject as it appears on the ground glass of these cameras can be another advantage because the image is generally the same size as it will appear on the film, although on certain cameras it may be reversed left to right. Image size may be especially important when accurate composition or careful positioning of a subject for a multi-image slide presentation is required.

Large-Format Cameras

These cameras include press and view-type cameras in many sizes (generally $4'' \times 5''$ or larger). They use cut sheets of film rather than rolls, thus permitting single pictures to be taken and immediately processed. The larger negative permits extreme enlargement and ease of retouching. These are ideal cameras for copywork and close-up photography since accurate viewing and focusing take place on a ground glass surface directly in line with the lens.

These types of cameras usually are used in a studio or copy room. Process cameras (page 138) for precision copywork fall into this category. They are bulky and the film for use with them is expensive.

On page 185 the preparation of photographic overhead transparencies with a sheet-film camera is described.

Their main use is for preparing large-size negatives from paste-ups (page 135) and other art work. From a negative, a metal plate is made for offset printing (page 136).

Self-Processing Film Cameras

The outstanding feature of this type of camera (manufactured by Polaroid and Eastman Kodak) is that the film is developed automatically into a print in the camera, thus eliminating the need for darkroom work. The results of your picture taking are seen almost immediately. Prints in both black-and-white and color, and large size slides in black-and-white, can be produced in one minute.

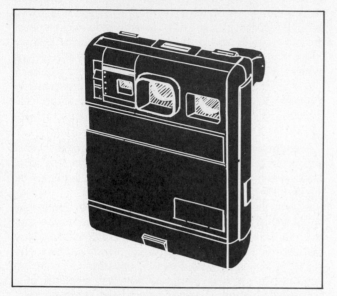

The film consists of light-sensitive negative film and a nonsensitive positive stock (paper or film). To the latter are affixed sealed pods, one for each picture, which contain developing chemicals. After a picture is taken, a pull on the tab (or motorized movement) advances the film, rollers squeeze the developer from a pod, allowing gelatinlike material to spread between the negative and positive stock.

Development takes place quickly. The result is a print or slide ready for use.

Instant 35mm color slides can be processed in a special unit after a particular film is exposed in a *regular* 35mm camera. See page 101 for further details of this latter procedure.

Motion-Picture Cameras

The same features and functions essential in the operation of other cameras are necessary in all Super 8mm and 16mm motion-picture cameras. The differences are in the threading of film and its movement through the camera, or within a cartridge, and the action of the shutter behind the lens.

The features of video cameras and their operating principles are described on page 222.

Review What You Have Learned About Camera Types:

1. Examine your own camera. Locate the seven parts as shown on page 73.
2. If your camera is a 35mm-type, is it single-lens reflex or one with a window-type viewfinder?
3. What type of camera would best fit each situation?
   a. Used to prepare photographs quickly and easily immediately after the pictures are taken
   b. Used as a roll-film camera for accurately viewing a subject in same size and appearance as it will be filmed
   c. Used to produce a quantity of color slides
   d. Used to make greatly enlarged close-up pictures of small objects on pieces of film

---

CAMERA LENSES

If you have an inexpensive still-picture camera, you probably have only the standard lens that came with it. But many of the higher-priced cameras permit removal of the original lens and its replacement with other lenses for special purposes. There are advantages in using different lenses to film various subjects under certain conditions.

The major feature of a lens is its **focal length.** This term refers to the distance measured from the center of the camera lens to the film plane within the camera when a subject at a far distance (infinity) is in focus.

Lenses of different focal lengths form different-size images on film (when the camera is used from the same position). As you substitute a lens for another one having

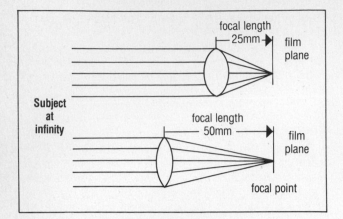

a *longer* focal length, a *larger* image is projected to the film. Thus, on the film you record only a portion of the image you had with the first lens. If you use a lens of *shorter* focal length, a *smaller* image reaches the film and you record a *greater* area of the subject.

The focal length of a lens is measured in millimeters (mm). A **normal** focal length has been established for various cameras. For the lens of a 35mm camera the normal focal length is 50mm. A lens with a *larger* or *longer* focal-length number than that of the normal lens is called a **telephoto** lens. A lens with a focal-length number *smaller* or *shorter* than the normal lens is called a **wide-angle** lens. For a 35mm camera the relationship illustrated below holds true.

By selecting from lenses with different focal lengths, you can take pictures more easily under difficult conditions. For example, a wide-angle lens is useful when you cannot move far enough away from a subject to shoot with a normal lens. When you cannot get close to a subject or do not want to, a telephoto lens may be helpful.

Also, recognize that some unusual optical impressions can be caused by both wide-angle and telephoto lenses. A person viewing a wide-angle picture will get the impression that the camera was farther away from the subject, an exaggerated feeling of depth. With extreme wide-angle lenses, objects close to the camera will appear proportionately larger than they really are. Also, near the edge of a picture straight lines are bowed, resulting in a distorted picture.

A telephoto lens may give the impression of compressing the distance between objects in a scene so that the foreground and background elements appear very close together.

A **zoom** lens, which in effect is a series of lenses within one housing, allows you to rapidly change the lens focal length along a predetermined range. Such a lens may be marked 35–70mm, 75–200mm, and so forth.

In using this lens, you choose your subject by zooming in or out without moving the camera with respect to the subject. With this lens on a video or motion picture camera, you can also continuously change the size of a subject by zooming while filming. A slow zoom on occasion to center on action or to show relationships between elements of a scene can be effective as well as dramatic. However, there

Field with Wide-angle Lens (28mm)

Field with Normal Lens (50mm)

Field with Telephoto Lens (135mm)

Wide-angle Lens

Normal Lens

Telephoto Lens

Varying Perspectives of a Subject with Different Focal Length Lenses

is a tendency to overuse the zooming feature, which can be distracting to the viewer.

When using a zoom lens, always focus your subject with the lens set at the *longest* focal length (close-up view of the subject) and with the lens aperture at its *smallest f/* number (largest opening). By following this instruction you will be sure to keep the subject in focus for all positions within the zoom operation.

In summary, as compared with a normal lens, a lens with a *longer* focal length has a *larger* millimeter number and is called a *telephoto* lens; while a lens with a *shorter* focal length has a *lower* millimeter number and is called a *wide-angle* lens. A *zoom* lens has focal lengths that can be varied over a certain range.

Special lenses, called **macro** lenses, are available for close-up and copywork. They are designed to reproduce details more sharply than do other lenses (page 97).

By setting a zoom lens at various focal lengths, you have access to different perspectives. The zoom lens will change both the **magnification** and **depth-of-field** (page 76) relationship between foreground and background objects in a scene.

Review What You Have Learned About Camera Lenses:

1. Make a sketch to illustrate the meaning of focal length for a lens. Label the parts.
2. What numerical measure is used to indicate focal length?
3. Into what three groups are lenses placed?
4. Which lens would you select for each situation?
   a. General picture taking of a group of children on a playground
   b. General scene in a factory showing as much of the equipment as possible
   c. As close a view as possible of a track start on the athletic field when filmed from the grandstand
   d. Two by two inch slide of a bird in a nest which is in a bush at some distance
5. Into what group(s) would you place the lens(es) you have with your camera?

CAMERA SETTINGS

If you have an automatic-setting camera, you may feel you have little need to recognize and understand the purposes for the settings made on an adjustable camera. But your camera may allow you to override the automatic feature and, for special situations, make settings to ensure correct exposure. For example, when filming a *backlit* subject (sun or other light source *behind* the subject, with much of the subject in shadow), or when making close-ups of an object on a light-colored background, you face situations in which the general exposure determined by the camera may be incorrect.

There are three essential settings on adjustable cameras —lens diaphragm, shutter speed, and focus for subject distance.

The amount of light that enters a camera is controlled by the lens diaphragm and the shutter, working together. The *lens diaphragm* controls the *amount* of light that can reach the film; the *shutter* controls the *length of time* the light can reach the film.

Lens Diaphragm

Light enters a camera by passing through the lens. The intensity of light entering is controlled by a metal diaphragm, which is located either directly behind the lens or between two elements of the lens. The diaphragm acts somewhat like the iris of an eye. It is always open, but its size can be changed to control the intensity of light passing through the lens.

Lens-diaphragm settings are indicated by a series of

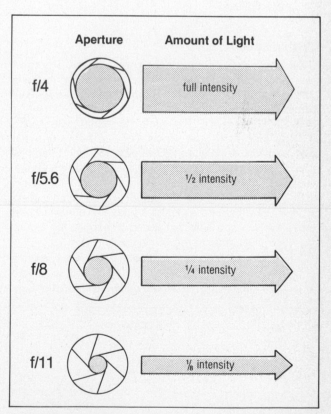

Aperture	Amount of Light
f/4	full intensity
f/5.6	½ intensity
f/8	¼ intensity
f/11	⅛ intensity

numbers—4, 5.6, 8, 11, 16, . . .—called **f/numbers** or **f/stops.** *The larger the f/number the smaller the opening.* A lens setting of *f/11* admits only *half the amount of light* passed by an *f/8* setting.

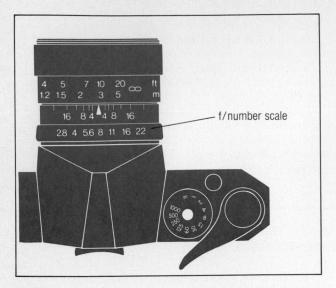

f/number scale

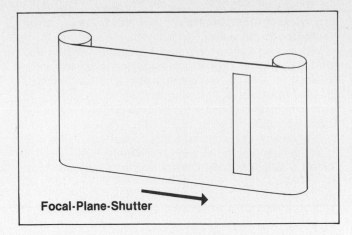

Focal-Plane-Shutter

Thus, adjacent numbers in the series admit light in the proportion of 2 to 1 (permitting the passage of *twice as much light* or *half as much light*).

An f/number expresses the relationship between the focal length (page 76) and the diameter of the lens's aperture. Thus, for f/11 the opening has a diameter one-eleventh that of the lens's focal length. In commonly used terminology, a lens with a low f/number (f/2 or f/1.4) is considered a "fast" lens. This means that the camera equipped with such a lens is capable of picture taking under low light conditions.

Shutter Speed

A second camera setting is the shutter speed. The camera shutter is similar to the eyelid as it closes and reopens rapidly. In the camera the shutter remains closed until opened to permit the lens to "see."

The most common type of shutter is placed between the elements of the lens and consists of thin pieces of metal that can be moved to allow light to pass for various lengths of time. (See illustration below.) The other method of permitting light to reach the film is with a cloth or metal curtain located just in front of the film plane. The curtain has a slit in it, and at the instant the picture is taken the curtain moves across the film at a selected speed to expose the film to light. (See illustration at top of next column.)

Generally, shutter speeds are measured in fractions of a second: $\frac{1}{2}$, $\frac{1}{4}$, $\frac{1}{8}$, 1/15, 1/30, 1/60, 1/125, and so on. On the camera they are printed as whole numbers instead of frac-

tions, the number shown being the denominator of the fraction. A shutter speed of 1/60 second is *slower* than a speed of 1/125 second, admitting light for *twice as long a time.*

Thus, adjacent speeds are in the proportion of 2 to 1 (permitting the entrance of light for approximately *twice as much time* or *half as much time*—1/60 is twice as much as 1/125; but 1/60 is half as much as 1/30).

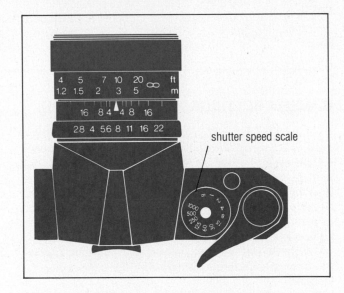

shutter speed scale

On your camera you may have many speeds from which to select. For general scenes, a shutter speed of 1/60 or 1/125 is suggested. But when a moving subject is to be filmed, the choice of a shutter speed is dependent on the speed of movement, the distance from camera to subject, and the direction of movement relative to the camera.

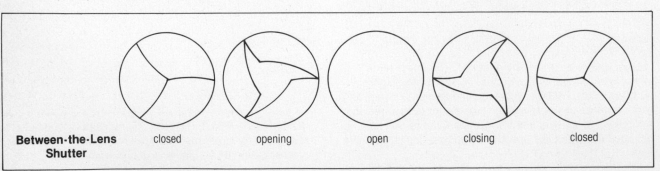

Between-the-Lens Shutter   closed   opening   open   closing   closed

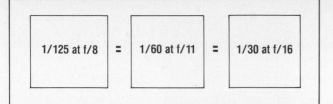

| 1/125 at f/8 | = | 1/60 at f/11 | = | 1/30 at f/16 |

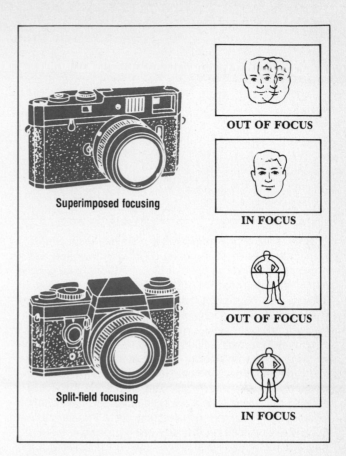

Superimposed focusing

OUT OF FOCUS

IN FOCUS

Split-field focusing

OUT OF FOCUS

IN FOCUS

Since both the adjacent lens-diaphragm settings and the adjacent shutter speeds are in the 2-to-1 proportion, they may be used in various combinations to allow the same amount of light to reach the film. As you will see, the selection of such combinations is important to obtain specific effects.

Focus

The third setting on many cameras is for focus. With your camera you may be able to focus the image on the ground glass. Or your camera may include a built-in **rangefinder** coupled to the lens, which, upon proper adjustment, automatically sets the lens for the correct subject-to-camera distance.

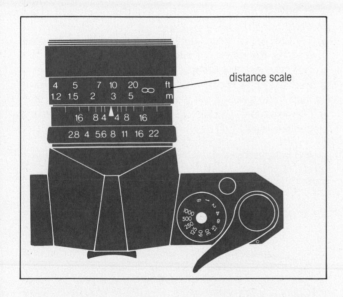

distance scale

The rangefinder may be either of two kinds: **superimposed image** or **split-field.** The former will show two images unless the focus is correct, at which point the two images are superimposed to make a single image. The split-field rangefinder will show two half images, one below the other, unless the focus is correct, at which point the two halves are matched together to make a complete image. See the illustration at the top of the next column for a visualized explanation.

Some cameras now include an *automatic* focusing feature. When the focus is adjusted, an electronic pulse, like a small sonar beam, is emitted from the camera. It strikes the subject toward which the camera is pointed. The time for the pulse to reach and then reflect from the subject is measured. This period of time is translated into a distance (feet or meters), and the camera lens is set automatically.

Depth-of-Field

While lens-diaphragm openings and shutter speeds work together to admit various amounts of light into the camera, lens-diaphragm openings and distance settings can also be coordinated to get sharp pictures.

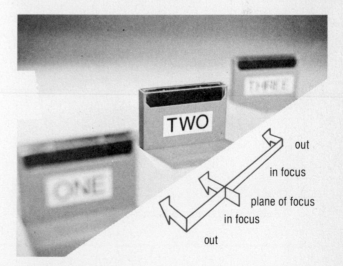

out
in focus
plane of focus
in focus
out

In this scene the camera is focused at 3 feet, but points both closer and farther away than 3 feet also appear sharp. This distance from the closest sharply focused point to the farthest spot in focus is the **depth-of-field** of the lens at the f/number used. To get a sharp picture, a photographer must have the subject in this field.

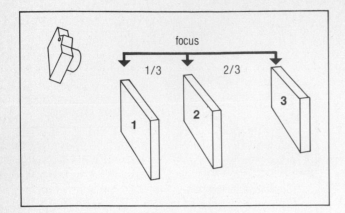

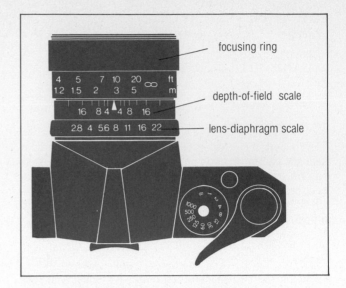

Of the total depth-of-field within a scene, about one-third is included ahead of the point of actual focus and about two-thirds beyond the point of focus. Therefore, to get the maximum value from the depth-of-field factor, focus your camera lens on a point that is about one-third of the way into a scene. If your camera gives you more exact information, use it.

Your camera may include a depth-of-field scale adjacent to the focusing ring. Refer to this scale to determine the depth-of-field for any combination of lens setting and distance to the plane of focus. The illustration that follows shows how this scale is to be read in combination with the distance scale on the focusing ring. Note that its graduations are like those on the lens setting scale: 4, 8, 16, 22 (the 2.8, 5.6, and 11 points are omitted for legibility in the scale used in the illustration). It should be easy to observe the two important facts. First, whatever the lens setting, it provides greater depth-of-field at far distances than at near distances. Second, whatever the distance, it provides greater depth-of-field at higher-numbered lens settings than at lower-numbered ones.

In summary: On adjustable cameras three settings must be made—lens-diaphragm opening (f/number), shutter speed, and distance. Study their relationships carefully and learn how to use them. Also, remember to ensure sharp pictures:

- Try to select a lens f/number in the middle of the scale (11, 16) for maximum picture sharpness.
- Hold the camera steady.
- Use a sturdy tripod whenever shutter speeds slower than 1/30 second will be used.
- Squeeze, don't punch, the shutter release, or use a cable release.

Scene with Limited Depth-of-Field — f/4

Scene with Maximum Depth-of-Field — f/16

Review What You Have Learned About Camera Settings:

1. If a lens setting of f/8 permits a certain amount of light to pass into a camera, then how much light does a setting of f/5.6 admit?
2. What shutter speed would you select to "stop" the action of (a) a person diving off a board, (b) a car driving past at 25 mph, or (c) a child walking by you?
3. If f/11 and 1/125 second are correct exposure settings, but you want to increase the depth-of-field by two f/

80

stops, what camera setting would you now use?

4. What is the depth-of-field on the lens on page 80?

5. Does a lens setting of *f*/4.5 permit *greater* or *less* depth-of-field as compared to that of *f*/11?

THE FILM

Selection of film is determined by a number of factors:

- Kinds of subjects, such as general scenes, action shots, or close-ups of fine details
- Lighting conditions, such as daylight, floodlights, flash, or low-level room light
- Use for materials, such as enlarged photographs, slides, overhead transparencies, or motion pictures

In addition, choices are based upon characteristics of different films, including:

- Degree of light sensitivity (film speed or exposure index, see explanation following)—*slow* films, such as Kodachrome 25 (index 25), Kodalith Ortho Type 3 (index 8); *moderate-speed* films, such as Ektachrome 160 (160) and Plus-X (125); *fast* films, such as Tri-X (400) and Kodacolor VR 1000 (1000)
- Reproduction quality of colors in a subject
- Useful exposure range to reproduce a range of tones from highlights to shadows

The terms **film speed, exposure index,** and ASA (American Standards Association), or ISO (International Standards Organization), *speed* refer to the degree of light sensitivity of a film or the degree of speed at which the film will create an image. These expressions are used interchangeably and are scaled by a number assigned to each film. The number is a relative one; the higher the ASA/ISO rating the less time and/or less light is required for proper exposure. A film with a speed of 100 requires less light for proper exposure than one with a speed of 64, and vice versa.

Information about a few types of black-and-white and color films is summarized in Table 10-1. The data are correct at the time of writing; but changes and new developments are to be anticipated. Check carefully the data sheet packaged with your film for the latest assigned exposure index and other details.

Film is available in various length rolls for 35mm cameras. Also, many types of film can be purchased in rolls of up to 100 feet in length. This permits you to spool off small amounts or full rolls and load your own film cassettes. To do this you will need a bulk film loader and reusable 35mm film cassettes.

A Bulk Film Loader

Using bulk film has certain advantages: cost of film is reduced appreciably; waste is lower, because you can measure off only what will be needed for a job; and film consistency is assured, because you can use the same film emulsion lot for a project.

Table 10-1 **CHARACTERISTICS OF SOME WIDELY USED FILMS**

FILM	EXPOSURE INDEX (ASA)	LIGHT SOURCE	SUGGESTED USE
PANATOMIC-X	32	any	Copying fine detail for extreme enlargements (black-white)
KODALITH ORTHO TYPE 3	8	flood	Copying high-contrast black-and-white print and artwork
TRI-X	400	any	Subjects under low light conditions (black-and-white)
KODACOLOR VR	100	daylight	General purposes to prepare color prints and/or slides in quantity
KODACHROME 40 TYPE A	40	flood	Close-up and copy work for color slides
EKTACHROME 200	200	daylight or fluorescent	Fast action or dimly lit subjects for color slides

Black-and-White Film

Black-and-white films are inexpensive and simple to process and print. Their primary use is in making enlargements from negatives onto paper or film, the latter for overhead transparencies.

Black-and-white films are available in both high-contrast and continuous-tone emulsions. The former is used for photographing black line material on a white background, while the latter reproduces subjects in all tones of gray as well as black and white.

Color Negative Film

Color negative film is versatile since it may serve as negative for color prints, for black-and-white prints, and for positive color slides. On a color negative the colors of the subject are complementary to their normal appearance (yellow in place of blue, magenta for green, and cyan or blue-green for red). Kodacolor VR100, VR200, VR400, and VR1000 are the principal color negative films available. The negative color films, available in sizes for most cameras, are more expensive than black-and-white, but they can be processed with prepared kits, thus reducing the cost when a number of rolls are handled.

Color Reversal Film

After exposure and processing, color reversal films become positives—slides, filmstrips, or motion pictures. In processing, the image on the film is *reversed* to make a positive picture.

2 × 2 inch (35mm film)

2 × 2 inch (126-size film)

Color reversal films are supplied for 35mm and other standard size roll-film cameras. They are available in a range of film speeds and in various types, each type designed for a specific light condition—daylight or artificial light (tungsten). The light supplied by these sources differs in "color temperature," a characteristic measured in de-grees Kelvin (°K). Daylight contains more blue wave-length rays, while artificial light contains more rays from the red end of the spectrum. Therefore, each film type has an emulsion balanced for a specific color temperature, such as 6000°K for daylight film or 3400°K for Type B film (used under photoflood lights).

Correction filters are employed to permit use of a film under lighting conditions that differ from its Kelvin rating, or when duplicating a slide to correct the overall color of a scene. A filter will pass light of its own color and block out light of its complementary color. For example, if you use daylight color film indoors under artificial (tungsten) light, place an 80A filter (blue) over the camera lens to correct the excess red color of the artificial lighting by filtering out yellow.

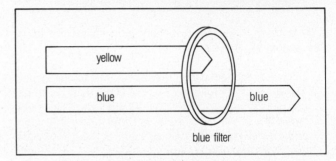
yellow

blue   blue

blue filter

Because a correction filter reduces the overall light reaching the film, the exposure must be increased two *f/* stops if a meter, separate from the camera, is used. If the camera has an internal meter, set the ASA/ISO normally and the meter will compensate for the loss of light. Thus, it is better to use the film appropriate to the light conditions so as not to lose effective film speed.

Color prints can be prepared directly on color printing paper from color slides (page 103). It is not necessary to make a color negative from a slide before a color print can be made.

All film packages are stamped with an expiration date. For best results, use the film before this date since film quality deteriorates with time. The changes are accelerated with high temperature and high humidity. Therefore, store unopened film to be kept for a long period of time under refrigeration (50°F or lower). To avoid damage from moisture that may condense on cold film, always allow the package to warm up to room temperature before being opened. This can vary from 2 to 8 hours. Then process exposed film as soon as possible after exposure to avoid deterioration of the latent image on the film.

Review What You Have Learned About Film:

1. Most reversal color films are available in two types. The selection of the type to use depends on what major factor?
2. If you wanted to have both color slides and enlarged color prints from the same subject, what film would you select to use?
3. What does the number 125 mean when referred to as the ASA or ISO rating?
4. What is the reason for selecting a film with an exposure index of 200 in preference to one with an index of 64?
5. On what principle of light does the selection of a filter depend?

Black-and-White negative

Enlarge Black-and-White Print

Black-and-White Slide

Color Reversal Slide

Color Print from Color Slide

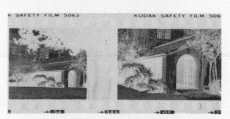

Color Negative

Color Slide

Print in Black-and-White or Color

CORRECT EXPOSURE

How do you put together information about f/numbers, shutter speeds, and film characteristics to get correct exposure? The simplest method is to refer to the data sheet packaged with the film, on which a table gives you *general* guides to proper exposure.

KODACHROME 25 Film (Daylight)

DAYLIGHT EXPOSURE: Cameras with automatic exposure controls—Set film speed at ISO (ASA) 25. **Cameras with manual adjustments—**Determine exposure setting with an exposure meter set for ISO (ASA) 25 or use the table below.

Bright or Hazy Sun on Light Sand or Snow	1/125 sec f/11
Bright or Hazy Sun (Distinct Shadows)	1/125 sec f/8*
Weak Hazy Sun (Soft Shadows)	1/125 sec f/5.6
Cloudy Bright (No Shadows)	1/125 sec f/4
Open Shade or Heavy Overcast	1/60 sec f/4

For general pictures to be taken under bright sunlight, you can use the *f/16 rule.* Set the lens at f/16 and the shutter speed at (or close to) the ASA/ISO rating of the film. For example, using Ektachrome 200, the exposure would be f/16 at 1/250 second.

But what about situations involving particularly dark or unusually light-colored subjects or backgrounds? What corrections should you make when the sun is behind the subject rather than over your shoulder? How do you determine camera settings when doing copy work or when using floodlights? These are common problems and their solutions may require more information than that provided by the data sheet tables or a simple rule.

Using a Light Meter

The most accurate method for determining exposure is with the use of a photographic light meter. Your camera may have a built-in meter that automatically sets the lens f/stop, or gives you an indication of where to set the aperture. Also, you can use a separate, hand-held light meter.

Light is measured by an exposure meter in either of two places: at the place where the subject is or at the place where the camera is. The **incident-light method** measures the light where the subject is, with an **incident light meter** held at or near the subject's position and pointed toward the camera. The **reflected-light method** measures the light where the camera is, with a **reflected light meter** held at or near the camera's position (or as part of the camera) and pointed toward the subject. Another way of describing the two methods is to say that the incident-light method measures the light that *falls on* the subject whereas the reflected-light method measures the light that reflects *off* the subject.

Some meters measure light by only one of these methods; many have attachments or components that permit measurements to be taken in either way.

If your camera has a built-in meter it is used as a reflected light meter. A photoelectric cell is located at a separate window, or, as with a 35mm single-lens reflex camera, behind the lens. With the latter type, the amount of light that actually reaches the film is measured.

When you use light meters that have been described above, overall or average brightness is measured to provide a general exposure. Should you want to expose for a small, critical part of a scene that may be lighter or darker than the overall area, then a selective reading should be taken from it—for example, a piece of stainless steel equipment on a light-colored background, or the face of a person dressed in a white uniform. A reflected-type meter can be held close to the important part of the subject (avoid reading a shadow cast by the meter), or you can use a special reflected-type, a **spot meter.** This meter has an optical viewing system with a narrow angle of light acceptance (1 degree as compared with 30 degrees for a regular reflected-light meter). It can be used from the camera position to measure accurately the light from a small portion of a scene.

An exposure meter measures light through its photoelectric cell as shown by a needle on a light level scale. With a hand meter, you then use this measurement to compute

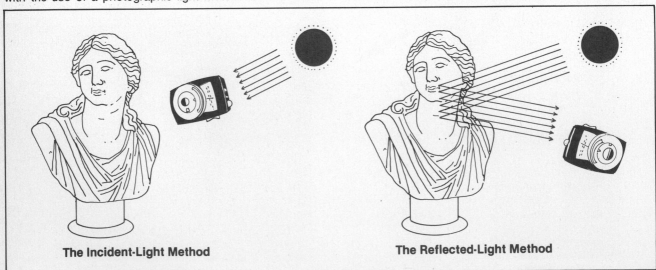

The Incident-Light Method **The Reflected-Light Method**

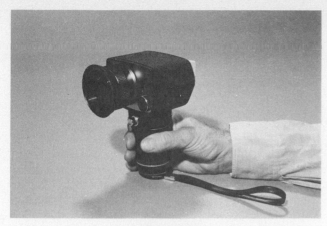

A Spot Meter

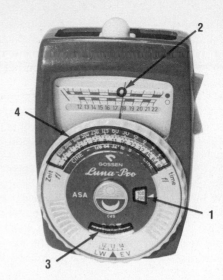

f/number and shutter speed on a dial scale. With a built-in meter, you directly set the lens or shutter speed to center the needle on the light level scale. For some cameras, an automatic setting for either the f/number or shutter speed is made that correlates with either an f/number or shutter speed you have selected already.

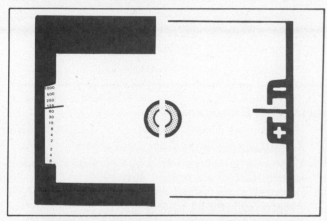

Two Kinds of Exposure Indications Appearing in Viewfinders of Single-lens Reflex Cameras.

To use a hand-held light meter, follow these steps:

1. Note the exposure index of the film you are using (as 160 for Ektachrome 160) and set the meter's exposure-index scale at this number.
2. Take your light level reading and note the light level indicated by the needle.
3. Adjust the movable scale until its pointer points to this light level.
4. You will now find lens openings and shutter speeds matched on two dials. Select the pair you will use.

The illustration at the top of the next column presents an example of the appearance of an exposure meter after it has been set as directed above.

Because shutter speed and lens opening work together, as has been explained, to admit the proper amount of light to the film, correct exposure is shown by any paired values for f/number and shutter speed. Can you read the paired figures in the illustration? If you find f/16 paired with 1/30,

then other pairs will be f/11 at 1/60, f/8 at 1/125, f/5.6 at 1/250, and so on.

Now which pair should you choose? Your selection is based upon answers to two questions:

1. How much movement is there in the scene? (Recall the examples of shutter speed selection on page 78; the faster the motion, the higher the necessary speed.)
2. How much depth-of-field is desired? (Recall the discussion of depth-of-field, page 79; for greater depth-of-field, use a setting with larger f/number.)

Now apply some of these relationships. Can you explain why the particular exposure settings were selected for these two examples?

f/4 at 1/500 second

f/22 at 1/15 second

A good exposure meter is a worthwhile investment. Use it *carefully* for correct exposure determinations.

● When using an incident light meter, hold it in the center

of the scene and aim the white cone toward the camera.

- When using a reflected light meter, aim it at the subject, especially for exterior scenes. Do not tip the meter and record too much light from the foreground or from the sky.
- Average the readings of a reflected light meter taken from various objects in a scene; avoid taking readings with a reflected light meter from very bright or very dark parts of the scene, unless you wish to expose especially for such a part.
- With an incident light meter, when a subject is back-lighted (that is, when the sun or other light source is behind the subject and parts of the subject facing the camera are in shadow), open the lens to one additional f/number beyond that indicated on the meter.
- Use an incident light meter to determine the evenness of illumination in a scene when photoflood lighting is used.
- When filming under photoflood lights, follow the additional suggestions for using a meter that you will find on page 89.
- When doing close-up and copy work, make use of the exposure information on page 98.
- When using color reversal film, an underexposed scene will appear darker overall, while an overexposed scene will appear thin and washed out.

- Follow other suggestions found in the instruction manual accompanying your light meter.

Keep a careful record of light and subject conditions, choice of exposure, and the quality of resulting pictures. From this record you can judge how well your meter is serving you and establish the modifications you must make in using it.

A final word about automatic-setting cameras. On such cameras an exposure meter is coupled directly to the lens diaphragm or shutter, and as light strikes this meter it automatically sets the lens (f/number) to correspond to a preselected shutter speed, or sets the shutter speed for a chosen lens setting. This diaphragm-shutter setting will be satisfactory *provided two requirements are fulfilled:*

- The light must come over your shoulder as you take the picture.
- The meter must not be measuring any unusually bright or unusually dark large areas (such as a white shirt or a background) that are unimportant to the picture (these could cause underexposure or overexposure of the main subject).

Automatic-setting cameras are almost foolproof—but your own experience should guide you to vary the camera setting under certain conditions.

Review What You Have Learned About Determining Exposure:

1. What three numbers, relating to exposure, can be determined from a film data sheet? Which two can be used directly to make camera settings and which one is for a setting on a light meter?
2. What camera setting would be made under the "f/16 rule" when film with an ISO of 100 is used to take pictures under bright sunlight?
3. What two settings are required on a light meter before determining exposure?
4. When these two settings are made on the meter, what

pair of numbers results?
5. When may a camera with a built-in meter *not* give a reliable reading or automatic setting?
6. According to the setting illustrated on the incident light meter on page 85, if a subject requires extreme depth of field what camera shutter speed would you use? (First, would you select f/2 or f/32?)
7. The type of light meter on which the measurement of light intensity is not affected by the color or other characteristics of the subject itself is the _____.

ARTIFICIAL LIGHTING

The purposes of artificial lighting are to provide sufficient illumination for satisfactory exposure and to create either a natural or special appearance of the subject to the viewer. The latter may mean that there is an illusion of depth or a 3-dimensional impression created by shadows on a 2-dimensional surface.

Indoor Available Lighting

Offices, classroom, laboratories, and many other areas are usually well illuminated by daylight entering through windows or by fluorescent lights. Frequently this available light is sufficient for many filming purposes, with high-speed color film or with moderate-speed black-and-white films.

But there are limitations to overhead fluorescent lighting. It causes a very flat type of illumination with few shadows, and is uneven as you measure light levels from ceiling to the floor. For these reasons, it is often necessary to add supplementary lighting to a scene under fluorescent lighting.

Also, because color films respond differently to various light sources, fluorescence may cause unusual effects. Test your film to determine any variations in color rendition and film speed (for example, tests show that Ektachrome 200 daylight-type film, under certain classroom fluorescent lights, at an exposure index of 80 and using a FLD filter over the lens, gives highly acceptable results).

An alternative is to add artificial light to the scene with one or more photoflood lights equipped with *blue bulbs.* With this lighting, skin tones will appear natural.

Exposure for Subject

Exposure for Background

Exposure Balanced with Flash

Electronic Flash Lighting

Notice the difference in brilliancy and in shadow detail between otherwise identical photographs above, two taken with the natural available light and one with added flash light.

By using electronic flash units, you create your own light. Even in sunlight, flash lighting can be used to add light to shadow areas. A flash is most useful for lighting relatively small areas or for lighting larger ones that have light-colored backgrounds. The light falloff from flash is so great that the background, if it is too far behind the subject (over 10 feet or so), will appear undesirably dark. Conversely, if flash is used too close to a subject, the subject may appear too light or washed out. (See illustrations below.)

The exposure for taking pictures with a manual-type electronic flash unit is determined by the following formula, using a guide number for the flash unit and the type of film in your camera.

$$\text{lens setting} = \frac{\text{guide number}}{\text{flash-to-subject distance}}$$

Many electronic flash units have scales on the unit from which correct exposure can be determined directly. Note the data with the sample picture in the next column, and the use of these data in computing the f/number.

Ektachrome film, exposure index 64
Guide number 45
Shutter speed 1/60 second
Distance 10 feet (flash to subject)

$$\text{lens setting} = \frac{45}{10} = 4.5$$

$$\text{lens setting} = f/4.5$$

(*Note:* f/4.5 would be set between the 4 and 5.6 "click stops" on your lens.)

An automatic flash unit contains a light-sensitive cell that reacts to the light being reflected by the subject and automatically controls the flash intensity to produce correct exposure. A predetermined f/stop is used for all settings. Some flash units can be adjusted to accommodate a number of f/stops to create different effects.

Flash with Distant Background

Flash with Near Background

These units usually are quite accurate, but can be fooled under certain conditions. For example, a subject standing in front of and very close to a white wall will often be underexposed and appear dark. If the subject is in front of a very dark background, or if the background is far behind, the subject will often be overexposed. In order to work effectively with an automatic flash unit, you should learn to recognize these conditions and adjust your camera to correct for them.

Here are some other points to keep in mind when preparing to use flash lighting:

- Guide numbers are for determining exposure with average subjects under average conditions. Darker or lighter subjects or the color of walls can affect light reflections. Modify the guide number accordingly.
- Because the duration of an electronic flash is so short, the guide number is the same at any shutter speed. But use a slow speed, especially with a camera that has a focal plane shutter (1/60 second).
- Aim the flash unit toward the ceiling to **bounce** the light before it reaches the subject. Soft, even lighting will result. (Be alert that a ceiling color will be picked up by the reflected light and will change the color tone of the scene.) For bounce lighting, open the lens setting at least two additional f/stops as compared to direct flash used at the same distance.
- When using an automatic flash unit to bounce light, make certain that the electric eye is pointed at the subject.
- Test your own equipment with the films you use to determine the best settings that give you proper exposure.

Photoflood Lighting

If controlled lighting is necessary, consider using photographic floodlamps. They may be essential for video recording and motion-picture photography; even for still photography floodlights are often better than flash since you see exactly what effect the lights are creating (reflections, heavy shadows, or uneven lighting) and can make corrections before taking pictures.

Two major types of photoflood lamps are available. The traditional incandescent-filament lamp, with wattages of 250, 500, or 1000 in a hemisphere-shaped reflector, is widely used. Lamp technology has made available small tubular units consisting of a quartz envelope containing a tungsten filament and filled with a halogen gas, commonly iodine vapor. These quartz-iodine lamps provide a very bright, narrow ribbon of light and project an extremely even illumination. People appearing in scenes in which quartz iodine lamps are used should be warned not to look directly into the light as the brightness of the source can cause eye discomfort.

When it is desirable to use available light, but a supplement is needed to raise the light level, "bounce" light is beneficial. From near the camera position, aim a floodlight at the ceiling or floor. It will "snap up" the scene without lights being reflected directly off the subject.

If floodlights or a flash are aimed directly at the subject,

Available Light   **Bounce Light Added**

sharp dense shadows are created somewhere in the scene. They must be controlled. Floods placed close to the camera will light the subject only from the front side. Such lighting results in a flat, shadowless subject with heavy background shadows, which is usually undesirable.

It is better to place one light about 45° to the side of the camera, somewhat closer to the subject, and 30° above the

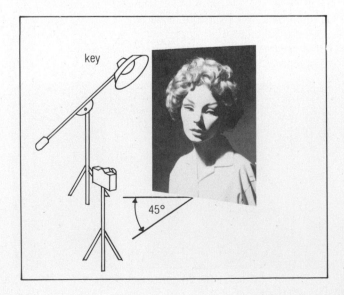

camera (as shown in the illustration at the bottom of the previous column). This becomes the **main**, or **key, light.** It substitutes for the sun, which shines on an outdoor scene or represents light from a window or a lamp. Therefore it should be the brightest light source (either by wattage or closeness to the subject).

Place a second light (or two) beside the camera (on the side opposite the main light) and at camera height. This light serves to fill and soften shadows created by the key light, thus bringing out more detail in the subject. It is called the **fill light.**

Some light from the key may fall on the background, but a third light (or two), aimed evenly at the background, will illuminate it, thus separating the subject from the background and giving the scene some depth. This is the **background light.** Always keep the subject at least 2 or 3 feet away from the background to minimize heavy shadows created by any lights.

These three lights—key, fill, and background—form the basic lighting pattern for good indoor lighting. With color film, the key, as the brightest light, is placed so that it illumi-

nates the subject with *twice* the intensity of the fill light. Intensity is measured by holding an incident light meter at the subject and aiming it at the light source. When the intensity of the key light is *twice* the intensity of the fill light (meter readings may be: key—100, fill—50) we say the "key to fill ratio" is 2 to 1. This will result in soft shadow areas being recorded on the film.

For close-up scenes, to reduce shadows, the intensity of the fill may be equal to that of the key light—key to fill ratio of 1 to 1. (Set a light for the intensity you want by moving it closer to or farther from the subject.) When black-and-white film is used, the key to fill ratio may be as high as 5 to 1.

One or more background lights may be set to illuminate the background evenly with a meter reading one-half to one stop lower than the general reading within the scene. Sometimes a spotlight (or photospot) is used as an **accent light** to highlight a person or an object in the scene.

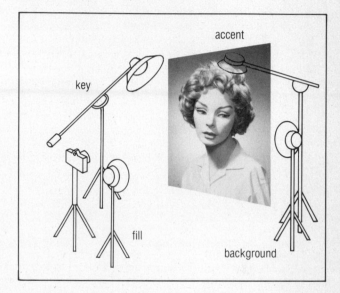

For most subjects, whether portrait, table top setups, or in a large area, this lighting pattern can be applied.

Lighting Metal and Glass Objects

When shiny metal surfaces or glassware are illuminated with the normal lighting pattern just described, the result usually is undesirable reflections and dark or light areas of the subject. See the first picture in each set below.

Indirect or diffused lighting should be used with such subjects. Place objects with metal surfaces in an enclosure consisting of tracing paper or other translucent material sides. Aim lights through the sides to illuminate metal sub-

jects. Light is softened as it passes through the translucent paper. By following this procedure, hot spots and uneven reflections are avoided.

To avoid reflections on glassware, place the objects to be photographed on a sheet of glass, raised above the table. Set a sheet of white cardboard on the table, under the objects. Aim a floodlight at the cardboard. Light will reflect from the cardboard through the glassware, providing an even, nonreflecting illumination of the subject.

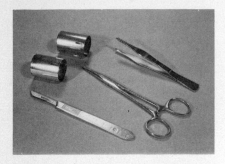
Result with Direct Lighting

Diffused Lighting Setup

Result with Diffused Lighting

Result with Direct Lighting

Reflected Lighting Setup

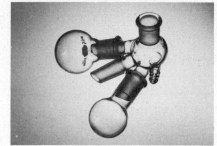
Result with Reflected Lighting

Review What You Have Learned About Lighting:

1. Is electronic flash effective for lighting *small* or *large* areas?
2. Apply the correct formula to determine *f*/number: flash guide number for film is 100, distance of camera to subject is 6 feet.
3. In a scene including people and large objects, what are some disadvantages of using lights placed only right beside the camera?
4. What is meant by the expression, "Key to fill ratio is 4 to 1"? Is this an acceptable ratio to create pleasant shadows on color film?
5. What method would you use for determining exposure under floodlighting?
6. Explain the positioning and purpose served by each light—accent, background, key, and fill. In what order is each set for use?
7. Why is it necessary to give special attention, when lighting, to metal surfaces and glassware?

CAMERA SHOTS AND PICTURE COMPOSITION

The effectiveness of instructional media is strengthened by giving attention to two aspects relating to picture composition—framing of the total image within the camera's viewfinder and arranging the necessary elements that comprise the image. These matters are important for all photographic visuals—still scenes in slides or motion scenes for video and motion pictures.

Long Shot

Medium Shot

Close-up

Long Shot

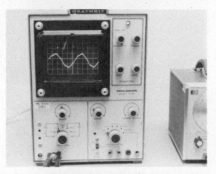

Medium Shot

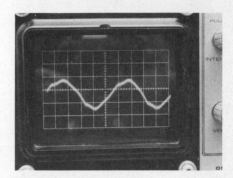

Close-up

Framing the Picture

The composition of a scene should be carefully decided when viewed through the camera lens. Therefore, the camera must be placed in the best position for recording all necessary elements in a scene. In deciding on where to place the camera, two questions should be asked:

- What is the best **viewpoint** from which the picture should be taken?
- How much **area** should be included in the scene?

The description of each scene (or "shot") in the script or on the storyboard provides information which helps you to answer these questions.

In doing this, you give attention to the various kinds of camera shots. Each one refers to the relation between the camera position and the subject.

Basic Shots

Three types of scenes are common in photography:

- The **long shot (LS)**—a general view of the setting and the subject. It provides an orientation for the viewer, by establishing all elements in the scene, and if important, shows size proportions relating to the subject.
- The **medium shot (MS)**—a closer view of the subject, eliminating unnecessary background and other details
- The **close-up (CU)**—a concentration on the subject, or on a part of it, excluding everything else from view

When the subject is the same, three successive shots assume a relation to each other.

LS, MS, and CU do not mean any specific distances. A long shot of a building may be taken from a distance of hundreds of yards, whereas a long shot of equipment may be taken from a distance of only a few feet. You may be close-up to a building when you are across the street, but you may need to get within a foot or so of the equipment to take a close-up.

Although LS–MS–CU is a fundamental sequence, it is not to be rigidly followed in successive sequences, and there is no set rule for the use of these three basic shots. The visual effect desired should determine the sequence.

Sequences need variety; without variations, your media may become monotonous. A still-picture sequence can include a number of relatively similar successive shots, while a video recording or film requires more diversity so it does not lose its pacing or impact.

For straightforward explanation the LS–MS–CU sequence may be satisfactory. For a slower pace, gradually increasing interest, LS–MS–MCU–CU may be used. For suspense or drama, consider CU–CU–CU–LS.

Note the differences in the shots that comprise the two sequences on the next page.

Although the subject does limit the kinds of shots that are called for in the script, two photographers covering the same subject may film the three basic scenes differently, each imparting his or her own interpretation and emphasis. To say that one version is right and the other wrong would most likely depend on personal preference (see page 93).

At the two ends of the LS–MS–CU sequence you can introduce **extremes** if they are important to your story—**extreme long shot (ELS)** and **extreme close-up (ECU)**. Also, there are situations in which you may designate a scene between two basic shots—a **medium long shot (MLS)** or a **medium close-up (MCU)**.

Sequence 1

LS MS MCU CU

Sequence 2

MS CU LS MS

Angle and Position Shots

Variety, emphasis, and dramatic effect can be accomplished through the use of *camera angles*. The normal or neutral camera position is at about eye level for a person standing. A camera in a higher position, looking down on the subject, makes a *high-angle shot* that gives the illusion of placing the subject in an inferior position, reducing its size and slowing its motion. A camera in a lower than normal position, looking up at a subject, makes a *low-angle shot* that seems to give the subject a dominant position, exaggerating its height and speeding up movement. High- and low-angle shots can be used to eliminate undesirable background or foreground details. (See bottom of page.)

The camera may be placed in the position of the eyes of the observer, that is, of an audience. This is an **objective camera position.** Or, the camera may be placed in the position of the subject's eyes to see the performance of an operation or the behavior of an object as the subject sees it. This is a **subjective camera position.** In the latter, the photographer may shoot over the subject's shoulder, with the camera at a high angle.

In summary there is a variety of filming shots at your disposal:

Objective Scene

Subjective Scene

- Basic shots—long shot, medium shot, close-up
- Extremes—extreme long shot, extreme close-up

- High-angle and low-angle shots
- Objective and subjective camera positions

High Angle Shot

Neutral Shot

Low Angle Shot

LS, Photographer A

LS, Photographer B

MS, Photographer A

MS, Photographer B

CU, Photographer A

CU, Photographer B

Extreme long shot (ELS)

Extreme close-up (ECU)

To these shots, when using a video or motion-picture camera, **moving** camera shots like **panning** and **tilting,** can be added. These are treated in the video and film chapter on page 227.

Each scene requires placing the camera in the best position for viewing the setting and action. Thus, camera placement determines image size and angle. These factors contribute to the communication value of any photographic media, whether still or moving.

Arrangement of Elements

From studies of perception and artistic design a number of composition principles have been stated so that the content of a picture will be clearly recognizable, easily understood, and convey a certain feeling. Those principles appropriate to the design of graphic-type materials are considered on page 108. Refer to them since they also relate to photographic images. The following principles apply when photographing pictures:

93

● Most visual materials normally have a horizontal format. If possible, plan your content for this format. Try not to mix vertical photographs or slides with horizontal ones in a series if there is a possibility that they might be used in a filmstrip, a motion picture, or on television.

● Have only one major subject or center of interest in a scene. Do not clutter a picture or make it tell too much. Eliminate or subordinate all secondary elements and focus attention on the main one.

● Because viewers have no way of judging the size of unknown objects in pictures, it is important to include some familiar object for comparison.

● Keep the background simple. Eliminate confusing background details by removing disturbing objects, by putting up a screen to hide the background, or by throwing the background out of focus (using a smaller f/number, thus controlling the depth of field).

● Place the center of interest near to but not directly in the physical center of the picture area. By making the picture slightly unsymmetrical you create a dynamic and more interesting arrangement.

● Include some foreground detail to create an impression of depth (principally in long-shot exterior scenes). Foregrounds help to balance the picture and to make it interesting.

● Try not to be static from one scene to another by shooting from the same relative camera position or angle. Plan to vary camera positions. Changing angles creates a dynamic impression and gives variety to composition.

● In black-and-white photography similar tones may blend together. Have the color of the center of interest contrast with the background and surrounding objects.

If action or movement is implied in a picture allow more space or picture area in the direction of the action rather than away from it.

Finally, use common sense in composition. Ask yourself, What am I trying to accomplish with this picture or scene? Then pick what appears to be the best angle and the best distance for the camera. If necessary, view the scene from two or three positions and make pictures from each one for future selection.

We have examined a number of suggestions and guidelines that should be helpful in creating good composition. Yet, there is personal choice when framing and arranging picture elements because there are no hard rules to follow mechanically. Each situation is different, requiring a unique decision. Use these principles for guidance, along with your own experience to insure that the images will best serve the scripted purpose.

Review What You Have Learned About Camera Shots and Picture Composition:

1. Following is a group of six pictures (scenes); each one composes the same subject differently. Which one do you prefer? Why?

2. Relate the shots—high-angle, low-angle, objective, subjective—to the following situations:
   a. Exaggerate subject height

(a)

(b)

(c)

(d)

(e)

(f)

95

b. Slowing subject movement
c. Over-the-shoulder filming
d. Filming from an audience viewpoint
e. Eliminating a disturbing background

3. Label the types of camera shots illustrated by each scene in the sequence that follows. In what order would you arrange these scenes to make a meaningful sequence?

(a)   (b)   (c)   (d)

(e)   (f)   (g)   (h)

CLOSE-UP AND COPY WORK

In photography it is often necessary to photograph subjects at a very close range, such as for titles, reproductions of charts and pictures, and for close-ups of subject details.

Your camera may be unsatisfactory for such close-up work unless you can make adjustments and allowances in two respects: **viewfinding and focusing.**

Viewfinding and Parallax

In some cameras the viewer and the taking lens do not see exactly the same area. This is the phenomenon of **parallax.** The different areas that the taking lens and the viewfinder see are illustrated in two pictures below.

Single-lens reflex and view cameras permit through-the-lens viewing. The subject is observed directly through the taking lens and focused on a ground glass. These cameras are preferred for close-up and copy photography. (See illustration top of facing page.)

If your camera does not have built-in features to deal with the parallax error, it is difficult to do close-up and copy work.

Focusing and Exposure

Cameras are adapted for close-up photography (picture taking close to the subject, often under 2 feet) by one of four devices:

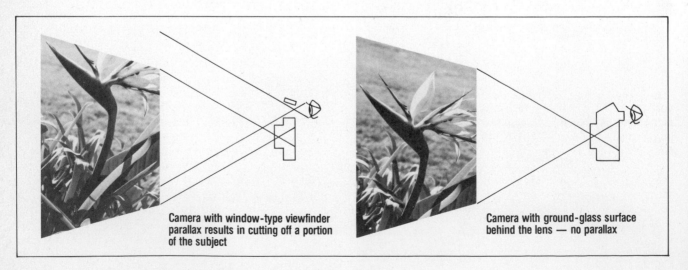

Camera with window-type viewfinder parallax results in cutting off a portion of the subject

Camera with ground-glass surface behind the lens — no parallax

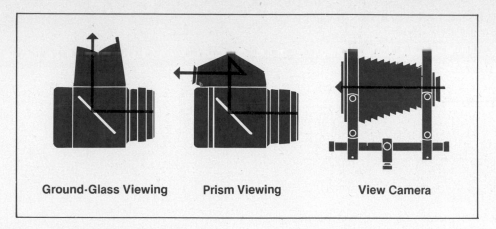

Ground-Glass Viewing   Prism Viewing   View Camera

- A lens specially made for close-up focusing, which maintains a flatness of field and does not distort lines in a subject (a **macro lens**).
- The camera is built with bellows that can be used to lengthen the lens-to-film distance.
- The camera permits the use of separate extension tubes or bellows that lengthen the lens-to-film distance.
- A close-up attachment can be mounted on the lens to change the optical character of the lens system.

Macro Lens

Bellows

Extension Tubes

Close-up Lens

When you use bellows or extension tubes for close-ups, you are using the same lens that you would use to take pictures at normal distances. But exposure must be adjusted because of the increased distance between lens and film. A through-the-lens meter accommodates the exposure for any extension. When a separate meter is used, the exposure for close-ups under these conditions is found by the usual exposure-meter procedure plus an additional computation. The additional computation takes into account the focal length of the camera lens (which is printed or engraved on the lens housing) and the amount of extension of the bellows or tubes. The formula can be used for measurements either in inches or in metric units. It is:

$$\text{increased-exposure factor} = \frac{(\text{length of extension})^2}{(\text{focal length of lens})^2}$$

The application of this formula to a specific problem is illustrated:

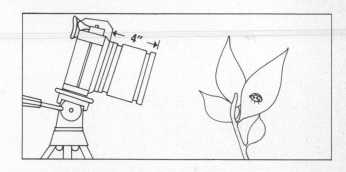

Focal length of lens; 50mm or 2 inches
Length of extension; 100mm or 4 inches
Normal exposure; f/8 at 1/60

Using the formula with the distances in inches gives:

$$\text{increased-exposure factor} = \frac{4^2}{2^2} = \frac{16}{4} = 4$$

Therefore exposure must be 4 times the normal exposure. This increase can be accomplished by opening up the diaphragm of the lens 2 f/stops (to give exposure of f/4 at 1/60 second) or by reducing the shutter speed one-fourth (to give exposure of f/8 at 1/15 second). To insure best focus and maximum depth-of-field when working close, use a larger f/number.

Review the explanation of f/number–shutter speed relationships on page 79 and the selection of a lens setting to increase depth-of-field on page 80.

The majority of cameras can be adapted for close-up work with one or more supplementary lenses (close-up attachments) placed in a retaining ring and attached over the regular lens. With this method no compensation for exposure is required as with tube or bellows extension.

The effectiveness of close-up attachments with some adjustable cameras is shown in Table 10-2 on the following page.

Table 10-2 **CLOSE-UP ATTACHMENTS AND CAMERA-SUBJECT SETTINGS**

Width of Subject (Inches)	Distance Camera to Subject (Inches)	Power of Close-up Lens	Distance Setting on Camera (Feet)
26½	39	1+	infinity
11½	15½	2+	6
6	8	3+	2

Using a Copy Stand

A copy stand, which holds your camera at various positions, is useful to accommodate materials of different sizes. Vertical stands are more serviceable than horizontal ones because of the difficulty of securing books in a vertical position.

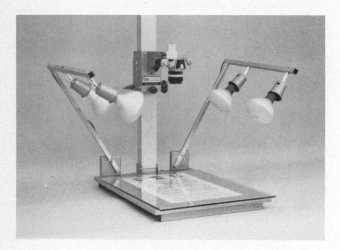

Commercial Copy Stand

Suggestions for Close-up and Copy Work

- Use a sturdy tripod or a sturdy stand to steady your camera.
- Use either photoflood lamps in reflectors or lamps with built-in reflectors. For copying set the lamps evenly at 45-degree angles to each side of the camera and at a sufficient distance to avoid uneven lighting, reflections, or "hot spots." Check for evenness of lighting with your meter.
- When copying from a book or other source that does not lie flat, hold the material in position with a sheet of nonreflecting glass (available from photo-supply, art-supply, or hardware stores).
- For close-ups of three-dimensional objects, avoid most shadow areas by using **flat lighting** (that is, use a key to fill light ratio of 1 to 1 as explained on page 89).
- Use a meter to determine exposure. An incident light meter gives a direct reading when placed in the position of the subject or material. If you use a reflected light meter, read either from a gray card (Eastman Kodak's

neutral test card) or from a sheet of white bond paper held against the main portion of the subject. Because such paper reflects a large portion of the light, a reading from it will be *5 times too high.* To correct for this high reading, divide the exposure index of the film by 5 and set your meter at the closest value (example: Ektachrome 64, exposure index 64, set meter at 64 ÷ 5, that is, at 13).
- Select camera settings of *f*/11 or *f*/16 to insure adequate depth of field (but note the exception on page 78 in order to keep the shutter speed faster than 1 second).

Remember: You must have the copyright holder's permission to reproduce copyrighted materials (see page 57).

Filming Titles

Titles are handled as is other copy work.

If a mask was used to frame the original art work, align the camera on the copy stand to take in the open area of the mask (see page 106). Remove the mask. Then proceed with routine copying of materials made to its size and format.

If a mask was not used to prepare the lettering, adjust picture size by raising or lowering the camera. Keep in mind that acceptable legibility requires lettering size to be a minimum of one-fiftieth the height of the projected area. (See page 121 for description of legibility standards.)

When a title is prepared on a sheet of film (using a procedure described in Chapters 11 and 15) and placed over a background picture or art work, cover the total area with a piece of nonreflecting glass. This should press the film to the background, eliminating shadows of the letters and reflections from the film. Since most 35mm single-lens reflex cameras photograph a larger field than is seen through the viewfinder, adjust the camera distance so the title almost fills the viewfinder. Generally this will provide large words and sufficient background bleed on the resulting slide.

Titles to appear over a special background as **white letter overprints** may require double exposure (see page 119 for preparation of art work). First film the background slightly dark (½ to 1 *f*/stop underexposed). Then, on the same negative without advancing the film, expose the lettering. Many cameras permit double exposures, but check yours before trying this technique.

98

Review What You Have Learned About Close-up and Copy Work:

1. Of the cameras described at the beginning of this chapter (page 73), with which ones would you expect to have a parallax problem in viewing?
2. If a close-up lens is used over the regular camera lens, need there be a calculation for change in exposure? Would this also be true with a bellows extension?
3. What is the name of a lens that may be directly used on the camera in place of the regular lens for close-up and copy work?

4. In close-up work is a higher or lower *f*/number desirable? Does this mean that a *slower* or *faster* shutter speed is used? Therefore, a tripod or stand *is* or *is not* essential?
5. Explain how a reflected light meter can be used with a sheet of white paper to determine exposure for copying. In such a situation, what setting for exposure index (film speed) is made on the meter for Kodachrome Type A film having a rated exposure index of 40?

PROCESSING FILM

You may choose to send exposed film to a commercial film processing laboratory for developing and even for printing. But in recent years the processing of both black-and-white and color film has been greatly simplified.

● Graduated or other calibrated measuring container and a funnel
● Thermometer
● Timer or watch with second hand
● Three to six stoppered bottles, preferably of brown-tinted glass or plastic

Facilities and Equipment

● Light-tight room for film loading (a daylight-loading tank eliminates this need)
● Sink with running water and countertop working area
● Clean, ventilated area for film drying
● Roll-film developing tank with one or more reels for the film size being used or one or more tanks and film holders for cut film
● Prepared chemicals for processing film

Black-and-White Film

Practice loading an old roll of film onto the reel of your tank (see suggestions on the instruction sheet with the tank) until you can do it smoothly. Then, *in the dark,* load and thread the film to be processed. Each turn of the film on the reel must fit into a separate groove so film does not touch. Place the reel in the tank and cap it. From here on do all processing under normal room light.

The purpose served by each step in the process is:

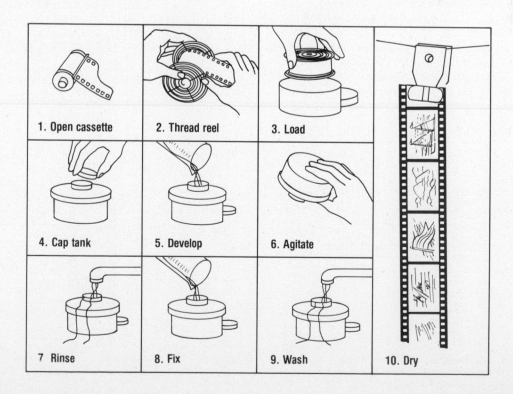

1. Open cassette	2. Thread reel	3. Load	
4. Cap tank	5. Develop	6. Agitate	
7. Rinse	8. Fix	9. Wash	10. Dry

- **Developer**—acts upon the exposed silver chemicals in the film that have been affected by light during picture taking, depositing the silver as tiny grains to form the black silver image of the negative.
- **Rinse**—removes excess developer from the film.
- **Fixer**—sets the image by changing the remaining undeveloped silver chemical so that it may be removed.
- **Washing**—removes all chemicals that may cause discoloration of the negative or deposits on its surface.

Refer to the information sheet packaged with your film for recommended developer and for specific processing instructions. Follow all directions, especially those for time and temperature controls.

Judge the quality of your negative by these points:

- A good negative will have a considerable amount of detail, even in its very darkest and lightest portions, unless these portions represent parts of the picture which were themselves entirely lacking in detail.
- A good negative will be transparent enough, even in its very blackest areas, so that you can read a newspaper through it.
- A good negative will have no part of the picture quite as clear as the borders of the film.

Color Negative Film

Kodacolor VR 100, 200, 400, and 1000 films can be processed by a film laboratory or with a color processing kit (Process C-41). Processing time in a tank is under 30 minutes with careful timing and critical temperature control (the first step requires a constant temperature of 100°F $\pm \frac{1}{4}°$ F and the remaining 6 steps should be within a range of 75° to 105° F). See the illustration below.

Mix the chemicals according to instructions with the processing kit. Store each solution in a tightly closed bottle. Write the date of preparation on each bottle. The useful life of the developer is 6 weeks and that of other solutions is 8 weeks. Before use, place the bottles in a tray of running water maintained at the proper temperature.

Color Reversal Film

All color reversal films, except Kodachrome, can be "home" processed. Kodachrome is handled only by authorized processing laboratories. As with color negative films, processing kits allow handling of one or more rolls in normal room light with the film in a light-tight tank during the initial part of the process. Here again, temperature (100°F $\pm \frac{1}{4}°$ for the first developer and 92°–102°F thereafter) and timing must be carefully maintained. The procedure for Ektachrome roll film (process E-6) is illustrated.

In some filming situations, the light level is too low for normal exposure. It is possible to increase the effective film speed (ASA/ISO rating) and then adjust the time of the first developer to obtain satisfactory color slides. For example, by increasing Ektachrome 160 (Tungsten) from an ASA/ISO of 160 to 320 you are in effect underexposing one f/stop. To compensate, increase the first developer by two minutes. Film speeds up to four times normal can be used with modified processing times.

If you will be processing many rolls of negative or reversal color film, or lengths longer than 36 exposure rolls, consider using automated equipment. In such a machine, film moves

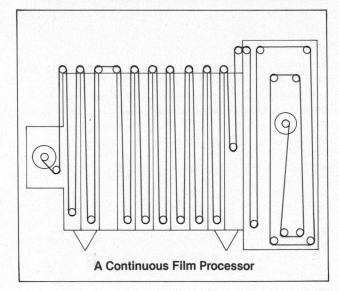

A Continuous Film Processor

1. Develop
2. Bleach
3. Wash
4. Fixer
5. Wash
6. Stabilizer
7. Dry

Color Negative

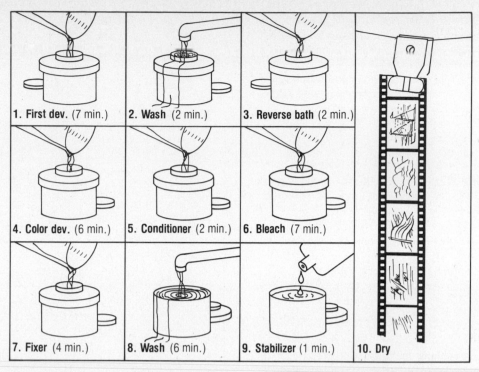

1. **First dev.** (7 min.)	2. **Wash** (2 min.)	3. **Reverse bath** (2 min.)
4. **Color dev.** (6 min.)	5. **Conditioner** (2 min.)	6. **Bleach** (7 min.)
7. **Fixer** (4 min.)	8. **Wash** (6 min.)	9. **Stabilizer** (1 min.)

Color Reversal

10. **Dry**

continuously through separate tanks containing chemicals for each processing step. Development time is controlled by the length of time film remains in each tank.

A reversal color film (ISO 40), available from Polaroid, can be exposed in a regular 35mm camera and then immediately processed in an "autoprocess" unit. Thus slides can be shot and the results seen in a few minutes. Continuous-tone black-and-white and high contrast 35mm films are also available in this process.

MAKING PRINTS

For successful contact printing and enlarging you need to know about:

- The selection of contact and enlarging papers (printing papers are classified by speed, weight, finish, contrast, color of image, and base material)
- Exposure—length of time for contact printing; lens diaphragm opening and length of time for enlarging
- The selection of chemicals—developer, stop bath, fixer
- Using filters for contrast control and color correction
- Processing time and procedure with each solution
- Washing, drying, and finishing

The negative is used to prepare a positive print on paper or film. *If the print is to be the same size as the negative,* the process is **contact printing,** but *if the print is to be larger than the negative,* the process is **enlarging.**

Negative

Contact Print Sheet

Enlarged Print

101

Facilities and Equipment

The standard equipment and materials for a darkroom consist of the following:

- A darkroom 6×8 feet or larger, equipped with running water (temperature controlled), countertop workspace, storage, and electrical outlets
- A contact printer or printing frame
- An enlarger with easel and timer
- A print washer or tray siphon
- A print dryer
- One or more sets of trays (three to a set) in various sizes (8×10 inches, 11×14 inches, and so on) or a processing machine
- Clock, tongs, and miscellaneous small items
- One or more safelights (with color filter based on printing paper to be used)
- Photographic contact and enlarging paper
- Prepared chemicals for developer (or activator), stop bath, and fixer

Contact Printing

This method is particularly useful for rapid preparation of **proof sheets** from negatives. A whole roll of negatives (12 to 36) can be printed at one time on a sheet of contact paper (8×10 inch). From these contact prints, negatives can be selected for enlargements.

Place the negative (emulsion, or *dull side,* down) on top of a sheet of photographic contact paper (emulsion, or *shiny side,* up); cover them with glass and expose the pack to light. Or use a contact printer with a pressure platen and a built-in lamp for exposing. Develop the paper as illustrated in the next section. The resulting print will contain positive images the same size as each negative.

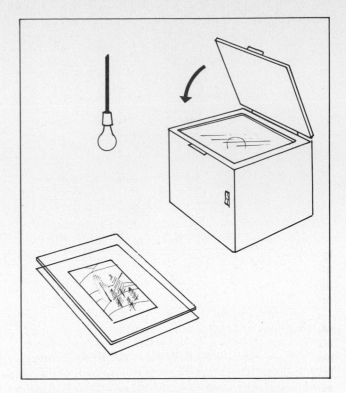

Enlarging from Black-and-White Negatives

Select printing paper (or a contrast filter) to match the contrast of the negative. For example, a contrasty negative requires number 1 paper or polycontrast filter 1.

Place the negative in the enlarger and project it through the lens onto a sheet of enlarging paper. Make tests on strips of paper to determine the correct combination of enlarger lens setting and exposure time before preparing the final prints.

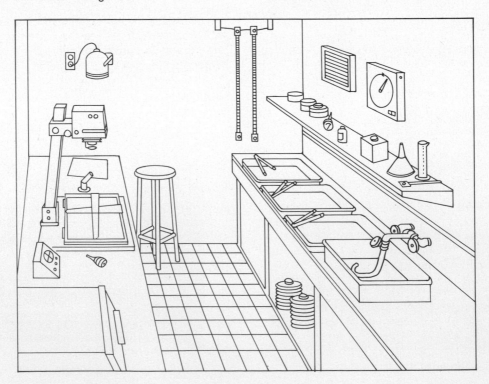

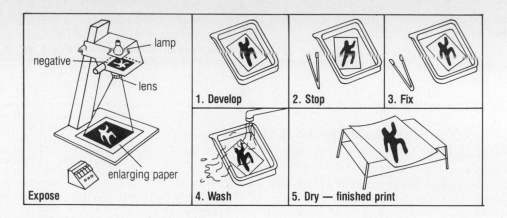

After the black-and-white photographic paper is exposed to light, processing follows. Use the same general chemical treatment as for film—develop, stop, fix, and wash (for a paper print, a paper developer is used in place of the film developer).

A two-step **rapid-processing** method for black-and-white processing is widely used. This method takes only a few seconds for developing and fixing exposed paper. It also eliminates the need for an extensive darkroom with a large sink and trays. It does require the enlarger and processing unit. In this method, known as **photostabilization,** the developing agents required are incorporated in the

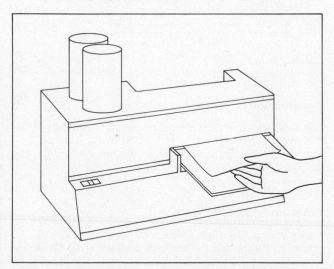

paper emulsion. The paper is carried automatically in timed sequence by a system of rollers, first through the **activator** and then through a **stabilizer** bath. The paper emerges damp, especially if RC (resin coated) photographic paper is used. The stabilizer arrests development and stabilizes the image (the chemistry is similar to, but not exactly the same as, fixing with hypo). Thus, processing is automatically accomplished in a matter of seconds. Because the print will fade in time, it is recommended that for greater permanence, a print processed by this method be fixed in regular hypo and then thoroughly washed and dried. In addition to various kinds of photographic papers, high contrast and continuous-tone sheet films are available for use in the photostabilization process. Developing PMT paper and film employs a process similar to this one (page 138).

Enlarging from Color Negatives

In the past, the procedure for printing from color negatives was complex and time consuming. Now, rapid, simple processes are available that result in high quality color prints.

As an example, with Kodak Ektaflex products a sheet of PCT film is exposed to a color negative as you would for conventional enlarging. The film is then placed in a processor containing one chemical solution. After 20 seconds the film is laminated to a sheet of Kodak Ektaflex PCT paper and the color image is transferred. After a few minutes, the film and paper are peeled apart. The finished color print is ready for use.

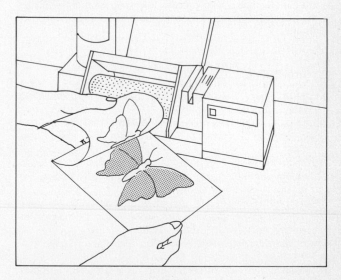

Enlarging from Color Slides

Formerly it was necessary to make a color negative from a positive color slide and from it prepare a positive color print. Today materials are available to produce a positive color print directly from a color slide. Various processes may be used:

- With the Kodak Ektaflex process described above, using Ektaflex PCT Reversal film
- With Ilford Cibachrome color print material requiring a 12 minute processing period in three chemical steps
- With a Polaroid fully automatic Polaprinter preparing instant color prints

● With Agfachrome Speed requiring one activating chemical and a five minute processing time

In the past, because of time requirements and expense, photographs for instructional purposes generally were prepared in black-and-white. Now, with the simplified processing available and unit costs being reduced, consideration should be given to the preparation of photographs in color.

Review What You Have Learned About Processing Film and Making Prints:

1. What are the four steps necessary to develop black-and-white film and the purpose of each step?
2. What are some characteristics of a good black-and-white negative?
3. In what ways and with what materials do the four steps in tray processing of black-and-white paper differ from those in developing of black-and-white film?
4. What purpose is often served by making contact prints of a roll of black-and-white negatives?
5. What are two advantages of using the photostabilization process for processing paper over the regular tray process?
6. What is the major difference in film processing procedure between color negative and color reversal processes?
7. If you plan to prepare a series of color photographs to explain how to operate a piece of equipment, what procedure would you follow to make the photographs?

REFERENCES

(Many references in the following sections are available from Eastman Kodak Co., 343 State Street, Rochester, NY 14650, or from a local photographic supply dealer. Following each of these titles is the Eastman publication code number without any other source indication.)

Picture Taking

Goldsmith, Marc. "Lens Perspective." *Technical Photography* 14 (May 1982):18–21.
Kodak Guide to 35mm Photography. AC-95S, 1981.
Lynch, David. *Focalguide to Better Pictures.* New York: Focal, 1981.
"Photographic Filters Part I." Slide/tape program X-104. In *Fundamentals of Photography Series.* Bloomington, IN: Indiana University, Audio-Visual Center, 1982.
"Photographic Filters Part II." Slide/tape program X-105. In *Fundamentals of Photography Series.* Bloomington, IN: Indiana University, Audio-Visual Center, 1982.
Photography for Audiovisual Production. Slide/tape program V10-70, 1982.
Sutherland, Don. "Zoom Lenses and Fixed Lenses." *Technical Photography* 14 (May 1982):24–26.

Films

Kodak Color Films for Still Cameras. AE-41, 1982
Kodak Ektachrome Duplicating Films (Process E-6). E-38, 1982.
Kodak Films—Color and Black-and-White. AF-1, 1981.

Lighting and Exposure

Accurate Exposure with Your Meter. AF-9, 1981.
Adventures in Existing Light Photography. AC-44, 1980.
Carraher, Ron, and Chartier, Collene. *Electronic Flash Photography: A Complete Guide to the Best Equipment and Creative Techniques.* New York: Van Nostrand Reinhold, 1980.

Exposure with Electronic Flash Units. AC-37, 1982.
Lynch, David. *Focalguide to Exposure.* New York: Focal, 1978.
Millerson, Gerald. *Lighting for Television and Motion Pictures.* New York: Hastings House, 1981.
Morton, Harry. "Spot Meters . . . Who Needs Them and Why?" *Functional Photography* 17 (September/October 1982):30–33.
Ulrich, J. D. "Lighting for Color Balance." *Technical Photography* 14 (February 1982):21–23.

Composition

The Beginnings of Photographic Composition. Slide/tape program AV-0008, 1979.
Bruck, Alex. *Practical Composition in Photography.* Woburn, MA: Butterworth, 1981.
Composition. AC-11, 1981.

Close-up and Copying

Basic Copying. Slide/tape program V10-29, 1981.
"Close-up Photography." Slide/tape program X-106. In *Fundamentals of Photography Series.* Bloomington, IN: Indiana University, Audio-Visual Center, 1982.
Goldsmith, Marc. "Industrial Macro Photography." *Technical Photography* 15 (April 1983):44–46.
O'Neill, Jerry. "Basics of Copying." *Photomethods* 27 (March 1984):48–50.
Roy, Sidney. *Focalguide to Close-ups.* New York: Focal, 1978.
Using Polarized Light for Copying. S-80-19, 1979.

Film Processing and Printing

Basic Developing, Printing, Enlarging in Black-and-White. AJ-2, 1979.
Basic Developing, Printing, Enlarging in Color. AE-13, 1981.
Making Color Prints with Kodak Ektaflex PCT Film and Paper. E-172, 1982.
Printing Color Negatives. E-66, 1982.
Selecting Slides for Color Prints and Duplicate Slides. AE-92, 1982.

Handbooks and Journals

Functional Photography. PTN Publishing Co., 101 Crossways Park West, Woodbury, NY 11797.

Industrial Photography. United Business Publications, Inc., 475 Park Avenue South, New York, NY 10016.

Kodak Professional Photoguide. R-28, 1981.

Photomethods. Ziff-Davis Publishing Co., One Park Avenue, New York, NY 10016.

Technical Photography. PTN Publishing Co. (see above).

Chapter 11

GRAPHICS

- Planning Art Work
- Using Graphic Tools and Basic Materials
- Illustrating Techniques
- Visualizing Statistical Data
- Using Computer Graphics Systems
- Coloring and Shading
- Backgrounds for Titles
- Legibility Standards for Lettering
- Lettering Techniques
- Mounting Flat Materials
- Protecting the Surface (Laminating)
- Making Paste-ups
- Duplicating Line Copy
- Reproducing Printed Matter

Visualization for instructional media can take place in either one or a combination of two forms—photographic or graphic. Photography provides illustration through pictures that closely represent the reality of a subject or situation. Graphic materials are symbolic and artistic representations of a subject.

The success of many instructional media can be attributed in large measure to the quality and effectiveness of the art work and related graphic materials. These are achieved through organizing preliminary thoughts, careful planning, and applying the techniques outlined in this chapter.

Many persons who develop instructional media have little or no professional art background. They need not, however, prepare amateurish and poor-quality graphic materials. First, they can consider a number of common-sense practical suggestions and guiding principles, then apply them as the need arises. Second, there are a number of easy-to-use manipulative devices that, with little practice, will ensure professional-quality results.

PLANNING ART WORK

Your art work must be planned with consideration for the size and dimensions of the working area, for the proportions of your visual materials, for design and layout features, for backgrounds, and for the resources, skills, techniques, materials, and facilities that you can employ.

Size of Working Area

It is important to select a cardboard or paper size that can be handled easily. Give attention to these requirements when deciding on a working surface:

- Lettering and drawing can be done easily.
- Sufficient margin *(safe* or *bleed area)* can be allowed on all sides of the visual.
- Parallax and close-up difficulties in the camera, if there are any, can be easily overcome when copying (see page 96).
- The art work is easy to store.

The minimum dimension that is likely to meet these requirements is 10×12 inches; therefore use cardboard of this size or larger. You can cut boards 11×14 inches without waste from standard-size 22×28 inch sheets, which are sold in 8-ply or 14-ply thickness. Commonly used working areas, on boards of either size, are 6×9 inches and 9×12 inches; minimum lettering sizes for these areas are suggested on page 124. Compose within the proper proportions (see below) of your selected instructional media. Provide generous margins around the sides of all work so the camera does not inadvertently film beyond the background.

The end papers inside the front cover of this book contain recommended mask sizes for filmstrips and slides. The end papers inside the back cover have diagrams for masks to use with overhead transparencies and video/motion picture formats.

If many scenes require art work and lettering, standardize your size and prepare a mask with a cutout of the proper working area. The mask will serve as a margin and as a frame when you view the prepared art work and will also be useful as a guide for positioning art work and camera during copying.

If titles, labels, or diagrams must be placed one over the other or over a background, you need equipment to hold them in **alignment,** or **register.** If you need to make only a few such graphics, you may be able to work on an ordinary drawing board with masking tape. If you have any quantity of work, a commercial *register board* will save time and allow accurate alignment. Use either prepunched paper and film, or obtain a punch that corresponds to the location of the pins on the register board. The register board can be used both during preparation of the materials and also when filming the final assembly.

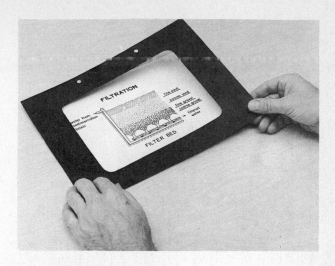

Proportions of Instructional Media

The proportion of each type of media, referred to as the **aspect ratio,** determines the shape of the area within which each visual should be composed.

In the following illustration proportions for the most common forms of instructional media are shown:

Visual Design and Layout

Examine some of the graphic materials that are a common part of your everyday world—magazine advertisements, outdoor billboards, animated cartoons, television titles and commercials, and so on. You can find many ideas for designing your own materials by studying the arrangement of elements in such commercial displays.

The process of combining the various elements that comprise a visual (picture or art work, lettering, texture, color, and so forth), in order to create a pleasing entity, is known as **layout.** The layout must be understandable, legible, and aesthetically pleasing in terms of the purpose it is designed to serve.

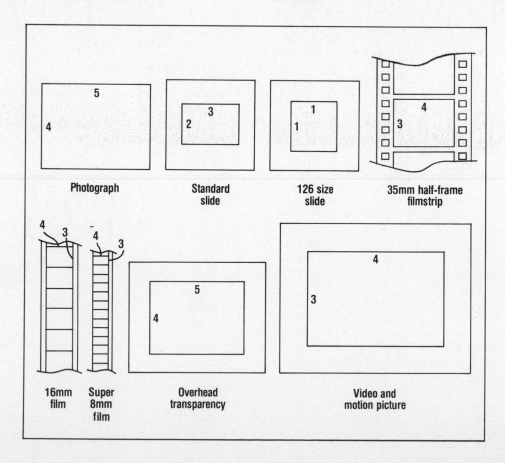

You may be planning a title for a slide series, a diagram, a transparency, a cartoon for a video recording; or your plans may deal with art work for a chart, a diagram, a poster, or even an instructional bulletin board. In these and other planning situations you should be aware of certain design principles and visual design tools. Then be prepared to apply those that can help you.

Design principles include: *simplicity, unity, emphasis,* and *balance*.

Simplicity

Charts, graphs, and diagrams suitable for page printing may not be suitable for projection. They may include large amounts of information and be acceptable in a printed report or for a manual, but these permit detailed, close-up study, which is not usually possible with projected materials. A cutting from a publication, used in a slide, might be so complex that it would be confusing. Therefore, evaluate the suitability of all items you consider for inclusion in your visual materials and try to limit your selection or design to the presentation of one idea at a time.

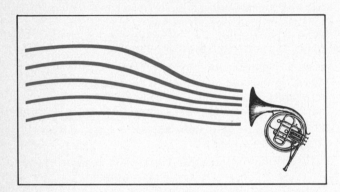

Generally speaking, the fewer elements into which a given space is divided, the more pleasing it is to the eye. Subdivide or redesign lengthy or complex data into a number of easy-to-read and easy-to-understand related materials. Limit the verbal content for projected visuals to 15 or 20 words.

Drawings should be bold, simple, and contain only key details. Picture symbols should be outlined with a heavy line. The necessary details can be added in thinner lines since they should appear less important. Many thin lines, particularly if they are not essential, may actually confuse the clarity of the image when viewed from a distance.

Finally, for simplicity use simple, easy-to-comprehend lettering styles and a minimum of different styles in the same visual or series of visuals.

Unity

Unity is the relationship that exists among the elements of a visual when they are perceived as all functioning together. This is particularly appropriate in display materials and other items to be viewed with little or no direction or guidance. Unity can be achieved by overlapping elements, by using pointing devices like arrows, and by employing the visual tools (line, shape, color, texture, and space) described on the facing page.

Emphasis

Even though a visual treats a single idea, is simply developed, and has unity, there is often the need to give emphasis to a single element—to make it the center of interest and attention. Through the use of size, relationships, perspective, and such visual tools as color or space, emphasis can be given to the most important elements.

Balance

There are two kinds of balance—formal and informal. Formal balance is identified by an imaginary axis running through the center of the visual dividing the design so that one half will be the mirror reflection of the other half. Such a formal balance is static.

Informal balance is asymmetrical; the elements create an equilibrium without being static. It is a dynamic and more attention-getting arrangement. It requires more imagination and daring by the designer. The informal balance may have an asymmetrical or a diagonal layout.

For titles, a symmetrical balance of lettering is formal in effect and is desired for many uses. It requires accurate positioning of letters and extra care when filming to insure even margins (equal side margins, but a somewhat greater area at the bottom than at the top).

Informal arrangements, when appropriately combined with sketches or pictures, make attractive titles. Such arrangements eliminate the problem of centering but not the problem of accurate positioning.

Try various arrangements before doing the final lettering.

The **visual tools** that contribute to the successful use of the above design principles include *line, shape, space, texture,* and *color.*

Shape

An unusual shape can give special interest to a visual.

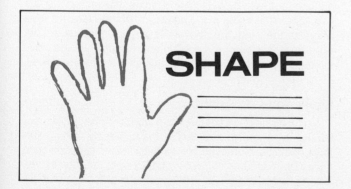

Space

Open space around visual elements and words will prevent a crowded feeling. Only when space is used carefully can the elements of design become effective.

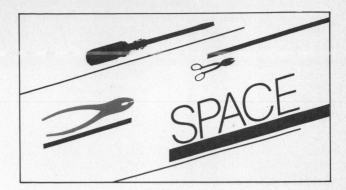

Line

A line in a visual can connect elements together and will direct the viewer to study the visual in a specific sequence.

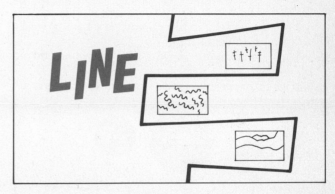

Texture

Texture is a visual element that may serve as a replacement for the sense of touch and can be used in much the same way as color—to give emphasis or separation, or to enhance unity.

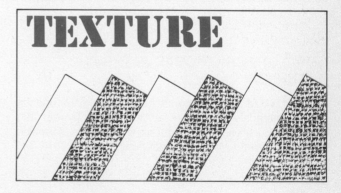

Color

Color is an important adjunct to most visuals, but it should be used sparingly for best effects. Apply it to elements of a visual to give separation or emphasis, or to enhance unity. Select colors that are harmonious together because colors that are dissonant (of equal intensity and complementary on the color wheel, like orange—blue and red—green) create annoyance in the audience and consequently interfere with a clear perception of the message.

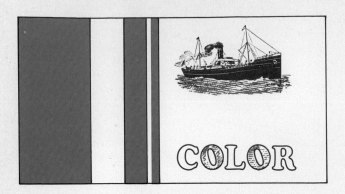

Other emotional impacts of specific colors have been identified. Some common ones are: *red*—danger or action; *orange*—warmth or energy; *blue*—aloofness or clarity; *green*—freshness or restfulness; *violet*—depression; and *yellow*—cheerfulness.

When selecting color for visual materials, attention should be given to three matters. First the *hue,* the choice of a specific color (red, blue, and so on). Second, the *value* of the color, meaning how light or dark the color should appear in the visual with relation to other visual elements. Third, the *intensity* or strength of the color for its impact or coordinated effect.

Review What You Have Learned About Planning Art Work:

1. What is a satisfactory size for a working surface on which to prepare art work?
2. Why is it important to leave a wide margin around art work?
3. For which instructional media would these pictures or art work have satisfactory proportions (dimensions in inches): 8×8, 8×10, 12×18, 9×12?
4. When is a register board used?
5. Which *principles* of design and *visual tools* for design are applied in each illustration?

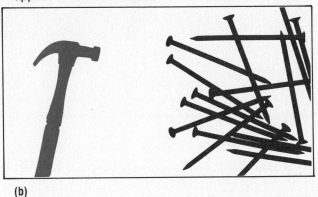

(a)   (b)

USING GRAPHIC TOOLS AND BASIC MATERIALS

In order to obtain professional results when preparing graphic materials, it is necessary to become familiar with and to properly use the basic tools of the trade. Attention should be given to the following items.

Working Surface

A smoothly finished table or a drawing board, preferably with a metal edge to guide a T-square, is required as the surface upon which all work is performed. A drafting table that can be slightly tilted is advisable for convenience when working. Along with the table, a high chair or stool, properly adjusted for comfort, is necessary.

T-square

A T-square rides along the edge of the table or drawing board. It provides an accurate horizontal axis when posi-

tioning art work, drawing horizontal lines, and it serves as a guide for triangles and other devices. When using an ink pen, tape small coins to the underside of the T-square. This will raise the T-square above the surface and prevent ink from smearing under the edge when making lines. Be careful not to move the T-square over lines that have not dried.

Paper and Cardboard

A high-quality bond paper has a smooth finish, will accept ink evenly, and erasures can be made cleanly with no surface damage. Some graphic artists prefer to work on a translucent surface like tracing paper or higher quality vellum. A layout design or a rough sketch placed under this paper can be seen and followed. Cardboard is designated by thickness or "ply" as well as the finish. The cardboards suitable for media graphics range from inexpensive 6–8-ply "railroad" board to 14-ply display or illustration boards, all of which are available in numerous colors. The smooth surface accepts inks very well.

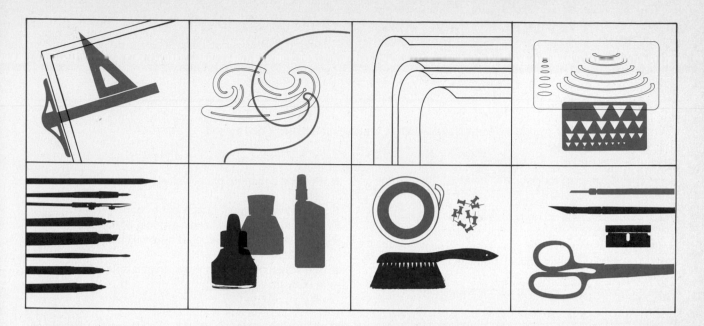

Triangles

Plastic triangles are used against a T-square when drawing vertical and diagonal lines. There are 30°–60° and 45° styles. Other triangles are adjustable for angles from 0° to 90° positions. The same use of coins taped under a triangle applies as described for the T-square.

Curved Surface Tools

The most common curved shape is a circle. A compass is used to form circles. In addition, plastic templates containing cut-out circles, ellipses, or other shapes in various sizes are available. Again, raise the template above the surface with coins as described above. For special shapes, french curves and flexible rods or "snakes" can be used. The latter can be adjusted to form any shape, becoming a guide for a pencil or ink pen.

Cutting Tools

Paper can be cut with a scissors, a single-edge razor blade, or an X-acto knife. The latter consists of a pointed blade in a holder. Be sure to protect the table surface with heavy cardboard before doing any cutting. As a straight-edge guide for cutting, use a ruler with a metal edge. When cutting cardboard, use a trimmer or mat knife. For circles, a compass with a cutting blade is useful.

Pencils

Drawing pencils are coded by their degree of hardness

Soft:	6B 5B 4B 3B
Medium:	2B HB F 2H 3H
Hard:	4H 5H 6H 7H 8H 9H

For sketching, a soft pencil may be preferred, while a harder pencil would be used for final layout over which ink lines will be drawn. A sharp pencil is a much better tool than a blunt one. A light blue pencil, whose marks do not reproduce photographically on high contrast film, is used for layout and writing instructions on a visual.

Pens

There are many types of pens available for drawing lines and filling in areas. Narrow-tipped and broad-tipped felt pens, ruling pens, and technical drawing pens are all used. The technical drawing pens are preferred when crisp, even-edged lines are required. They have a hollow tip fed from an ink reservoir. As the pen is held vertical and the tip is moved lightly over the surface, ink is drawn from the tip onto the writing surface to form uniform lines. Select an ink formulated for technical pen use. Felt pens may bleed on many surfaces. Ruling pens must be handled carefully so ink does not run too freely and form a blot. Keep all pens capped when not in use to prevent the writing tip from drying and clogging.

Erasers

A soft, white rubber eraser will remove pencil marks cleanly. It is available as a block or as a pencil to be sharpened for use on small areas. When marks are close together, place an erasing template with cut-out area over the mark to be eliminated, and use the eraser. An ink eraser is hard and abrasive; rather than using it, consider covering the error with "white-out" correction fluid.

Materials for Securing Art Work

Masking tape most often is used to adhere the corners of paper and cardboard to a drawing surface. Be sure the tape is applied outside of the visible area of the visual as it may damage the surface when removed. By rolling a small piece of tape on itself (with the adhesive side on the outside), it can be placed under the paper or cardboard and will not harm the surface or interfere with a T-square or other tools. Pushpins and thumbtacks can also be used for securing corners of art work. Do not use clear cellophane tape as it damages the art work surface.

Drafting Brush

Keep the work surface and art work clean by sweeping them with a drafting brush. Its soft bristles prevent smearing and scratching.

Review What You Have Learned About Using Graphic Tools and Basic Materials:

Which type of graphic tool or material would you use for each situation?

1. A high quality material as the surface for a display.
2. Drawing lines that would be visible but should not reproduce on photographic film.
3. Locating and drawing horizontal lines across a sheet of paper.
4. Drawing numerous long, smooth lines with ink.

5. Drawing a number of circles, all having the same diameter.
6. Cutting an irregular shape out of paper.
7. Drawing lines to designate vertical columns.
8. Drawing lines on cardboard over which a felt pen will be used.
9. A material on which to draw when it is necessary to see a diagram under it.

ILLUSTRATING TECHNIQUES

In addition to photographed subjects, your script may require illustrations made as original drawings or as copies of available pictures. If you have an art background, you will have little difficulty in preparing such illustrations. If you do not have this ability, you can resort to a number of easy-to-apply methods.

Using Ready-Made Pictures

Pictures from magazines, from free or inexpensive booklets, or from similar sources can serve your needs for some illustrations. If you maintain a file of clipped pictures *(tearsheets)* on various subjects, you may have suitable pictures as called for in your script. At times, part of a picture or combined sections of two or more pictures may serve a need. Mount pictures on cardboard (see page 128) and add lettering if it is appropriate (see page 124). And remember, always, that such pictures may be copyrighted and to use them you need the permission of the copyright holder (see page 57).

For certain general uses **clipbook** pictures are ideal. Clipbooks on many subjects are available commercially (see the list on page 143 for sources). Each book contains a variety of black-and-white line drawings, on paper. These may be cut from the page or, more frequently, duplicated; pictures or copies are then combined with suitable lettering in paste-ups (see page 135) to make titles or visuals. To use these, reproduce on diffusion-transfer material (page 138), copy them on a high-quality copy machine (page 141), or photograph them using high-contrast film (page 139) and print the negative on photographic paper (page 102) for use in the visual.

In addition to line drawings, printed sheets containing arrows, circles, stars, and other symbols in multiple sizes are available. They are used directly or can be reproduced.

Another type of ready-made pictures, similar to clipbooks, is **dry-transfer** art. Pictures are printed on the underside of an acetate sheet. A picture is transferred to the

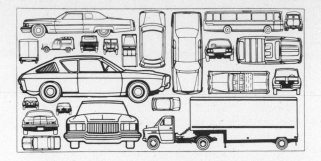

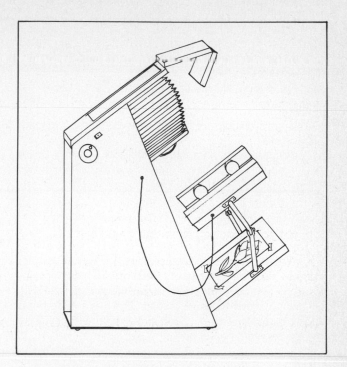

working surface with a burnishing tool in the same way that dry-transfer letters are used (see page 125).

If a picture cannot be used directly or easily reproduced, you can place a sheet of translucent tracing paper over the picture and outline the main lines with a pencil. Then transfer the tracing to a cardboard or other material by backing the tracing paper with a sheet of carbon paper and tracing over the lines.

Enlarging and Reducing Pictures

The most acceptable way to adjust the size of a picture is to photograph it. Both the diffusion-transfer process (page 138) and high-contrast film (page 139) can be used to make a copy of a black-line illustration. Also, certain office copy machines permit a percent of enlargement or reduction to be made.

A device, especially designed for enlarging and reducing art work, is the **photo modifier.** It resembles a large view camera (page 74) with a ground-glass back against which tracing paper can be taped. (See illustration at top of next column.) The size of the original picture can be reduced or enlarged in accurate proportion by moving the device and then focusing the image by adjusting the bellows. Perspective can be changed and distortion created by tilting the ground-glass surface or the front lens.

In addition, a small picture on a single sheet or in a book can be enlarged by using an **opaque projector.** Place the paper or book on the holder of the projector and attach a piece of cardboard to a wall. Adjust the size of the projected picture to fit the required area on the cardboard by moving the projector *closer* to the cardboard *(to be smaller)* or

farther away from the cardboard *(to be larger)* and focusing as necessary. Then trace the main lines of the projected picture with pencil. After completing the drawing, ink in the lines using pen and ink or a felt pen. This is one of the easiest and quickest ways to enlarge a picture.

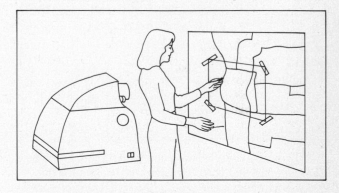

If a transparency or a slide of the original diagram is available or can be made by one of the processes to be described later in this book, an overhead projector or a slide projector can be used to make an enlargement.

But also with the **overhead projector,** large pictures can be *reduced* to fit $8\frac{1}{2} \times 11$ inch or other formats. This technique uses the overhead projector in reverse fashion as compared to its normal enlarging use. The original, large diagram is attached to a wall and a light (floodlight or a slide projector) is aimed at it. Sufficient light must be reflected from the diagram through the lens of the projector to be visible on a white sheet of paper placed on the projection stage. A cardboard light shield, placed on the edge of the projector stage, will block out stray light and make the image more visible on the paper. Move the lens up and down to focus the image on the paper. Control the size by moving the whole projector closer to the wall or farther from it. Sketch the visual over the image on the sheet of paper.

Review What You Have Learned About Illustrating Techniques:

Which method of illustrating would you use for each situation?

1. Enlarge a diagram of a map without having a camera available.
2. Locating pictures on a current topic for a bulletin board display.
3. Tracing a map after changing the size from the original six inches to nine inches.
4. Line drawings of persons in vocational trades, copied without damaging the original illustrations.

VISUALIZING STATISTICAL DATA

Illustrating the numerical relationships among various factors and visualizing tabular information are important applications of graphic methods. These are accomplished by designing various types of graphs. There are three basic kinds of graphs—*line, bar,* and *circle.* A different purpose may be served by each one.

Line Graph

A line graph usually relates a factor of changing quantity to successive time periods. Points, representing quantities at each time period, are joined together as a line or a smooth curve.

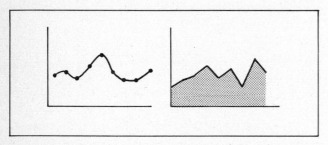

If the space under a line graph is shaded, it becomes a *surface* chart and gives emphasis to this area under the curve. The effect will dramatize the data being presented.

Bar Graph

A bar graph makes simple quantitative comparisons. The length of the bars represent amounts, and when drawn side by side, relationships and changes can be observed.

When a bar chart is arranged vertically it becomes a *column*

chart. Such a graph can make comparisons at various points in time.

Circle or Pie Graph

A circle graph illustrates the proportion of the whole that each element of a subject represents. The elements are numerically shown as percentages. In total, they add to 100 percent.

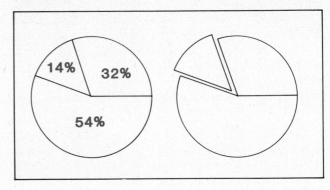

When preparing a graph, in addition to the actual graph lines, bars, or circle segments, give attention to these elements:

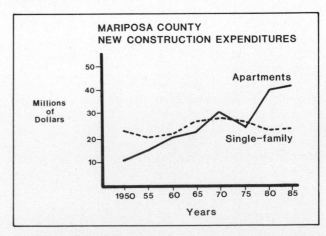

- Title—briefly describes the topic treated in the graph; preferably placed at top of the visual.
- Scales—shown as vertical and horizontal axes and section lines for line and bar graphs; spacing between adja-

114

cent sections should permit a fair interpretation of data (not too compact or exaggerated); numerical values are written for each section.

- Axes captions—each scale (horizontal and vertical) requires a brief descriptive phrase or caption.

- Labels—as necessary for identification, label graph lines, bars, and circle segments; if parts are visually distinguishable (solid—as opposed to dashed, dotted, or hatched; different colors; and so forth), identify components in a boxed legend.

Review What You Have Learned About Visualizing Statistical Data:

Which type of graph would you use for each situation?

1. Relating product sales to the three divisions of a company for a three-year period.
2. Showing proportionate uses of energy in the average home for heating and to run appliances.
3. Illustrating the consumption of gasoline in a state for a long series of months.
4. Comparing the amounts of calories found in ten different foods.
5. Indicating the extreme change in the population of an endangered species for the last ten years.

USING COMPUTER GRAPHICS SYSTEMS

Computers are now being used to produce media such as camera-ready art (charts, graphs, cartoons, and illustrations), 2×2 inch slides, overhead transparencies, video graphics, and animation. Computer graphics systems offer many advantages to organizations producing large quantities of visual materials. These advantages include:

- **Increased speed of production.** Generating a typical word slide (a few lines of text on a colored background) by conventional graphic techniques requires such intermediate steps as typesetting, paste-up, high-contrast photography, and color photography. The entire process can take from 15 minutes to several hours, depending on the procedures used and the complexity of the slide. Producing a word slide on an efficient computer graphics system, eliminates almost all handwork other than entering information through a keyboard—a process that typically takes a trained operator just a few minutes.
- **Ease of revision.** With conventional production methods, even a "minor" revision can result in many hours of work. But since the information used to construct a computer-generated visual is stored electronically, it takes only minutes to retrieve the stored information in the computer, use the keyboard to make revisions, and generate the revised visual.
- **Increased emphasis on design.** Because the color, size, location, or style of the elements within a computer generated visual can be altered rapidly, a computer graphics operator is able to explore alternative layouts before selecting the final design. Thus, time that might normally be spent on more mundane production tasks (lettering, paste-up, etc.) can now be devoted to design considerations.
- **Reduced storage space.** Dozens, or even hundreds of visuals can be stored electronically in the space normally occupied by a single piece of conventional art.

Components

While computer graphics systems vary somewhat in appearance and operation, they tend to have components with similar functions.

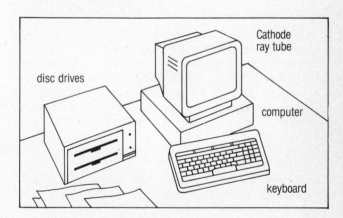

The heart of any computer system is the computer itself, where information is processed. However, to be of use, a complete computer system must also have a variety of devices which can input, store, and output information.

As the computer processes information, it displays the results in the form of text or graphics on a video monitor or cathode ray tube (CRT). This display often prompts the user to supply additional information (input).

Input devices

The most common input device is the keyboard. It may contain special keys for computer functions. However, a variety of additional devices are available which enable a user to manipulate elements on the screen.

Graphics tablets or digitizers have specially designed pens which are connected to the computer by a wire and touched to a sensitized surface. A symbol, called a **cursor,**

115

then appears on the screen and follows the movements of the pen. These devices are used extensively to trace a picture into the computer memory, or to select the picture element that is to be manipulated.

A light pen acts much like the pen of a graphics tablet except it is touched directly to the screen of a CRT.

Some computer graphics systems have video digitizing cameras, called *image grabbers,* which input existing art by photography. Once in the system, the art can be modified by eliminating unwanted elements, manipulating color, inserting text, or adding special effects such as a glow or halo.

Storage devices

In order to function efficiently, a computer must be able to store and later retrieve information and/or instructions. A variety of storage devices is available. The simplest and least expensive is a tape recorder, which often uses a standard audiocassette. However, tape systems are usually

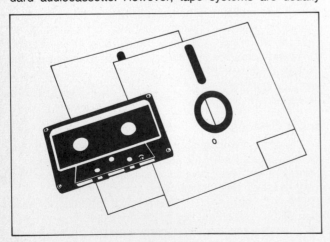

slow and unreliable, and are recommended for only the simplest of computer systems.

Disc drives record information on a magnetic floppy disc. They offer much greater speed and reliability. Larger, hard disc systems are available which further increase speed and greatly expand the storage space. However, their cost restricts them primarily to large mainframe computers or the more expensive microcomputer systems.

Output devices

Once a computer has finished processing information, the results can be fed to an output device. The most common is the video screen, as previously mentioned. Information from a screen, or directly from the computer memory, can be printed on paper through a printer attached to the output of the computer, Some printers have a **dot matrix imaging** system which allows them to produce visual materials such as graphs or drawings. Illustrations produced in this manner are fairly coarse, but are suitable as proofing copies.

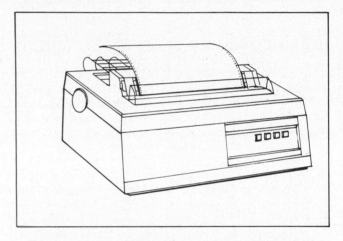

More expensive **ink-jet** printers work much like dot matrix printers except they create images by spraying colored inks in a very fine dot pattern onto paper (for reports or camera-ready copy) or on acetate (for overhead transparencies).

The **plotter,** another output device, uses a mechanically controlled ink pen to trace an image on paper or acetate. Multicolor images can be produced by changing the pen tip.

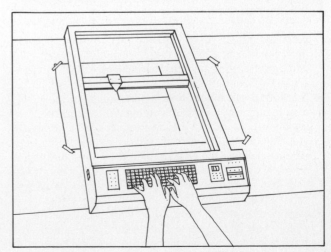

This device is excellent for reproducing line drawings, but illustrations with excessive detail or shading may require as much as an hour to reproduce.

Full-color graphic images can also be reproduced using special cameras called **film recorders** which photograph the images from a high-quality CRT. Some film recorders can photograph in a variety of formats ranging from 35mm slides to 8×10 inch overhead transparencies or 16mm motion picture film. Others incorporate sophisticated image enhancement equipment to improve the resolution of the visuals.

Some computer graphics systems output the image in a form which can be recorded directly onto videotape and used for video graphics or video animation. In this situation, the videotape recorder serves as an output device.

Software

Software provides the instructions a computer needs to perform a repetitive task. Programs can be purchased that will guide you in generating various kinds of graphics (graphs, charts, standard illustrations, and so forth). For example, a program designed to construct graphs will query the computer operator for the necessary information and then manipulate that information to produce a usable graph. Without such software, the computer is of little value for graphics work.

Operating the System

For most routine visuals, computer graphics systems work from standard formats. Often they use a **menu** which lists the options that are available at a particular point in the development of a visual. The operator simply selects the desired option from the menu.

For example, generation of a word slide begins with the computer asking the operator to specify several basic parameters—size and style of type, location and arrangement of text, desired colors for text and background, and so forth. The operator then types the text into the computer and it appears on the CRT for preview and correction. When the image is acceptable, the operator either routes the information to an output device or stores it for later reproduction.

Computer graphics systems are also good for producing charts and graphs since these are visual representations of numerical data. To produce a graph, the computer may ask the operator to specify the type of graph in addition to the basic information requested for word slides. It may also ask for other relevant information—the graph title, axes labels, scale ranges, and scale increments.

Drawings that are not mathematically based, such as a diagram of the heart, provide a greater challenge to a computer system. Such drawings are usually stored in the computer's memory by tracing the outline on a graphic tablet. They can be recalled and manipulated (enlarged, reduced, colored, modified) to produce an appropriate visual. This allows you to create an electronic clip art file of visual elements (maps, etc.) which can be rapidly modified and used in several different illustrations.

Image Resolution

A computer-generated image is composed of rows of tiny dots called **pixels,** which are the smallest manipulable portion of an image. The sharpness or resolution of the image is directly proportional to the number of lines (rows of pixels) comprising that image.

Visuals created on systems containing relatively few lines (for example, 200) appear to be extremely coarse. Curved lines drawn on such systems take on a sawtooth or ragged appearance. Visuals created on systems containing many lines (for example, 2000) have much higher resolution, and curved lines drawn on such systems will appear smooth. For comparison, a television image approximates the resolution of a computer graphics system with 500 lines.

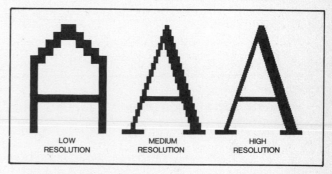

LOW RESOLUTION — MEDIUM RESOLUTION — HIGH RESOLUTION

Image resolution is the factor that most affects the price of a computer graphics system. A working 200-line system can be assembled for a few thousand dollars. However, the resolution of such a system limits the value of the material produced on it. A full-scale computer graphics system with 2000 to 4000 lines of resolution can cost much more. Such systems are capable of producing art work that only a trained person can differentiate from art work produced by conventional means. If your instructional needs do not require high-resolution graphics, then a low-resolution system might be adequate.

Typical Approaches

There are several approaches to the use of computer graphics. The approach you take will depend upon your needs, your budget, the volume of work, and the availability of outside services.

- **Complete systems.** Complete in-house computer graphics systems typically consist of a computer with keyboard and graphics tablet inputs, a floppy disc drive, a color CRT, and an output device such as a film recorder, a plotter, or a video recorder. Options might include a proofing device and a hard disc drive. The larger systems may provide multiple terminals so that more than one operator can work at a time. These systems generally provide 35mm slides, but with options can also include overhead transparencies, paper prints, and video images.
- **Partial systems.** In an attempt to save money, organizations frequently invest in partial systems which do not include output devices such as a film recorder. The operator uses the computer system to design the visual and

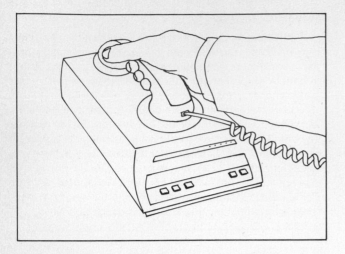

then transmits the information by telephone line from the computer, using a device called a **modem,** to a remote site having a film recorder where the visuals are produced. Production turnaround time is usually 1 to 3 days and a service charge is paid for each slide produced. Systems such as these have reduced the start-up costs for computer graphics systems. By using a microcomputer as the graphics design station, and transmitting by a modem to a remote film recorder, high resolution art can be created for word slides, charts, and graphs.

● **Computer graphics production services.** Computer slide services that will prepare slides on a per-item basis are available in many areas. You submit a job request to the service and a trained computer operator produces the slide to your specifications. Prices for such a service can vary from twenty to several hundred dollars per slide, depending upon the complexity of the visual. Average prices range from twenty-five to eighty dollars per slide or transparency.

Review What You Have Learned About Computer Graphics Systems:

1. In what final forms can computer-generated graphics be produced?
2. What are *three* advantages for using a computer to generate graphics over regular hand methods?
3. Categorize each of the following as an *input* device, *output* device, or *storage* device.
   a. disc drive
   b. graphics tablet
   c. video screen
   d. printer
   e. keyboard
   f. plotter
   g. film recorder
4. Define these terms:
   a. modem
   b. software
   c. menu
   d. pixels
5. When quality and price of a computer graphics system are important considerations, what matter should receive careful attention?
6. If your organization cannot afford a high-quality computer graphics system, what are the alternatives for obtaining computer-generated graphics products?

COLORING AND SHADING

In addition to using color to make visual materials more attractive, the choice of colors can make the communication process more effective (See the reference by Green.) Using certain color combinations or coloring selected parts will contribute emphasis and even clarification to a complex diagram. For example, since yellow and orange are colors of high visibility, black or blue lines on a yellow or orange background will command more attention than black on white.

Those who have art backgrounds may be able to use such techniques as wash drawing and air brushing; even those with limited training may consider using several simple techniques.

Felt Pens

Colored lines of various thickness can be made with felt pens. Both permanent and water-based inks are available

in a variety of colors. Felt pens are useful for coloring small areas. Since the colors are transparent, apply them carefully; each overlapping stroke deepens the tone and may produce uneven coloring in large areas.

Spray-Can Paints and Airbrushing

Paint in pressurized cans can be used to apply color to areas on a visual. This technique is also used to color cardboard around three-dimensional letters which are removed, after spraying, to reveal the unpainted image of the letters. By controlling the distance between the spray can and the surface to be sprayed, either a spatter effect or an even paint coverage can be achieved. Protect parts of the visual not to be colored by covering with paper attached with masking tape.

 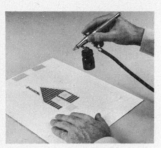

Carefully controlled color spraying can be done with an **airbrush** attached to a compressed-air line or to a pressurized-air can. By adjusting the airbursh nozzle and depressing the control knob the spread of spray is controlled. As with the spray-can paints, it is necessary to cover parts of a visual around the area to be colored.

Colored Art Paper

Various kinds of paper can provide color and emphasis for areas of a visual, can serve as backgrounds, or can direct attention when used for arrows and special shapes. Construction paper, available in many colors, can be used for any of these purposes. But its tones are not rich and its coarse surface, when photographed, reflects light poorly. These limitations result in dull colors on slides, filmstrip frames, and motion pictures.

A better material is color-coated art paper, which does not fade and accepts ink readily. It is available in a wide range of strong and vibrant colors and tones. Most colors reproduce reliably on film.

Color and Shading Sheets

Prepared color and shading sheets are excellent for use on areas of any size. They are available in a wide range of patterns and colors, both transparent and translucent. They will adhere to all surfaces.

Translucent color sheets have an adhesive wax backing, which makes them partially opaque. Such sheets should be used to color art work prepared on cardboard. Transparent color sheets are prepared with a clear adhesive backing so the color will project brilliantly when applied to an overhead transparency or to other visuals for projection.

The shading or color is printed on a thin plastic sheet, which has an adhesive backing. This in turn is protected with a backing sheet. To work with these sheets, use a razor blade or a sharp X-acto knife and follow the procedure illustrated on the next page.

1. Place a sheet of the selected material over the area to be shaded or colored.
2. Lightly cut a piece slightly larger than the area to be colored or shaded. (Try not to cut through the backing sheet.)
3. Peel the cut piece from the backing.
4. Place the cut piece of adhesive-backed material over the area and rub to adhere.
5. Cut to match the area, using lines in the diagram as guides.
6. Peel off the excess pieces of the coloring or shading material.

BACKGROUNDS FOR TITLES

Select backgrounds that are appropriate to the treatment of the subject and that do not distract attention from the title. Such backgrounds will be inconspicuous in color and design, yet will contribute to the mood or central idea of the topic. Cool colors (blue, gray, green) are preferred for backgrounds, and warm colors (red, orange, magenta) for titles and visuals over the background.

For backgrounds you may consider plain, colored, or textured papers; cardboards of various finishes and colors; cloth, wood, or other unusual materials like wallpaper samples, or pictures and photographs. Rich colors give maximum contrast in black-and-white and pleasing effects in color. But do not let the background design or extremes of color interfere with legibility or with the purpose of a title or diagram. Refer to the visual design principles on page 107 for additional suggestions.

Special Techniques

For most uses, prepare simple titles directly on the background material. But for special purposes, you can make an *overlay* and place the title over the background before filming. Overlays are particularly useful when several titles or diagrams must appear over the same background. Lettering for overlays can be in black, in white, or in color.

Illustrations of the three following overlay techniques can be found on page 121.

119

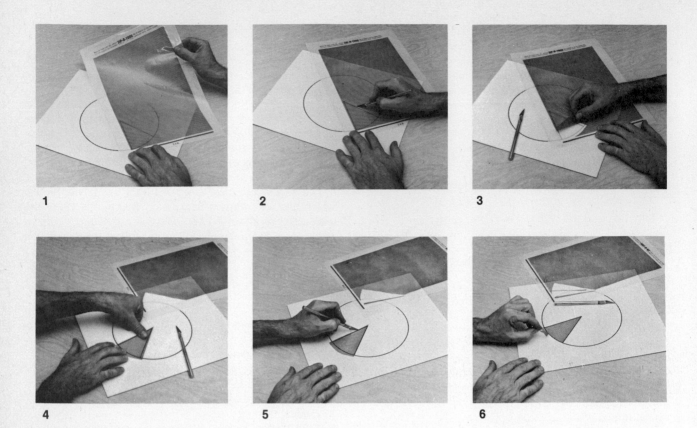

1

2

3

4

5

6

Black Overlay Lettering

- Adhere punch-out letters, cutouts, or dry-transfer letters directly to the background (page 125).
- Make regular black-line thermal (page 179), diazo (page 181), or photographic (page 185) transparencies and overlay them on the background.

White Overlay Lettering

Put white letters on black nonreflecting cardboard, or film black letters on a white background with high contrast film to make a negative (page 186). Then, during filming, dou-

ble-expose the film to record first the background and then the white title.

Color Overlay Lettering

- Use colored dry-transfer letters directly on the background or on clear acetate.
- Make single color diazo or thermal transparencies (page 181).
- When more than one color is needed on a title, prepare different color diazo or thermal transparencies.

For applications of these techniques in preparing slides, see the section on Titling for Slides on page 199.

Review What You Have Learned About Coloring, Shading, and Backgrounds for Titles:

1. How does the use of felt pens compare to the use of spray paints for coloring moderate-sized areas?
2. What two matters should be considered when selecting a paper for coloring an area on art work?
3. When using a color adhesive sheet why is it *not* proper to cut through the color sheet *and* the backing sheet when first cutting the piece for use? Also, why should you *not* cut the piece to be used the exact size at first?
4. Of the coloring methods described, which one might be selected to carefully tint a large, irregular area?

5. Which color combinations are preferable when preparing titles on backgrounds?
   a. gray background, orange title
   b. green background, magenta title
6. To which special category for preparing title backgrounds does each apply?
   a. red, dry-transfer letters on the background
   b. black thermal transparency over the background
   c. expose the background scene, then expose lettering on a black-and-white negative

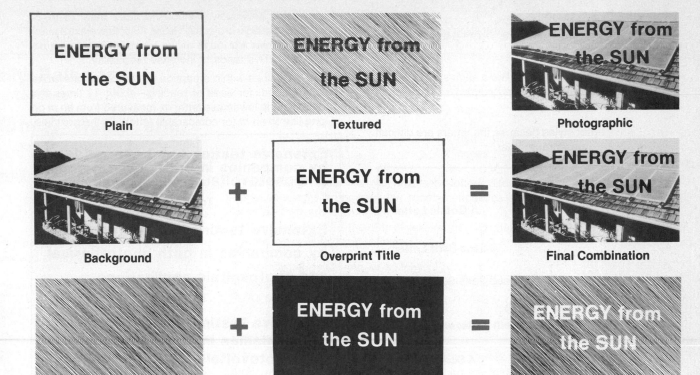

| Plain | Textured | Photographic |

| Background | **+** | Overprint Title | **=** | Final Combination |

| Background | **+** | Overlay Title | **=** | Final Combination |

Allow 1½ letter wid
3 widths betwee
space again mak

Hold lid fir

Hold lid

Hold lid fi

Contrast the lette
ground so that se

SOLAR
COLLECTORS

Red Lettering on Yel

Light letters again
visibility than do da

For satisfactory leg
background require
letters on a dark b

DRUG
REHABILITATI

Table

ME

SLIDE
FILMS
TRAN
MOTI
VIDEO

Sourc
Co., R
situati

LEGIBILITY STANDARDS FOR LETTERING

The legibility of the words, numerals, and other data that an audience is expected to read is frequently neglected during the planning and preparation of visual materials. This neglect is especially common since simple and quick methods have become available for duplicating typewritten and printed materials. An illustration with lettering copied from a book usually is not as easy to read when projected on a screen as it is on the printed page.

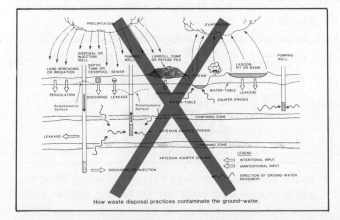

How waste disposal practices contaminate the ground-water.

It is very important that planners give proper attention to legibility—hence to methods of lettering, sizes of letters, and styles of lettering. These matters physically control the amount of information that can be presented in one visual unit. Reciprocally, the psychological limits on amount of information affect the choices in respect to lettering. Keep in mind, therefore, the methods for dividing lengthy or complex data into a sequence of visuals (page 175); see the discussion of layout and design earlier in this chapter.

The suitability of lettering is further complicated by a number of other factors—characteristics of the meeting room, such as its shape; the type of screen surface (rear-screen transmitted projection is not as brilliant as front-screen reflected projection and will require larger lettering for legibility); and the amount of ambient or outside light that cannot be controlled.

If your instructional media are designed for use in a specific room, then take into account as many of these factors as possible in deciding on lettering sizes. But no one can predict or be prepared for all eventualities in viewing situations. Often visuals must be presented under less than ideal conditions; therefore, it is advisable that *minimum standards* be recognized. As a general guide, select minimum lettering size for *all* materials so that any member of an audience, seated at an anticipated maximum viewing distance, can easily read titles, captions, and labels. If you do not heed this advice, you are likely to find members of the audience losing interest in your presentation because they cannot read the lettered information.

The information that is presented as the following guidelines applies equally to projected and nonprojected media types. They also need consideration as words, expressions, and sentences are designed for the computer CRT screen.

121

Guidelines

To assist with
recommended:

- Select a rea
  type, which i
  strokes.

- Avoid script
  to distinguis

MANY

MANY I

MANY BUY

MANY

Many I

𝔐𝔞𝔫𝔶 𝔅𝔲𝔶

- For a visual,
  verbal inform
  typefaces (tw
  monize with e

STRUCT

1 Amino
   (AA)

   Am

2 Pepti

3 Polyp

- Use capital le
  longer caption
  lowercase lette
  ercase letters

A HISTO
OF
OUR CHUF

Rear-screen projection does not permit as bright or as contrasty an image as does front protection. For suitable legibility, lettering half again as large as for front projection is required for rear-screen projected materials.

The maximum group viewing distance for video, in terms of the screen size, is greater—16H to 24H. As you might expect, minimum letter size for video is therefore greater.

In reviewing the research on legibility of projected information, Sharpe concluded that a suitable ratio for minimum letter size should be 1 to 30 rather than 1 to 50 and letter height should be $\frac{1}{4}$ inch minimum rather than $\frac{1}{8}$ inch. Thus, you should recognize the value of using lettering that is larger than the minimum shown in Table 11-1.

You can make a rough test of the legibility of lettered materials for projection by first measuring the width of the art work in inches, then dividing this number by 2 and placing the material that many feet away from a test reader. If the person reads the lettering easily, then for normal conditions the material, when projected, will be legible. But don't trust yourself as a test reader if you prepared the lettering

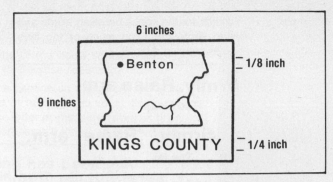

or know how it should read—your memory may help your vision too much.

The recommendations for insuring legibility of visual materials are only for your guidance. Be alert to special conditions in any situations—seating arrangements, light level, image brightness, and so forth. These may require larger images or bolder lettering to insure satisfactory legibility.

Review What You Have Learned About Legibility Standards for Lettering:

1. What is the single most important reason why legibility standards should be considered for projected materials?
2. How might you relate the degree of legibility to quantity of information possible in a visual?
3. What are five of the guidelines that will contribute to good legibility in lettering?
4. What is the minimum letter size for materials displayed at the front of an average-sized classroom—30 feet deep?
5. What is the minimum size for lettering used in the direct preparation of a transparency ($7\frac{1}{2}$ inches vertical dimension)?
6. Should visuals for video use be lettered larger, smaller, or the same size as materials for regular classroom use?
7. Does rear-screen projection require larger or smaller lettering than comparable front-screen projection?
8. At what distance should legibility be checked when lettering is prepared on a 9″ × 12″ working area?

LETTERING TECHNIQUES

However good the photography and picture content of visual materials, their effectiveness is enhanced by well-appearing titles, captions, and labels. Neat lettering, simple designs, and attractive colors or background patterns all add a professional touch to your materials.

Titles generally require large, bold letters. Since there are relatively few major titles, their letters may be hand drawn, or set individually in place by hand. But such methods may be too slow for preparing captions that consist of many words; here other lettering techniques are appropriate, adequate, even better. You need to know and select techniques with regard to the results needed and the time available for preparation of your materials.

Some remarks follow concerning nine specific lettering techniques. No one technique is necessarily the best for any lettering job. You need to evaluate as many of them as you can—for your own needs, in respect to availability, cost, ease of use, time required for preparation, and resulting quality.

Typewriter Lettering

Lettering prepared with a typewriter is good for captions requiring many words. This form of lettering is also satisfactory for preparing labels and simple titles. A boldface type (called Bulletin or Orator type) provides legibility that is superior to that of the pica or elite type on regular office typewriters unless the latter are photographed and enlarged. Use paper of good quality, whether white or colored. Make sure the typewriter has a good grade carbon ribbon for preparing crisp-looking letters.

Produced With
The Assistance Of
Dr. John Marsh, M.D.
Martin County Hospital

Felt-Pen Lettering

The use of the felt pen for coloring has been described on page 118. These pens can also be used for lettering. For successful results:

- Hold the beveled-tip pen firmly in a "locked" or set position in the hand.
- Make no finger or wrist movement. *All* movements should consist of arm movements.

Sharp-tipped nylon pens make a thinner mark than the beveled-tip felt pens. They are easier to use and are good for quick lettering on all surfaces. The inks in some make permanent marks, but most have water-based inks.

Always replace the cap on a felt or nylon pen as soon as you are done using it. Since their inks dry quickly, uncapped pens will dry out, resulting in a hardened unserviceable tip. If this happens, soak the tip in lighter fluid (permanent-ink pen) or in water (water-based-ink pen).

Dry-Transfer Letters

These letters have sharp, clean edges much like those printed from good type, and are easy to handle. They come in sheets of many sizes, styles, and colors. They are excellent for titles and labels—on many types of backgrounds. While suitable for short headings, their use for materials requiring many sentences would require a lengthy preparation time.

Dry-transfer letters are printed on the back of the sheet and each sheet is backed with a protective sheet of paper. It is important to keep this backing sheet behind the letters except when exposing a portion of the letter sheet for use.

Follow this procedure in using dry-transfer letters:

1. Slip the backing sheet below or above the row of letters having the letter to be used.
2. Position the letter by aligning the printed line under the letters over the guideline drawn on the mounting surface.

1

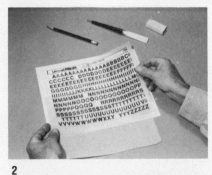
2

3

4

5

6

3. Burnish (rub) the entire letter to the mounting surface with a commercial burnisher, the round part of a pen, or other blunt object on which you can exert pressure without tearing the paper.
4. Slowly pull the sheet of letters from the mounting surface. The letter will remain transferred.
5. To secure the letters to the surface, place the backing sheet over them and burnish firmly.
6. Replace the backing sheet behind the letter sheet. Then erase the guideline from the mounting surface.

Dry-transfer letters are also available in transparent colors for direct use on transparencies. When they are adhered to acetate, a clean transfer results with no adhesive residue appearing around the letter.

If a dry-transfer letter must be removed from paper or cardboard, this can be done by firmly sticking a piece of masking tape on the letter. Then carefully pull the tape up. A new letter can be adhered in the same place.

Cut-out Letters

Inexpensive ready-to-use letters, cut out of construction paper or gummed-back paper in many styles, colors, and sizes, are easy to manipulate and are satisfactory for bold titles. They can be placed over any background. To align them neatly, use a T-square or lightly rule a guideline on the mounting material. Position the letters on guidelines. Attach construction paper letters with a small amount of rubber cement or other adhesive. Moisten the adhesive on the gummed-back letters with a sponge dipped in water. After the letters are in place, erase the guidelines.

Three-Dimensional Letters

Three-dimensional letters are manufactured in cardboard, wood, cork, ceramics, and plastics, and are available in plain backs or pin backs. They are excellent for main titles, and when photographed with side lighting they give shadow effects and three-dimensional effects. Costs vary widely according to kind and size. Surfaces can be tinted with paint or water colors. Position the letters against a T-square or on a guideline and adhere temporarily with rubber cement.

Stencil Lettering Guides (Wrico Signmaker and Wricoprint)

Stencil lettering guides are offered in a variety of styles and sizes. The better ones can be used, after a little practice, to produce neat lettering, even in lengthy captions.

These stencils are designed to be raised off the background and positioned against a metal guide. A special pen is used, which fits and follows the letter outline in the plastic stencil.

Lettering from $\frac{1}{2}$ inch to 4 inches in height can be made with the *Wrico Signmaker* unit. It consists of a *stencil guide,* a *brush pen* or special *felt pen,* and a *guide holder* to raise the guide from the paper.

For lettering $\frac{1}{2}$ inch tall and *smaller,* use the *Wricoprint* unit. It consists of three parts—a *lettering guide* a *pen,* and a *lettering pad.*

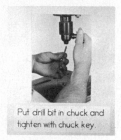

Put drill bit in chuck and tighten with chuck key.

Another stencil-type guide for smaller size lettering, similar to Wricoprint, is the *Koh-i-noor Rapidoguides* and pens. The unit consists of a plastic stencil guide having raised edges and a rapidograph pen, which is held vertically as the point follows the letter cutout in the guide.

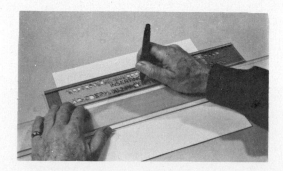

Template Lettering Guides

Lettering templates produce precise, uniform characters, especially for captions of many lines. Although more expensive than stencils, *Leroy* lettering guides permit faster work and give higher quality lettering. You can choose lettering in sizes from $\frac{1}{16}$ inch to 2 inches in height. An ink reservoir pen is used in a tripod scriber, one leg of which follows the letters grooved in the template.

WATER PURIFICATION
1. Aeration
2. Coagulation
3. Sedimentation
4. Chlorination
5. Filtration

Another manufacturer, *Varigraph,* makes a widely used lettering system that consists of a scribing device (called a mechanical typesetting instrument), a metal matrix (the template), and pen. By adjusting two scales on the instrument a wide range of letter sizes and shapes can be made.

Pressure-Machine Lettering (Kroy)

High-quality, professional lettering can be prepared with a table-top lettering machine. The unit includes a variety of interchangeable plastic discs, each with its own type style and letter size. The disc is placed in the machine and the selection of letters, numbers, or symbols is controlled manually by dialing or pushing a button. Pressure is applied to a character, causing an impression on carbon paper to transfer to film having a paper backing. The film is peeled from its backing and pressed on paper or cardboard for use.

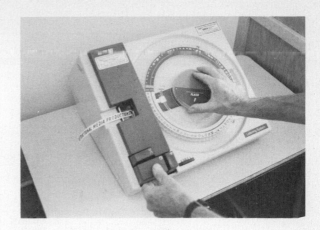

Phototypesetting

This equipment usually consists of a keyboard and a microcomputer which controls a typesetting device. The desired text is either entered through the keyboard or loaded from an attached storage device such as a disc drive. The microcomputer causes the typesetting device to project letters from a negative, through a lens system, onto photographic material, which must then be developed. By changing lenses, a variety of type sizes can be created. Some systems will even compose the text at the desired location on the page, eliminating some of the paste-up work.

For a summary chart comparing the various lettering techniques, see the next page.

Review What You Have Learned About Lettering Methods:

1. What rules guide the holding and using of a beveled-tip felt pen?
2. How do you line up and space punch-out letters?
3. What purpose is served by the backing sheet with dry-transfer letters?
4. What kind of tool can be used with dry transfer letters?
5. What method of lettering might you select to prepare a caption of 12 words?
6. Which type would you select when a great amount of carefully formed lettering must be done with ink?
7. What is the name of a machine that can be used to impress letters from a disc, through carbon, onto film?
8. Compare the lettering methods available to you in terms of best uses, time and skill to use, relative cost of equipment if required, and quality of results.

Table 11-2   **Summary Comparison of Lettering Methods**

Type	Main Use	Quality	Relative Equipment Cost	Relative Materials Cost
Typewriter	Credits Captions Sentences	Good with carbon ribbon	High	None
3-D letters	Titles	Excellent with care in lighting when photographed	Low	None
Construction or Gummed-back paper cutouts	Titles	Good to superior if aligned carefully	None	Low
Dry transfer	Titles Short headings	Excellent if handled carefully and surface protected	None	Moderate
Felt pen	Titles Labels	Good if lettering performed with care	None	Low
Stencil	Titles Captions	Fair with careful use of pen	Low	None
Mechanical	Titles Captions Sentences	Good with careful use of scriber and pen	Moderate	None
Pressure	Titles Captions Labels Sentences	Excellent and fast to use	High	Moderate
Phototypesetting	All uses	Excellent	High	Moderate

MOUNTING FLAT MATERIALS

A variety of techniques can be considered for mounting art work and for preserving finished visual materials and commercially available pictures, maps, charts, and so forth. These techniques have various characteristics: Final results are temporary or permanent; special equipment may or may not be required; heat and or pressure may or may not be required; and sealing may be on cardboard, cloth, or other surfaces. Because of the variety of these methods, you should study them carefully. Evaluate them in terms of materials, and your needs.

The methods to be described on the following pages include:

- Rubber cement mounting
- Dry mounting on cardboard
- Pressure-sensitive mounting
- Dry mounting on cloth

Rubber Cement Methods

Mounting with rubber cement is a simple procedure that requires no special equipment. It will accomplish temporary or permanent mounting.

Temporary mounting

Temporary mounting is useful for making paste-ups (page 135) of line drawings and accompanying lettering that are to be photographed rather than used directly:

1. Trim the material to be mounted.
2. With rubber cement, coat the back of each piece to be mounted.
3. Place the coated pieces cement-side down on the cardboard, while the cement is wet. They can be moved as necessary to get exact position and alignment, or picked up and repositioned.
4. Allow the cement to dry before using the paste-up. Rub away any visible cement.

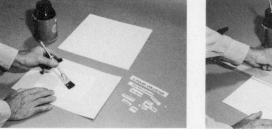

1   2

3

Permanent mounting

Permanent mounting is not truly permanent, but materials thus mounted with rubber cement will adhere for long periods. Cement is spread on back of the picture *and* the cardboard surface. There are twelve steps:

1. Trim the picture or other piece to be mounted.
2. Place the picture on the cardboard backing; make guide marks for each corner.
3. Coat the back of the picture with rubber cement.
4. Coat the marked area on the cardboard with rubber cement.
5. Allow the cement to dry on both surfaces.
6. Overlap two waxed paper sheets on the cement covered cardboard after the cement is dry.
7. Align the picture on the guide marks as seen through the waxed paper.
8. Slide out one sheet of waxed paper.
9. Smooth the picture to the cardboard on this exposed cement.
10. Remove the second waxed sheet.
11. Smooth the remainder of the picture to the cardboard.
12. Rub excess cement away from the edges of the picture.

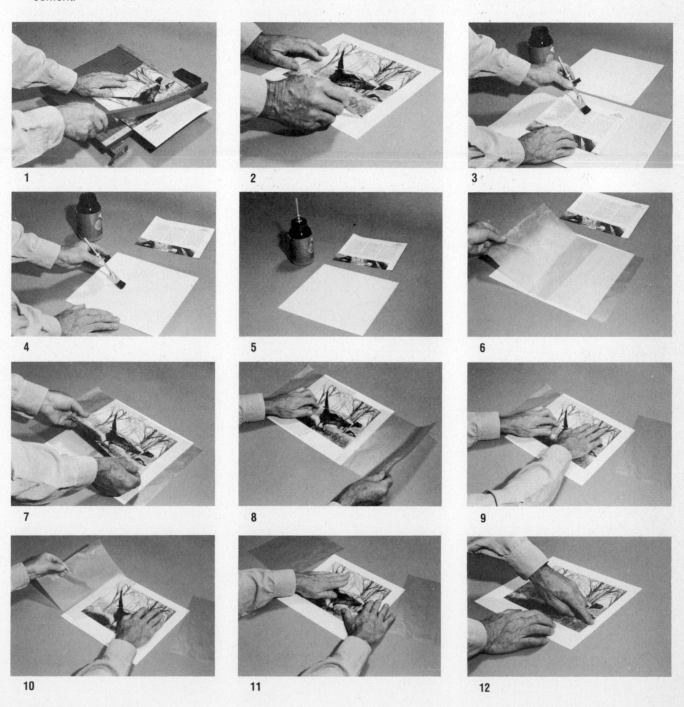

129

Consider these additional details when using the rubber cement methods of mounting:

- Make guide marks in corners on mounting board lightly in pencil so they are not noticeable after the mounting is completed. Erase the obvious ones.
- Make sure the brush is adjusted in the lid of the cement jar so the bristles are below the cement level in the jar.
- Cement should flow smoothly from the brush. If it thickens it will collect on the brush and fall in lumps. Add a small amount of rubber-cement thinner and shake the jar well.
- If colored cardboard is to be used, test the cement on a sample, as rubber cement may stain the surface.
- Apply cement with long sweeping strokes of the brush. Move moderately fast as cement dries quickly.
- Keep the lid on the dispenser jar tightly closed when not in actual use.
- In the permanent method, the cemented surfaces can be considered dry when they feel slightly tacky.
- The sulfur in rubber cement may react with the silver in a photograph to stain the face of the picture a yellowish-brown color.

Dry Mounting Methods on Cardboard

This is a fast method, resulting in permanent and neatly mounted materials. It is particularly useful when a number of pictures, photographs, or other visual items are to be mounted. The procedure requires the use of a heat-sensitive adhesive and attention to three variables—temperature, pressure, and time—to insure satisfactory bonding of the picture to the backing surface. Temperature and pressure are controlled in a **dry-mount press,** and you select the time.

In the dry mounting process, a tissuelike paper, coated on both sides with the heat-sensitive adhesive, is placed between the picture and the cardboard or other backing material. When heat and pressure are applied, the adhesive is activated. Upon cooling the adhesive forms a strong bond between the picture and the cardboard.

Thus the cooling phase of the process is particularly critical when preparing a successful dry mount. To insure a satisfactory result, immediately after the heating and pressure phase, place the mount under a metal weight to allow cooling to take place undisturbed. The liquified adhesive hardens to form a firm, even seal between the picture and its backing.

1

2

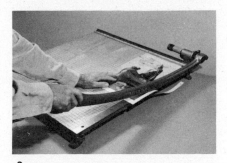

3

4

5

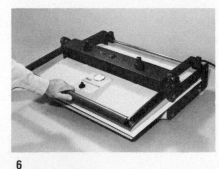

6

7

The Completed Mount

Either an electric hand iron or a dry-mount press is used to provide heat and to exert pressure.

A shortcoming of the dry mounting method is the possibility that bubbles of steam may form under a picture when heat is applied if there is moisture in the paper or cardboard. Most bubbles can be eliminated by predrying the cardboard and picture. If bubbles do appear after mounting, puncture them with a pin and then reapply heat and pressure. Unfortunately, when bubbles do form, the paper may stretch, resulting in a wrinkle.

The dry mounting method requires seven steps:

1. Set the dry-mount press at the suggested temperature for the mounting product to be used. Preheat both the picture and cardboard for 30 seconds to remove moisture from them.
2. Adhere the dry-mount tissue to the back of the picture by touching the iron directly to the tissue. Always protect the table top with paper.
3. Trim the picture and the tissue together on all sides.
4. Align the picture on the cardboard.
5. Tack the tissue to the cardboard in two corners.
6. For protection, cover the picture with a sheet of silicon-treated release paper (Seal product). Seal the picture to the cardboard with heat and pressure for 30 seconds.
7. Cool the mounted picture under a metal weight for one minute or more.

Consider these additional details when dry mounting:

- Every item placed in the dry-mount press should be covered with paper for protection, preferably with Seal Release paper. (Replace the carrier paper when it wrinkles.)
- In order to minimize any color change on color photographs, skip the preheat step when mounting color prints.
- If more than one sheet of tissue must be used, butt the edges together; do not overlap them.
- If a mounting is too large to be sealed in the dry-mount press at one time, seal it in sections. Make sure successive areas placed in the press are overlapped so none of the picture is missed.
- Be alert to temperature-sensitive materials such as resin-coated photo papers. Temperature of the press must stay below the melting temperature of the paper!
- When pressing is completed, *quickly* place the mount under a weight for cooling. This is when actual sealing takes place. A metal weight, which absorbs heat quickly, is preferred.

In addition to mounting complete rectangular pictures, cutouts around pictures can be made to eliminate unnecessary details. Be sure to tack tissue to back of the picture before cutting it out. Then follow the same procedures as described for regular dry mounting to mount a cut-out picture. (See example next column.)

Pictures that extend across two pages in a magazine require special handling. The pages must be spliced together so neither a clear line nor picture separation is visible where they are joined. Follow these steps:

1. Remove the picture from the magazine by cutting or take out the staples.
2. Trim the edge to be joined on each part of the picture. Remove as little paper as possible.
3. Tack dry-mount tissue to the back of one part, leaving part of the sheet of tissue overhanging the edge to be joined.
4. With the picture face up, align the edges to be joined and touch them together tightly. Make sure there is protective paper under the picture and tissue.
5. With a small piece of paper over the joined splice, tack the second part of the picture to the tissue in two or three places. Remember, the picture is face up.
6. Tack additional tissue if necessary to the back of the picture to cover it entirely. Always butt pieces of dry-mount tissue together, do not overlap them.
7. Then finish the mounting as with a regular dry-mounted picture—trim, tack to cardboard, seal with the press or iron, and cool immediately under a weight.
8. If a slight white line shows at the joined splice, darken it lightly with soft lead or colored pencil.

Completed Two-Page Mount

Various commercial products are available for use in dry mounting. Most are manufactured by Seal, Inc. Each one is designed for a special use:

- MT5—general use dry-mount tissue which bonds while it is heated in the press at standard 225°F.
- Fotoflat—low temperature (180°F) dry-mount tissue for mounting heat-sensitive materials; bonds as it cools, therefore can be used for temporary displays; picture can be removed from the backing by reheating in the press and then immediately peeling; electric hand iron (set at rayon or low) can be used with this material; start the iron from center of picture and move outwards with slow movements and a moderate amount of pressure.

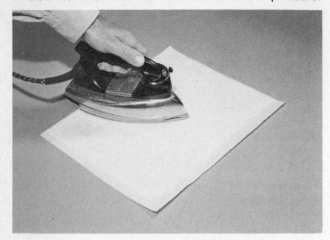

- Colormount—designed for mounting photographic papers at a temperature of 205°F; with glossy resin-coated (RC) papers use Colormount Coversheet to preserve glossy finish of the print.
- Fusion 4000—low temperature adhesive (180°F) that can be pieced together, overlapped, and used on irregular shaped items (like cut-out picture described previously) since it contains no tissue core; suitable for all low temperature uses; allows for removal of material mounted as with Fotoflat.

Pressure-Sensitive Methods

Another category of material useful for mounting includes products that consist of an adhesive which will seal the picture to the backing by only applying pressure. Such a procedure is useful when heat should be avoided due to potential damage to a delicate surface or to a photographic print.

Sheet mounting adhesive

Use either Print Mount (Seal product) or Positionable Mounting Adhesive (3M product) with a roller-type press to insure even, sufficient pressure for mounting. Follow these steps (for specific details see directions with product to be used):

1. Cut a piece of adhesive to size of picture to be mounted.
2. Transfer adhesive surface to back of picture by passing both through the press.
3. Position picture on mounting surface; move it around as necessary.
4. Pass picture and backing through press to seal securely.

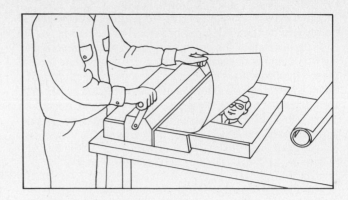

Print spray

A spray-mounting adhesive (Seal product) can be applied to the back of a picture. After the adhesive sets for one minute, the picture is adhered to the backing with uniform hand or applicator pressure.

Dry Mounting Method on Cloth

Large materials, like charts and maps, or items that need to be pliable, can be backed with a dry-mount cloth (Chartex). This product will provide durability while maintaining flexibility.

The adhesive is a coating on one side of the cloth, which is ironed on the back of the materials or applied with a dry-mount press. Large materials can be mounted so as to be rolled or folded. Follow these steps:

1. Set the dry-mount press at 180°F and the tacking iron at *medium.*
2. Dry the chart in the press. Besides removing moisture from the paper, this treatment will flatten folds in the chart.
3. Cover the back of the chart with sufficient cloth. Place the adhesive coating (smooth side, on inside of roll) against the back of the chart.
4. Tack the chart to the cloth in one large spot.
5. Cut the cloth to match edges of chart or leave excess cloth all around for a cloth margin.
6. If you have left a margin, fold the cloth and tack it to the edge of the chart along the full length of each side.
7. Cover the chart with release paper and seal in the dry-mount press. Cool under a weight.
8. Check for bubbles or wrinkles. Reiron or repress as necessary.

A chart can also be cut into sections and mounted with a slight separation between adjacent sections. It can then be folded for easy storage.

Dry-mount cloth is available in rolls up to 42 inches wide and 100 feet long. For wide subjects, pieces of cloth can be spliced and overlapped.

To facilitate displaying a cloth-backed chart, add gummed eyelets or grommets to the upper cloth margin. Place them 3 inches in from the outer edge. Add an additional one or two toward the center.

3

4

5

6

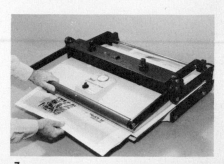
7

Completed Cloth-Backed Mount (rolled)

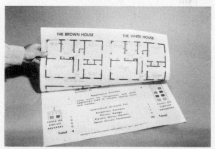

Completed Cloth-Backed Mount (folded)

PROTECTING THE SURFACE (LAMINATING)

The face of a photograph or other mounted material to be handled a great deal needs protection. A clear plastic spray can be applied, but an even better and more permanent protection is achieved by sealing a clear plastic laminating film over the face of the picture. A dry-mount press set at the recommended temperature for the product adheres this film to a mounted picture in a few seconds.

The following steps apply to laminating a previously dry-mounted picture (see next page).

1. Set the dry-mount press at 200°F (for Seal-lamin) and the tacking iron on high. (If laminating film thicker than 0.0015 inch is used, set the press at 275°F.)
2. Dry the mounted picture in the press for 10 seconds.
3. Extra pressure is required for laminating. Put a piece of heavy cardboard or masonite (smooth face to the mount) under the rubber pad of the press.
4. Cut a piece of laminating film (Seal-lamin) to cover the entire mount surface (front and back if desired) or use a two-sided laminating pouch.

5. The adhesive side of the film is on the inside of the roll. Tack the film to the mount in one spot with a piece of paper placed between the film and the tacking iron.
6. Trim excess film so none overhangs the mount edge.
7. Smooth the film over or around the mount. Cover it with a sheet of Seal Release paper. Seal the assembly in the press for one minute.
8. Immediately after removing the picture from the press, cool it under a metal weight for 1 or 2 minutes.

Consider these additional details when laminating:

- If bubbles appear under the lamination film they are due to moisture in the picture or cardboard expanding to form steam. Place the mount back in the press for about 45 seconds and cool again.
- It may be helpful, instead of cooling under a weight, to rub firmly over the affected area with a wadded handkerchief.
- If bubbles persist, break them with a pin and press again or rub by hand again. Because of bubbles, the film may stretch, resulting in a wrinkle.
- If unmounted pictures are laminated (to protect and dis-

133

3

4

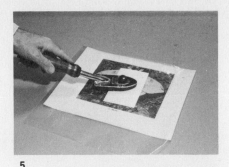

5

6

7

8

Completed Laminated Mount

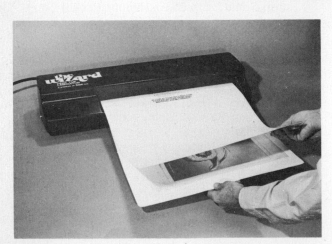

Wizard (Seal, Inc.)

SeaLaminator (Seal, Inc.)

play both sides of a sheet), seal one side with film, as above. Then repeat the process on the second side or use a two-sided laminating pouch.

- If materials to be laminated are wider than the roll of film to be used (roll widths up to 25 inches are available), butt adjacent pieces together. Seal each piece in turn, in the press.

- Thin three-dimensional objects, such as leaves, can be laminated easily to cardboard. Before starting, secure the object in place with rubber cement or a small piece of double-faced tape.
- Photographic prints cannot be successfully laminated consistently in a press because it is difficult to remove moisture trapped in the paper.

134

In addition to using the dry-mount press, other units are designed especially for laminating. Because of the increase and evenness of pressure, their use is preferred when photographs are to be laminated.

- A compact heat laminator (Wizard) accepts individual pieces or pouches of laminating film in which pictures are placed. Temperature and operating speed are preset. Both sides of a picture, or two pictures back to back, can be laminated at the same time.

- Another *heat* lamination unit (SeaLaminator) feeds thin film over the picture between heated, motor-driven rollers. With the use of two rolls of acetate, mounted on the machine, both sides of a picture, or two pictures back to back, can be laminated at the same time.

Review What You Have Learned About Mounting and Surface-Protection Methods:

1. What are two differences between *temporary* and *permanent* rubber-cement mounting methods?
2. For what two reasons is wax paper used in the permanent rubber-cement method?
3. What is the principle of mounting with dry-mount tissue?
4. Why is tissue tacked to the back of the picture *before* the latter is trimmed to size?
5. What procedure is used if bubbles appear under a completed mount?
6. Compare three methods—rubber-cement permanent, hand iron, and dry-mount press in terms of speed of process, ease, equipment, cost, and quality of mount.
7. For what major reason should special materials be used to mount colored photographs?
8. What two variations of the regular dry-mount procedure are used when a two-page picture is mounted on cardboard?
9. How do you decide whether to mount a map on cardboard, on cloth to be rolled, or on cloth to be folded?
10. Why should cloth be trimmed flush with the chart?
11. What is a variation of the regular dry-mount procedure when covering a mounted magazine picture with laminating film?
12. Is it possible to mount a picture by using pressure but no heat?

MAKING PASTE-UPS

When a layout is to consist of a number of elements that have to be prepared or gathered separately, such as headings, texts, illustrations, symbols, captions, and labels, they must be assembled for reproduction. Arranging and adhering all these parts to paper or cardboard is a procedure called **paste-up.** For example, before printing, a paste-up was made for each page of this book. In the printing trade this procedure results in what is called a **mechanical.** Upon completion, the paste-up or mechanical is **camera ready.**

The paste-up may be duplicated in single copies on an office copy machine or by means of the diffusion-transfer process (page 138). The result should be a clean copy, free of all paste-up marks. When a printer receives the camera-ready paste-up, first a high-contrast litho negative is made, and from it a metal plate for printing on an offset press.

There is also a direct plate-making procedure for preparing a paper plate from the paste-up.

Line Copy

In making a paste-up, start by trimming each piece to size. Then, paste-up is accomplished by putting rubber cement on the back of each piece, as in temporary mounting (page 128). This method permits moving a piece for alignment before the cement dries and the piece adheres firmly.

A special form of rubber cement called "one-coat" is very useful for preparing paste-ups. It remains tacky when applied to the back of a piece and allowed to dry, thus permitting pieces to be easily repositioned.

Stick-glue and heated wax can also be used as a backing adhesive for paste-ups. The latter is applied with a hand-operated roller or a mechanical coating unit. The glue or

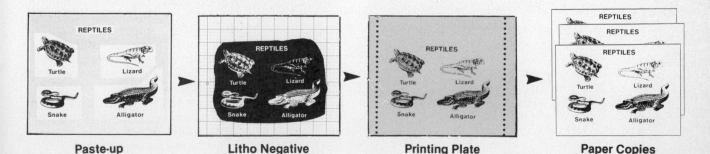

Paste-up   Litho Negative   Printing Plate   Paper Copies

135

wax backing permits ease of alignment and a tight seal between copy and paper. It also eliminates some of the cleanup problems that rubber cement presents.

Light blue pencil (nonreproducible blue) can be used for making guidelines and adding notes on the white surface. These blue marks will not reproduce when photographed.

The black areas on white paper are referred to as **line copy** or **high-contrast** subjects since there are no middle gray tones. When photographed on high-contrast film, the resulting negative will contain clear areas where the black marks were located on the original or the paste-up. Red is the only color that will also result in clear marks on the high-contrast film negative. Then, when the negative is printed, the original black (or red) areas again appear as black on photo paper or as visible images on a printing plate.

Be as clean as possible when working on the paste-up. Avoid smears, fingerprints, or unwanted spots. If any dirt, stray marks, or errors appear on the paste-up sheet, cover them with white paper or brush "white-out" correction fluid over the area. Edges of paper or any area covered with white-out (even though unevenly) usually will not show on final copies.

Continuous-Tone and Halftone Subjects

A black-and-white, **continuous-tone** photograph contains shades of gray, varying from white to black. A picture printed on paper, as in a newspaper or a magazine, is called a **halftone.** It consists of uniformly spaced black dots of different sizes, which blend together when normally viewed, thus conveying shades of gray. If a continuous-tone photograph is placed in position as part of a paste-up and reproduced as is described below, the gray tones will be lost, resulting in an undesirably contrasty picture. Therefore, it is

important to convert continuous-tone photographs (and art paintings) to halftone pictures before such materials can be used for reproduction, after the paste-up is completed.

As the paste-up is prepared, adhere proper size rectangular (or other shape) pieces of *black paper* or adhesive-backed *red acetate* (called rubylith) to the page at the location where each photograph is to appear. Then, when the paste-up page is filmed, the resulting high-contrast negative will have a clear "window" at each place where the black paper or red adhesive was placed.

The original, continuous-tone photograph is filmed separately to proper size through a **halftone negative contact screen.** This screen consists of a dot pattern. The greater number of dots per inch, the finer will be the tonal reproduction. (For this book a 133 dot-pattern screen—133 dots per inch—was used for reproducing each photograph.) The resulting negative, now called a **halftone negative,** is taped or "stripped" into the high-contrast, line negative at its proper clear window location. The final copies of the page, printed on paper, will show the photograph as a halftone picture with the surrounding line copy material. See the illustration of this procedure on the following page.

When the paste-up sheet and photographs are being prepared, make certain to number the black or red areas with corresponding numbers on the photographs. In this way, the printer will be certain to strip each halftone negative into its proper location on the litho negative.

An alternative procedure is to copy the photograph through a *positive* dot-pattern screen to prepare a positive halftone print, often using a direct Polaroid filming process. The resulting print can be pasted up directly as part of the black line-copy page. Regular line copy reproduction, as described above, follows. Because the continuous-tone photograph has been converted to a dot-pattern picture (the halftone) it now responds on the litho film as if it were line copy. One limitation to this procedure, using a positive dot pattern, is that only a coarse screen can be used (60–80 dots per inch). The result will not be as sharp a picture as is obtained from the previously described negative contact screen procedure.

Two or More Colors

When all elements of a page are pasted up on one sheet, final copies will be printed in only one color of ink. If more than one color is to be printed, then a *separate* paste-up (and separate negatives and printing plates) must be made for the headings, text, diagrams, labels, photographs, and any other items that will appear in each color.

The main, or base-color materials (usually to be inked in black) are pasted up on paper or cardboard first. Then for *each color* a *separate paste-up* containing the necessary materials is prepared. The elements for each color are usually pasted on a sheet of clear acetate or translucent paper or film, and then carefully *registered* over the base sheet so that everything fits accurately in position.

In order to align each layer properly, attach register mark symbols, printed on adhesive-backed acetate, in at least three widely separated locations on each sheet. They are set near the edge of the paper or acetate overlay outside

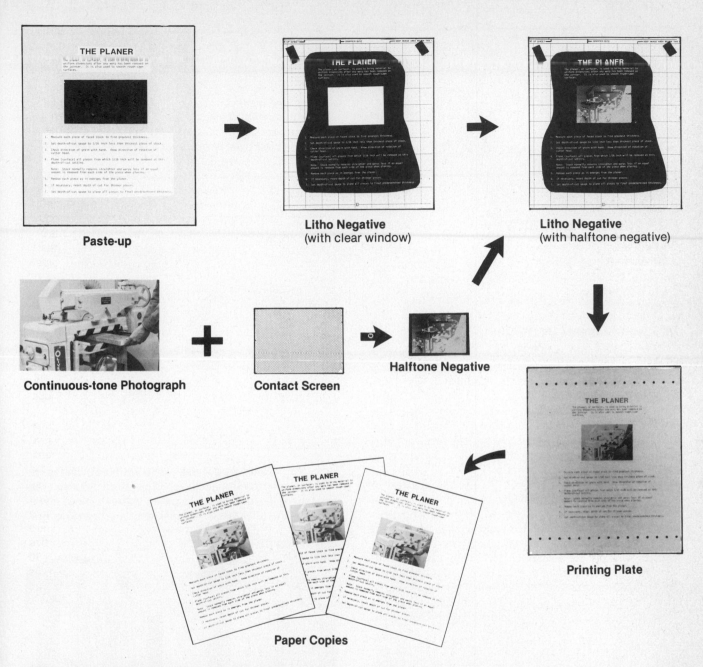

Paste-up

Litho Negative
(with clear window)

Litho Negative
(with halftone negative)

Continuous-tone Photograph

Contact Screen

Halftone Negative

Printing Plate

Paper Copies

the area that will be printed. When each of the three marks on an overlay sheet lies exactly over its matching mark on the base sheet, the two sheets are "in register." The same marks are used by the printer to align each plate on the press for printing each color in proper position onto the same sheets of paper. Be sure to tape each overlay sheet to its base page for ease of locating each one during the negative and plate-making processes.

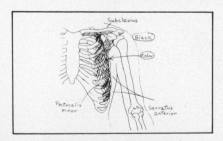

**Original Layout with Colors
Marked for Printing**

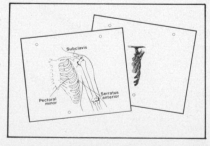

Base and Overlay with Register Marks

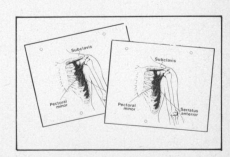

Printed Copies in Color

137

Review What You Have Learned About Making Paste-ups:

What is the meaning of each term?

1. Paste-up
2. Nonreproducible blue
3. Continuous-tone photograph
4. Halftone picture
5. Halftone negative
6. Clear window
7. Registered overlay paste-up

DUPLICATING LINE COPY

Besides using an office copier, line copy materials can be duplicated with or without size change by either of two methods. Each one requires the use of a camera to photograph the paste-up.

Diffusion-Transfer (PMT) Method

A process camera, commonly found in a graphic arts lab or printing shop, is used in the diffusion-transfer, or photomechanical transfer (PMT), method. It has a film holder or ground-glass surface, 8×10 inches or larger, often with a vacuum back and good alignment for accurate reproduction. The materials consist of a light-sensitive negative paper, a chemically sensitive positive receiver paper or transparent film (for overhead transparencies), and processing solution called the activator.

An image of the paste-up or other original is adjusted to the desired size on the ground-glass viewing surface of the camera. The negative paper, placed on the ground-glass surface (under a red safelight), is exposed. It is then placed emulsion to emulsion on the positive receiver paper or film and run through the processor containing the activating fluid. The solution acts on both emulsions, developing the image on the negative and then allowing the image to transfer to the positive paper or film. The processing time is 30 seconds. Because of this image transferring procedure, the process is termed "diffusion transfer."

In addition to the basic diffusion-transfer process just described, there are specialized applications of the proce-

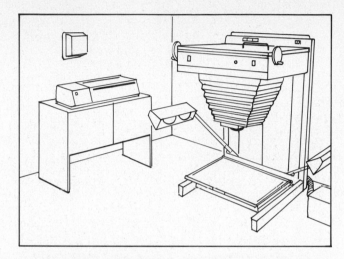

dure. For example, reversal (white on black) and direct halftone (gray tone) images can be prepared. Another creative use of these materials is called **posterization.** A continuous-tone photograph, consisting of a full-range of gray

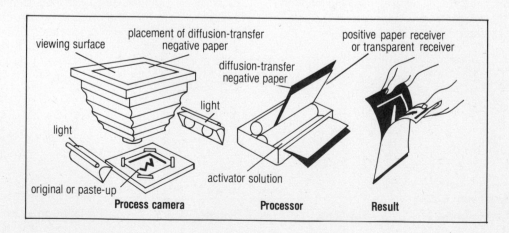

viewing surface | placement of diffusion-transfer negative paper | positive paper receiver or transparent receiver

diffusion-transfer negative paper

light

light

original or paste-up

activator solution

Process camera   **Processor**   **Result**

tones, is exposed to diffusion-transfer negative paper. On the resulting receiver paper is a reproduction of the photograph, now with *high contrast quality*. Such a reproduction can have an abstract artistic quality. It may be used as a visual itself or as a background for a title.

High-Contrast Film Process

In the high-contrast film process, Kodalith film (or the equivalent litho film from manufacturers other than Eastman Kodak) is used to photograph the paste-up or other line material.

The high-contrast film process has four steps:

1. Film the subject, using high-contrast film. (For exposure determination, film speed of Kodalith Ortho Type 3 is 8.)
2. Process the film to a negative under a red safelight.

3. Opaque the negative to eliminate clear spots, paste-up marks, or unwanted printed areas.
4. Enlarge the negative onto a sheet of photographic paper. Process the paper in a paper developer or through a photostabilizer according to instructions (see page 103). Dry the print for use.

Both the diffusion-transfer and high-contrast film procedures result in good quality duplicates of a paste-up or other original material in the required final size. The diffusion-transfer procedure is much faster to carry out than is the high-contrast film process. But recognize that it is only feasible to prepare one or a few duplicate copies with these methods.

The following page contains an illustration of the various results that can be obtained by applying both of the duplication processes just described.

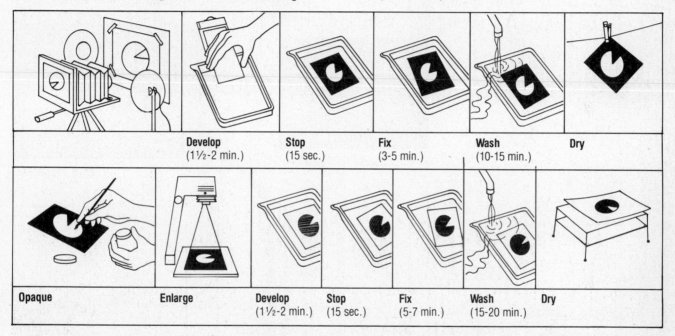

| | Develop (1½-2 min.) | Stop (15 sec.) | Fix (3-5 min.) | Wash (10-15 min.) | Dry |

| Opaque | Enlarge | Develop (1½-2 min.) | Stop (15 sec.) | Fix (5-7 min.) | Wash (15-20 min.) | Dry |

Review What You Have Learned About Duplicating Line Copy:

1. What are the *two* methods that may be used to duplicate line copy with a size change?
2. What simple-to-operate equipment might be used if *no* size change is required?
3. Explain how each of the two methods in your answer to question 1 is carried out.

REPRODUCING PRINTED MATTER

When instructional media are prepared, there is often the need for accompanying reading materials. In addition, various types of printed media (Chapter 13) may be designed and then prepared in multiple copies. When a few copies (less than 10) are made, we use the term **duplication.** If a larger number of copies is run, we refer to this as **reproduction.**

A number of duplication and reproduction processes can be utilized. Each one serves one or more particular needs and has certain requirements, advantages, and limitations. We will examine three of the most commonly used methods. They range from inexpensive duplication to more complex, costly reproduction procedures. Since this is a rapidly changing area in the graphic arts, further simplified and automated developments can be anticipated.

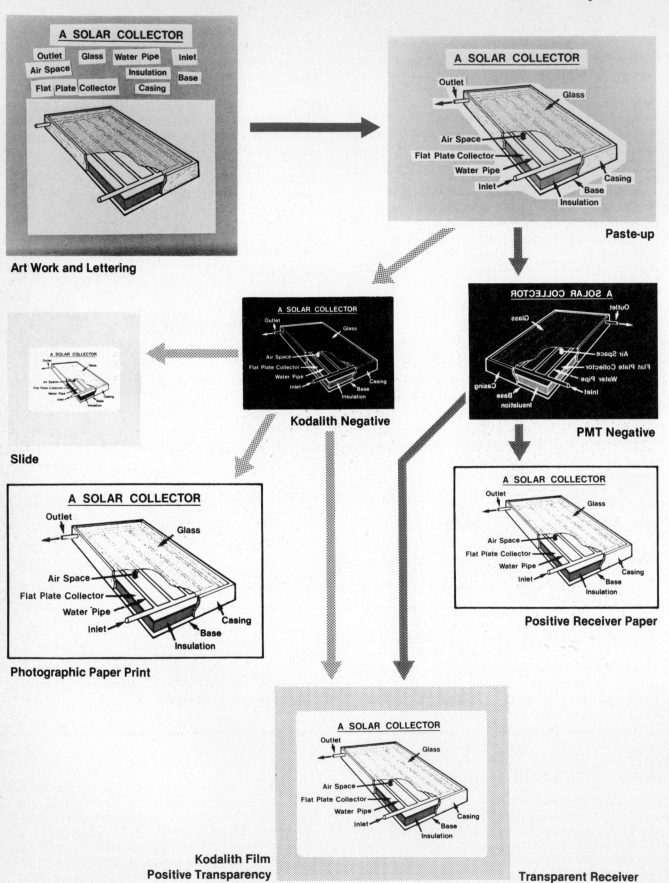

Spirit Duplication

For spirit duplicaton a master is prepared with colored carbon sheets placed in contact with the back side of the master paper. Impressions are made on the front side of the master by writing, typing, or drawing. At the same time a carbon impression is deposited on the back of the master sheet. Up to five carbon colors may be used to prepare one-color or multi-color paper copies.

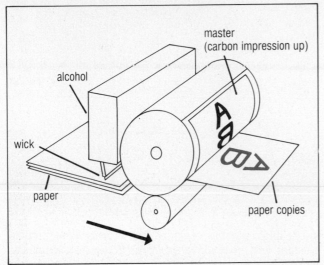

The carbon transferred to the back of the master is an aniline dye, soluble in methyl alcohol. In the spirit duplicator, alcohol is spread over the duplicator paper. When the master comes in contact with paper, some of the dye is deposited on the paper, producing a printed copy.

This is an inexpensive process resulting in fair quality copies. Printed sheets have a short life as the colored dyes that form the words and diagrams fade when exposed to light for a length of time.

Electrostatic Process

The electrostatic process is known as **xerography** and the copy machine used is termed a "plain paper" copier. The procedure requires a selenium-coated plate or drum, which is photoelectrically sensitive, to receive an electrical charge. The original to be duplicated is exposed to light, with the white area (and gray portions of photographs) reflecting light to the charged plate. The light that reaches the selenium-coated surface dissipates the electric charge, leaving a charge only in the image area. The copy paper (any ordinary bond paper) is charged as it is fed into the machine, and a toner (a fine black powder) is transferred to it from the selenium surface and appears wherever there was an image on the plate or drum. Finally, the copy paper is heated to fix the powder permanently to the image area and it exits from the machine.

1. Positive electric charge is placed on selenium-coated plate or drum.
2. Image of original is projected onto plate to form latent image.
3. Negatively charged powder *toner* is dusted onto selenium plate.
4. Sheet of paper is placed over plate and receives positive charge.
5. Final copy is heated to fuse image into paper.

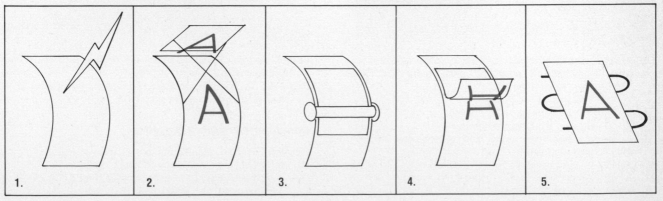

There is a wide range of copiers that utilize the xerography process. They range from single-sheet, table-top models to large, floor units that reproduce original material rapidly in quantity, on both sides of a sheet, that permit enlargement or reduction, and that collate final pages. In addition, *full-color* originals can be duplicated or reproduced in full color with certain models of xerographic equipment.

Offset Printing

The basic principle of offset is that grease (ink) and water do not mix. The printing plates have ink-receiving (greasy) image areas and ink-repelling (watery) nonimage areas. Plates are prepared directly on paper or aluminum with

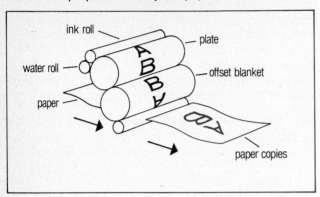

special pencils and typing ribbons, by thermal method, by the electrostatic process, or photographically using the previously described high-contrast film.

A water roll on the press coats the plate with a thin layer of water; where there is an image the grease in the image repels the water but allows the ink, which is oily, to adhere. The ink is then *offset,* or transferred, to a cylinder covered with a resilient rubber "blanket." The ink image from the blanket is then transferred to paper.

The offset process is extensively used for long-run educational and commercial printing. It requires expensive equipment and trained personnel.

Review What You Have Learned About Reproducing Printed Matter:

1. To which type of reproduction process does each refer?
   a. Electrically-charged areas are transferred from a selenium-coated surface to paper, thus allowing for retention of toner powder to form image.
   b. Ink adheres to greasy areas on plate, transferring image to paper.
   c. Master is made on carbon sheet, and image is transferred with alcohol to paper.

2. Which reproduction equipment should you select for each of the following jobs?
   a. Inexpensive preparation of 100 copies in two colors
   b. Single-sheet duplication; copies all colors
   c. Usual reproduction of printed materials in quantity, like pages for this book
   d. Reproducing 10 to 15 copies of pages for an instructional guide

REFERENCES

Design

Designing Visuals that Communicate Series. Slide/tape presentations. Bloomington, IN: Indiana University, Audio-Visual Center
   X-100—The Design Elements
   X-101—The Design Principles
   X-102—The Design Principles Applied
   X-103—Utilizing the Design Principles
Visuals for Projection Series. Slide/tape presentations. Bloomington, IN: Indiana University, Audio-Visual Center
   X-112—Designing Visuals for Projection
   X-113—Simple Title Slides
   X-114—Complex Title Slides

Illustrating

Bostick, George. "Graphing Statistics." *Performance and Instruction Journal* 20 (March 1981):11–13.
Basic Art Techniques for Slide Production. Publication VI-27. Rochester, NY: Eastman Kodak Co., 1978.

Computer-Generated Graphics

Bickford, Susan. "Computergraphics: Coming of Age. Part I." *Audio-Visual Communications* 17 (April 1983):16–19
———"Computergraphics: Coming of Age, Part II." *Audio-Visual Communications* 17 (May 1983):16–17,64–65.

Borrell, Jerry. "Computer Graphics: Replacement for the Dark-room." *Photomethods* 25 (September 1982):61–65.

Computer Graphics Today (monthly publication). Holmes, PA 19043, P.O. Box 656.

Knight, Kirk. "Slide into the Future with Computer-Generated Graphics." *Industrial Photography* 32 (July 1983):19–21.

Lewell, John. "The Dollar Guide to Computer Graphics. Part 1." *Audio Visual Directions* 4 (September 1982):24–37.

———"The Industry and Its Equipment—Part 2." *Audio Visual Directions* 4 (October 1982):43–47.

———"Plotters, Printers, and Photographic Recorders—Part 3." *Audio Visual Directions* 5 (January 1983):42–45.

———"Systems and Software—Part 4." *Audio Visual Directions* 5 (February 1983):31–35.

Coloring

Green, Ronald E. "AV Graphics: Communicating with Color." *Audio-Visual Communications* 12 (November 1978):14–18.

Standards—Legibility

Legibility—Artwork to Screen. Publication S-24. Rochester, NY: Eastman Kodak Co., 1980.

Sharpe, David M. "Can They Read It? Legibility of Projected Visual Information." *Performance and Instruction Journal* 20 (March 1981):15–16.

Lettering

Lettering for Instructional Materials Series. Slide/tape presentation. Bloomington, IN: Indiana University, Audio-Visual Center
X-107—Principles of Lettering
X-108—Cutout Lettering
X-109—Stencil Lettering
X-110—Dry Transfer Lettering
X-111—The Wrico System

Mounting

Maffeo, Thomas. *How to Dry Mount, Texturize, and Protect with Seal*. Naugatuck, CT: Seal, Inc., 1981.

Duplication and Reproduction

Many of the following references are available from Eastman Kodak Co., 343 State Street, Rochester, NY 14650. Following each of these titles is the Eastman publication code number without any other source indication.

Basic Photography for the Graphic Arts. Publication Q-1. 1982.

Basic Printing Methods. Publication GA-11-1. 1976.

Copy Preparation. Publication GA-11-4. 1979.

Copy Preparation and Platemaking Using Kodak PMT Materials. Publication Q-71. 1976.

Croy, Peter. *Graphic Design and Reproduction Techniques*. New York: Focal, 1982.

Graphic Design. Publication GA-11-2, 1976.

Halftone Methods for the Graphic Arts. Publication Q-3. 1982.

LeTissier, David. *Instant Graphic Techniques with Instant Art and Agfa-Gavaert*. Maidstone, Kent, England: The Graphics Communications Centre, Ltd., 1981.

Pett, Dennis W. *Copying and Duplicating Processes*. Bloomington, IN: Indiana University, Audio-Visual Center.

Photography and Layout for Reproduction. Publication Q-74. 1978.

Photoreproduction. Publication GA-11-5. 1980.

Stecker, Elinor H. *How To Create and Use High Contrast Images*. Tucson, AZ: HP Books, 1982.

SOURCES FOR EQUIPMENT AND MATERIALS

Many of the common items needed for the graphic preparation of your instructional media can be purchased from local art, stationery, photography, and engineering-supply stores. Other equipment and specialized materials are listed here. Although the headquarters or main-office address is given, you will find local offices or local dealers distributing most products. If you do not, write to the address given for further information.

General Art Supplies

Alvin and Co., P.O. Box 188, Windsor, CT 06095
Dick Blick, P.O. Box 1267, Galesburg, IL 61401
Arthur Brown and Bros., 2 West 46th Street, New York, NY 10036
Flax's Artist Materials, 1699 Market Street, San Fracisco, CA 94108

Illustrations

Board Reports, P.O. Box 1561, Harrisburg, PA 17105.
Clipper Creative Art Service, Dynamic Graphics, Inc., 6000 North Forest Park Drive, Peoria, IL 61614
Volk Art, Pleasantville, NJ 08232

Coloring and Shading Materials

Dr. P.H. Martin's Transparent Water Colors, Salis International, 4093 North 28th Way, Hollywood, FL 33021
Peerless Color Laboratories, 11 Diamond Place, Rochester, NY 14609
Thayer & Chandler, 442 N. Wells, Chicago, IL 60610
Zipatone Inc., 150 Fencil Lane, Hillside, IL 60162

Lettering Equipment and Materials

Kroy Industries, 1728 Gervais Avenue, St. Paul, MN 55164
Letterguide Co., P.O. Box 30203, Lincoln, NB 68503
Letraset Inc., 33 New Bridge Road, Bergenfield, NJ 07621
Lumocolor Markers, J. S. Staedler, Inc., P.O. Box 787, Chatsworth, CA 91311
Projectachrome Markers, Eberhard Faber, Inc., Crestwood, Wilkes-Barre, PA 18703
Koh-i-noor Rapidograph Inc., 100 North Street, Bloomsburg, NJ 08804
Stik-a-Letter Co., 3080 McMillan Road, San Luis Obispo, CA 93401

Mounting and Laminating Equipment and Materials

General Binding Corp., One GBC Plaza, Northbrook, IL 60062
Graphic Laminating Inc., 5122 St. Clair Ave., Cleveland, OH 44103
Laminex Inc., P.O. Box 577, Matthews, NC 28105
Seal Inc., 550 Spring Street, Naugatuck, CT 06770
Southwest Plastic Binding Co., 123 Weldon Parkway, Maryland Heights, MO 63043
3M Company, Visual Products Division, 3M Center, St. Paul, MN 55101

- Recording Equipment
- Magnetic Recording Tape
- Recording Facilities
- The Narrator and the Script
- Preparation for Recording
- Recording Procedure
- Music and Sound Effects
- Editing Tape and Mixing Sounds
- Variable Speed Recording
- Synchronizing Narration with Visuals
- Duplicating Tape Recordings

The audio portion of instructional media should receive as much attention as do the visual elements. Good sound quality will enhance the visuals and the effectiveness of a presentation; poor sound will detract and result in an ineffective program. For any type of material—whether projected or nonprojected, still or motion—the same procedures apply when preparing the accompanying recording.

RECORDING EQUIPMENT

Two basic pieces of equipment are required to make a recording. Narration can be recorded with a **microphone** and a **tape recorder.** If sounds from more than one source, such as from two microphones or a record player for music and a microphone, are to be combined, then a **mixer** and playback equipment are necessary.

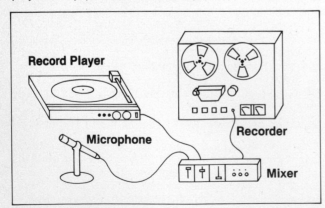

The Microphone

Small, portable cassette recorders include inexpensive built-in microphones that adjust automatically to the volume

of the sound source. But this feature can be a drawback because recording volume is increased automatically as if adjusting to a low sound level. Therefore, extraneous sounds like traffic noise or distant voices may be recorded.

The use of such a microphone results in a recording of poor to moderate quality. It should be used only to record a speech or a small-group discussion for documentary purposes. A separate microphone and manual volume control are necessary to produce a recording of acceptable fidelity without distracting noise.

Three common types of microphones are:

- **Crystal** or **ceramic**—contains crystals or granules that produce electricity when pressure from a sound wave is applied; picks up a limited range of sounds; is not rugged; does not provide good fidelity; is included with many inexpensive cassette recorders.
- **Dynamic**—has a diaphragm attached to a coil of wire placed within a magnetic field; produces electricity when the coil vibrates; picks up a wide range of sounds; is rugged; provides good fidelity.
- **Condenser**—has a plate that vibrates according to sounds received, causing variations in current carried on an adjacent fixed plate; picks up an even wider range of sounds; is rugged; provides high quality sound; can be very small and inobtrusively clipped to clothing.

In addition to the sound-generating feature of a microphone, each microphone has a sound pickup pattern. This refers to the way in which the microphone responds to sounds coming to it from different directions:

- **Nondirectional**—responds to sounds coming from all directions.
- **Unidirectional** or **cardioid**—heart-shaped pattern that

144

Nondirectional

Cardioid

Bidirectional

Shotgun

picks up sound with strength from one direction, some sound on the adjacent two sides, and almost none from the back side of the microphone.
- **Bidirectional**—picks up sound only on two opposite sides.

To record a group of people, as for an interview or a discussion, use a nondirectional microphone. Select a cardioid type for a single narrator or performer so that other sounds (ambient noise) can be held to a minimum. A bidirectional type is desirable when two persons are to be recorded as in an interview. When a speaker must be at a distance from the microphone, aim a unidirectional or "shotgun" microphone. It will accept only a narrow angle of sound and will reject most all other noises.

Use this information to select an appropriate microphone for each recording situation. A good microphone affects the quality of a recording just as a good lens on the camera relates to the quality of the resulting picture.

The Recorder

An audiotape recorder may be either a **reel-to-reel** (open reel) or a **cassette** type. An open-reel recorder is preferred when recording narration. It can be operated at a faster speed for higher fidelity sound ($3\frac{3}{4}$, $7\frac{1}{2}$, or 15 inches per second) as opposed to the speed of a cassette machine ($1\frac{7}{8}$ inches per second). The former is used when editing the tape. Although the principle of recording and playback of the two types is the same, the appearance and operating procedures are quite different.

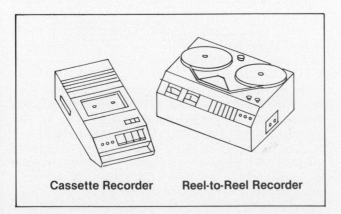
Cassette Recorder   Reel-to-Reel Recorder

Electrical signals from the microphone or other sound source enter the tape recorder, where they are initially amplified to a satisfactory recording level. Then the current

variations flow through the recording head, which is an electromagnet. As the recording tape passes the head, magnetic patterns are recorded according to the frequency of current variations and the intensity of the signals. The resulting patterns of the magnetic particles on the tape, when they are played back, create small electric currents through the playback head that vary in frequency and intensity. When these are amplified and fed into a loudspeaker, the recorded sound is heard. Prior to the recording operation, the erase head demagnetizes any existing pattern of signals on the tape.

Track configuration on open reel

Become familiar with the configuration of tracks (also called *channels*) on your recorder so that you will know its flexibility and limitations for combining sounds to create special audio effects, for adding synchronizing signals, and for permitting tape editing.

An open-reel recorder may have any of these configurations and possible uses:

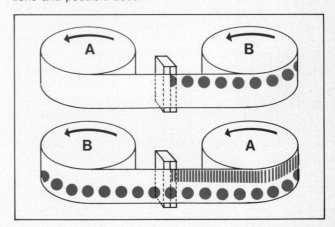

In **two-track monophonic** recording a single track is recorded on one half of the tape. When the tape is turned over, a second track is recorded on the other half of the tape. Each track is played back separately and is heard through a single speaker.

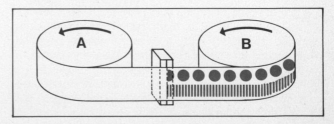

In **two-track stereophonic** recording, two separate tracks, each being one-half the width of the tape, are recorded simultaneously as the tape moves in one direction. On playback each track is heard at the same time through separate speakers. Although the main use of a recorder with this configuration is to record and play stereophonic music, it is possible to record a narration on one track and simultaneous music or an audible or inaudible control signal for changing visuals on the other track.

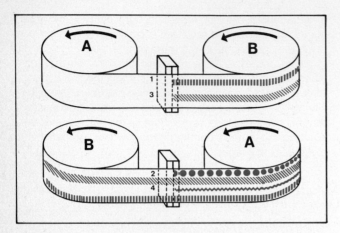

In **four-track stereophonic** two separate tracks (1, 3), each one-fourth the width of the tape, but with room for another track between them, are recorded simultaneously as the tape moves in one direction. When the tape is turned over, the other tracks (2, 4) are recorded on the unused fourths of the tape. Thus four tracks or, more correctly, two sets of stereophonic tracks are recorded and can be played back two at a time. This doubles the recording time on the tape from the previous two-track stereophonic configuration recorder. Most recorders of this type require that the recording be made on both tracks (1, 3 or 2, 4) at the same time. But some models permit separate track recording and erasing on each one. With the latter configuration, you can put narration on one track and add music, effects, or signals on the other one, in synchronization with the narration.

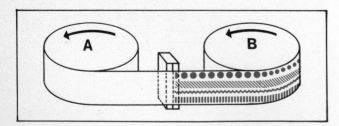

A professional recording is usually made on open-reel recorders with 4, 8, 16, or more tracks. Each track can be recorded separately with a different sound—narration, music, sound effects, or control signal. Each track can be played back separately, played together, or played in any combination. This configuration is the most flexible, because initial recordings and changes by rerecording on individual tracks can be made. Then when a tape prepared on

a machine with this configuration is dubbed to a cassette, any combination of tracks can be transferred together.

Track configuration on cassette

Audiocassette recorders can have either of two commonly used configurations.

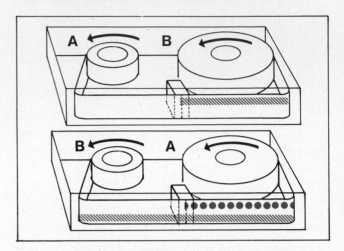

The **monophonic** configuration is similar to that of the open reel two-track monophonic recorder. A single track is recorded on one-half of the tape. When the cassette is turned over, a second track is recorded on the other half of the tape. Each track is played back separately.

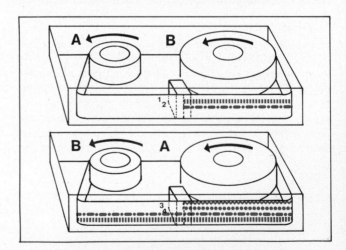

The **four-track stereophonic** configuration can be compared with the open reel four-track stereophonic recorder. The difference is in the location of the pairs of tracks. Two separate tracks are recorded simultaneously, side by side (1, 2) as the tape moves in one direction. When the cassette is turned over, two additional tracks (3, 4) are recorded at the same time on the other half of the tape. A stereo-recorded cassette can be played on a monophonic cassette player, because the side-by-side stereo tracks (1, 2 or 3, 4) are picked up by the same recording head. Thus narration and music or control signals, which had been recorded on separate tracks, are played simultaneously on the monophonic recorder.

Even though final audio material may be used with a cassette tape recorder (or on videotape), it is preferable to make the original recording on a high quality one-quarter-inch tape with a reel-to-reel recorder. The fidelity will be better. Most small, portable cassette recorders do not have the higher quality potential of better machines, but with careful placement of the microphone and volume level, satisfactory recordings can result. When playing back a recording from a small recorder use a large speaker. Sound quality will be appreciably better than with the small, built-in speaker.

When more than a single track is recorded on an open-reel machine, editing tape by cutting and removing sections is not possible without damaging other recorded tracks on the tape. Editing tape in a cassette is difficult and not practical.

Maintenance

After every 20 hours of use, two important maintenance operations on the recorder must be performed to avoid imperfections in recordings. First, remove specks of iron oxide that come off the tape as it passes over the heads, the capstan (the rotating cylinder that pulls tape across the heads at a constant speed), and associated rollers. Use a cotton swab wet with isopropyl alcohol or a recommended solvent to clean carefully all points that come in contact with tape. Turn the swab as you rub until no brown deposit is being collected, thus all oxide is removed.

Remove accumulated magnetism from the heads by moving an electronic *demagnetizer* across each head (without touching the head). Cleaners and demagnetizers that are enclosed in cassettes can be used for maintenance of a cassette recorder.

The Mixer

Frequently sounds from more than a single source, such as persons speaking into separate microphones, must be blended and fed into a single track on the recorder. This blending takes place through a **mixer.** An operator balances volume levels, adjusts tones, and fades sounds in and out as they are received.

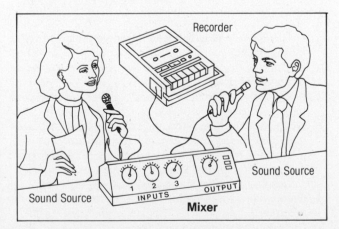
Recorder

Sound Source

INPUTS   OUTPUT

Sound Source

Mixer

During editing, a mixer also can be used to blend sounds smoothly onto the final tape from records and other tapes.

MAGNETIC RECORDING TAPE

Magnetic tape consists of a plastic base coated on one side with microscopic magnetic particles dispersed throughout a binding agent. These particles form the magnetic patterns that constitute the recording. Some details about audio tape include:

Base material—Most commonly a plastic, mylar, or polyester.

Base thickness—0.5, 1.0, 1.5 mil (a measurement in thousandths of an inch); the thinner the tape, the greater the amount that can be put on a reel and the longer the running time.

Coating material—particles of iron as ferrous oxide are most common, but a coating of chromium dioxide allows for higher quality recordings in terms of lower tape noise (hiss) level, a greater range of sound reproduction, and a higher sound output level. On many recorders a switch must be set according to the type of coating on the tape to be used. It adjusts circuits in the recorder so that the electronics will match the type of tape.

Your tape choice depends largely on how you plan to use your recording. The best guidepost is recording time per reel (at your chosen recording speed).

For a master recording, 1.5-mil tape offers the advantages of low cost, low print-through (signal transfer from layer to layer), and long life. It provides recordings of excellent quality.

For longer programs, a 1-mil tape is preferred. It provides 50 percent more recording time than the 1.5-mil tape on a reel of the same size.

The longest available playing time is provided by 0.5-mil polyester tape. Tape this thin usually is *tensilized,* a process which protects it from excess stretching during use. Some tape recorders will not handle this extra-thin tape; consult the instructions with your machine.

Table 12-1 on page 148 shows maximum recording time for two-track tape recorders *recording on both sides* at $7\frac{1}{2}$ and $3\frac{3}{4}$ ips. Use half of the time shown for a recording made on only one side of the tape.

Audiocassettes are designated according to their recording and playback length of time. The number indicates the *total* time, of which one half applies to each side of the cassette.

C-30—15 minutes per side
C-45—$22\frac{1}{2}$ minutes per side
C-60—30 minutes per side
C-90—45 minutes per side
C-120—1 hour per side

Exact tape lengths may vary in cassettes of the same size. Therefore it is advisable to record about 30 to 45 seconds less time on a side than the list indicates. Also allow 6 seconds at the start of the tape for the clear leader to pass. The C-120 size, because of its extremely thin base, jams many cassette machines and is not recommended for general use.

It may not be necessary to use an expensive tape solely

Table 12-1 **Recording Times for Various Tape Lengths on Open Reel**

TAPE	REEL SIZE (INCHES)	TAPE LENGTH (FEET)	RECORDING TIME MINUTES	
			AT $7\frac{1}{2}$ ips	AT $3\frac{3}{4}$ ips
STANDARD 1.5-MIL	5	600	30	60
	7	1200	60	120
LONG-PLAY 1-MIL	5	900	45	90
	7	1800	60	180
EXTRA-LONG-PLAY 0.5-MIL	5	1200	60	120
	7	2400	120	240

for voice recording. Buy a familiar brand name tape and avoid inexpensive tape; the latter not only produces lower quality sound, but it also can foul a recorder with oxide or cause excessive head wear.

RECORDING FACILITIES

Good quality recordings are made in an acoustically treated and soundproofed room. Where possible use a room hav-

ing some wall drapes, carpeting, and an acoustically treated ceiling. Do not try to "deaden" the room entirely, as it would be difficult for the narration to sound vibrantly alive in such a room. When desirable facilities cannot be found, improvise a recording booth in a corner of a room with some drapes or blankets to reduce sound reflections. Or use a carrel or study booth by covering the sides with blankets. Then record after normal working hours to eliminate extraneous noises.

When selecting a room in which to record, keep in mind the fact that many microphones cannot distinguish between the sound coming from the narrator's voice and any ambient noise—machinery, a ventilating fan, or sounds you can hear from outside the room. Any of these extraneous sources can cause undesirable background noise on the recording. Furthermore, an annoying hum, caused by 60 cycle electrical interference, may be picked up by the microphone and recorded on tape. This is often caused by the close proximity of fluorescent lights or high voltage (220 volt) power lines.

Keep these facts in mind when selecting a suitable recording facility.

Review What You Have Learned About Equipment, Tape, and Facilities for Recording Sound:

1. Which type of microphone would you select in each situation?
   a. To record a narrator in a room while outside noise can be heard through a wall on the far side of the room
   b. To record comments by any of five people seated around a table
   c. To record interviews at a sporting event where the microphone will be handled a great deal
2. What two kinds of maintenance should be performed periodically on a tape recorder?
3. What type of recorder would allow each of the following?
   a. Using the full tape to record narration and music on separate tracks in synchronization
   b. Using the full tape to record, on separate tracks, narration, music, sound effect, and a picture-changing audible signal

c. Using one half of the tape to record premixed voice and music
   d. Using a cassette to record two separate sounds at the same time
4. What four elements of sound control are handled through a mixer?
5. Which tape thickness and reel size would you select for recording voices and music in each situation?
   a. Average-quality tape for 46 minutes of recording
   b. Very durable and low print-through tape for a 35-minute recording that will be edited and spliced
6. Onto what size cassette would you copy a 25-minute recording?
7. Consider the recording facilities available to you. What arrangement would you make to ensure the best possible recording?

THE NARRATOR AND THE SCRIPT

Select a person who can read the script clearly and in a conversational tone. This means that he or she has the ability to raise or lower the voice for emphasis, to change the rate of speech according to the mood and intent of words, to provide pauses for variety, and to allow time for the listener to study the visuals. Evaluate a test recording before you decide on a narrator.

The narrator may be a friend or business associate having the necessary qualifications. Professional narrators are available in many communities. They may be listed in the yellow pages of the telephone book (one heading is "Recording Services"), or can be located through radio stations or media production companies. Even though there is a cost for the services of a professional, in addition to the quality of their work, there can be savings in recording and editing time.

Review the script with the narrator. Let him or her study it carefully and consider suggested changes in terms of the narrator's style and experience. The script should have markings to indicate points to be emphasized. Verify the pronunciation of proper names and special terms, and indicate by writing them phonetically in the script. The script should have all cuing places plainly marked. See the sample narration script on page 63.

PREPARATION FOR RECORDING

The first step in preparing an audio recording is to record the narration. The quality of a recording depends primarily on proper microphone selection, its placement, and on regulating volume level. Follow these practices:

- If possible, attach the microphone to a stand or clip it to the narrator's clothing about 10 inches below the mouth so it cannot be handled or moved during recording.
- If a stand is not available, set the microphone on a table with a sound-absorbing towel or blanket under the microphone.
- Determine by test the best distance from the narrator's mouth at which to place the microphone for good intelligibility (about 10 or 12 inches) and have the narrator speak across the front of it rather than directly into it to avoid popping sounds.
- If only voice is to be recorded, a tape speed of $3\frac{3}{4}$ inches per second (ips) is satisfactory. Music, however, requires higher fidelity than voice; if it is to be included, the tape for narration and music should be run at $7\frac{1}{2}$ ips.
- Avoid use of the automatic volume level control if there is one on the recorder (marked as AGC, ALC or AVL). See page 145 for reasons.
- Make a volume level check for each voice to be used. Select a moderately high volume setting, but one below the distortion level. This setting permits greater flexibility for controlling volume during playback and reduces background noise and tape hiss.

- Be sure to turn off fans and other apparatus that make noises which may be picked up by the microphone.
- Provide a comfortable yet alert working position for the narrator so that full energy can be put into reading the script. Allow him or her to stand or sit upright on a chair or stool. Standing is preferred as most people project their voice better when standing. Set the script, as separate sheets at eye level, on a music or other stand so the narrator will not have to handle the script too much or lower his or her head while talking. Change pages quietly to avoid possible paper rustle.
- Have a glass of water nearby for the narrator to "lubricate" the throat if necessary.
- Run a test recording of 2 to 3 minutes and listen to it to make sure all equipment is operating properly.

RECORDING PROCEDURE

Three people may be necessary to make a recording: the narrator, a cue giver (you or someone familiar with the timing of the narration in relation to the pictures), and a person to operate the recorder. The roles of the cue giver and recorder/operator may be combined for a person who is familiar with the script and the equipment.

You may wish to develop a standard lead-in for identification at the beginning of the tape. Allow 6 to 10 seconds of silence for the leader to pass, although cassette tape is now available without leader. Then identify the program by title, number, or both. Make certain that the volume and the tone levels are the same as those for the program content that follows.

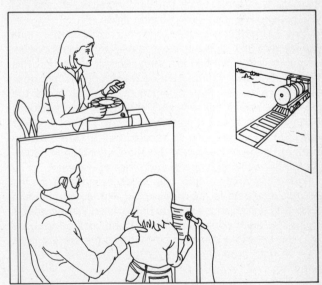

Stand the person to give cues behind the narrator. This person will indicate when each section of the narration is to start by tapping the narrator on the shoulder. Some narrators like to make gestures and move their hands or body while reading, as in face-to-face communication. This normal behavior should be encouraged, because these actions can stimulate good voice delivery.

Watch the sound level meter (VU meter) on the recorder and make adjustments to maintain a consistent sound level.

Experience shows that the best recording takes place in the first one or two tries. As the narrator repeats, there is a loss of spontaneity and errors can be made more frequently. Rehearse the presentation, and then make the recording. When the recording is complete, you and the narrator should listen to it together. Check it against the script to make sure that nothing has been left out, that no words are mispronounced, and that there are no extraneous noises. Your ear is the best judge of a successful recording.

Often when a tape is being edited, you will need to lengthen a pause to allow the listener time to study one or more accompanying visuals, to engage in an activity such as replying to a question on a worksheet, or to allow time for thinking about what has been presented. The length of tape that is added to provide the pause should have the same signal characteristics as the recorded tape, even though no sound is heard. Therefore, at the end of the recording session, ask the narrator to remain silently in position before the microphone. Record a few minutes of blank tape at the same volume setting as was used during the actual recording. Splice in lengths of this tape to adjust a pause period during editing.

Correcting Errors

It is unusual for a narrator to read through an entire script without making mistakes or mispronunciations. When a mistake is made catch it at that moment. It is easier to correct it immediately than to leave it until the end of the recording. At the end, the narrator might not match the intonations in context with the surrounding words and you must spend more time rearranging tape.

Two methods can be used to correct mistakes when they are made. First, stop the tape, rewind to the last correct sentence with a planned following pause in the script. This blank time allows you to start the recorder (in record position, which erases the unwanted material), and cue the narrator to reread the section. The narrator then proceeds normally on to the next part of the script. Second, make a correction by continuing to record. Call an identification such as "page 7, line 3, take 2." Then, after a short pause, repeat the material properly and proceed normally.

In the first method editing is unnecessary, because the correct recording is continuous. But with some standard recorders a click sound may be heard when the recorder is started at each correction. The second method requires that the unusable takes be cut out and the correct ones spliced, with proper spacing, to the previous part of the narration.

Once the recording has been completed, sit quietly with the narrator and playback the entire recording. Listen to it carefully for any errors—mispronounciations, extraneous noise, machine sounds, or anything else that might have been missed during the recording session. It is best to catch imperfections and correct them now rather than to redo later.

MUSIC AND SOUND EFFECTS

Research evidence indicates that background music is not essential to effective communications with instructional media. In some instances, indeed, it interferes with the message. But for other purposes it may help in creating a desirable mood and in setting the pace for the narration.

Music as background for titles will assist the projectionist or individual user to set the volume level for the narration that follows. Introduction music gets the attention of the audience, helping individuals to settle down for the presentation.

Here are some suggestions for selecting and using music:

- Always start music a few beats before the first image appears on the screen.
- When music is used under narration, maintain it at a low enough level so it does not interfere with the commentary or compete with the picture for the viewer's attention.
- Select music carefully. Avoid music with vocals unless the words are part of the message. Keep in mind the nature of the audience and the message to be communicated.
- A single musical selection may be suitable for the total production. Or, a number of brief musical cuts provide variety and support for the visuals. They must be carefully matched and mixed in keeping with the nature and pace of the visuals and narration.
- The music chosen for the conclusion of the presentation should receive special attention as it helps to set the emotional attitude with which viewers leave the program.

Three potential sources for music are:

- Selections from a commercial music library, which can be purchased for unlimited use or for a "needledrop" fee (meaning that a payment is made each time a selection is played). See page 155 for a listing.
- Original music recorded from good local talent—from a single guitar or organ, to a full orchestra.
- A professional music production company to score original music for a fee.

Sometimes it seems easier to select music from popular, semiclassical, or other entertainment records. But remember that when a listener hears background music with which he or she is familiar, it can distract attention from the narration. If you do plan to use such selections, permission must be obtained from the copyright holder (page 57). Obtaining music rights is complex and time consuming. Often a high fee is requested by such sources if your program will have wide use or commercial distribution. See the reference by Kichi for how to go about acquiring clearance for using a popular musical selection.

Natural sounds, when carefully used, help a production to sound professional. Sound effects, which add a touch of realism, also are available commercially. If you do not want these commercial effects or cannot find them you can record actual sounds on tape, or create sound effects (see

150

page 155 for a book on this subject). Sound effects on tape can later be transferred to the final audiotape or to videotape.

EDITING TAPE AND MIXING SOUNDS

After the narration recording is completed and other required sounds have been put on separate tapes, editing may be necessary to remove bad takes, to rearrange elements into a more cogent order, and to add tape for lengthening pauses. The resulting tape becomes the *master* recording.

Editing Tape

Audiotape may be edited by two methods. One is performed **electronically** and the other by **cutting and splicing** the tape physically. Electronic editing is a dubbing (duplicating) process in which a copy of the original recording is made, and the recorder on which the copy is being made is stopped while parts of the original recording are skipped or reordered. This requires the use of a high-quality recorder which can silently and easily pause, stop, and start again.

For physical editing, which requires cutting the tape, it is a good idea to make a quality dub and store the copy safely away. This will protect you if an incorrect cut is made while editing.

If the original recording is made on cassette equipment, it is necessary to make a copy on an open reel of tape. There is no practical way to cut and splice the thin, narrow tape in cassettes. Assuming that the recording is on a single track, follow these steps for successful tape editing:

1. Listen to the recorded tape, listing spots to be edited (use the index counter on the recorder to note locations on the tape).
2. Replay the tape and stop at the first spot.
3. Pinpoint the spot to be edited by moving the tape manually back and forth across the playback head.
4. Carefully mark cutting points on the base side of the tape. Use a fine-tipped felt pen or a china-marking (grease) pencil.

5. Listen to that part of the recording again to be certain that the mark is at the correct point.
6. Cut the tape; remove the felt pen or grease-pencil marks; then splice the ends together or add tape as necessary.
7. Repeat the same procedure at the next editing spot.

Tape Splicing

Splicing is best and most conveniently done with a tape-splicing block. This is a piece of metal that holds a short length of recording tape in precise alignment while you use a single-edge razor blade to cut the tape at a 45-degree angle, following slots cut in the block to guide the razor and then apply the splicing tape. The 45-degree cut makes a strong splice and provides smooth movement between spliced pieces as they pass over the playback head of the recorder. Use a sharp blade to make straight, clean cuts. Follow this procedure when splicing (see next page):

1. Set one piece of tape, with shiny (base) side up, firmly in the splicing channel so it just passes the cutting groove.
2. From the other side, do the same with the second piece.
3. Draw the razor blade across the 45-degree cutting groove to cut both pieces of tape at the same time. Remove the top waste end of tape.
4. Cover the cut with one 1-inch piece of splicing tape.
5. Rub firmly with a fingernail or nonmetallic burnisher.
6. Draw the blade along both edges of the splicing channel to trim any excess splicing tape extending beyond the edges of the magnetic tape.
7. Examine the splice for strength.

Mixing Sounds

There are various methods for mixing sounds. It is best to do this electronically rather than with a microphone. By placing the microphone in front of a loudspeaker to pick up the sound from a recording as it is played, the quality will not be the best and extraneous room sounds may be recorded.

A preferable way to combine sounds from two recordings (narration and music), on a single track, is to use a "Y"

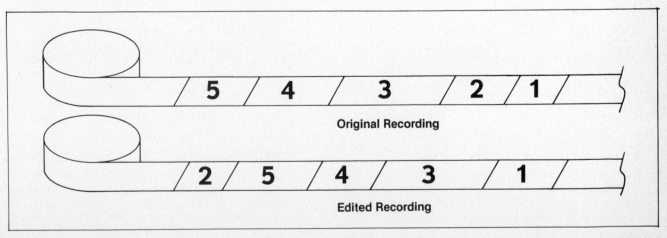

Original Recording

Edited Recording

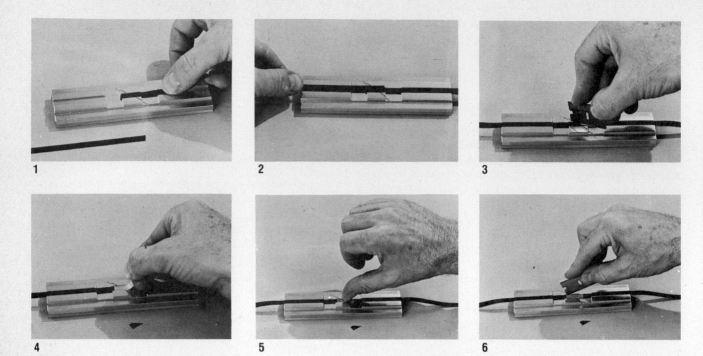

1 2 3

4 5 6

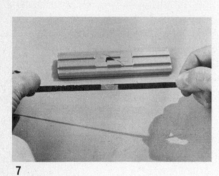

7

jumper cord (also called a **patch cord**). Connect an arm of the cord from the *output* of each source to the *line input* of the recorder for making the final master. After setting a volume level on the recorder, adjust levels for matching sounds at each source.

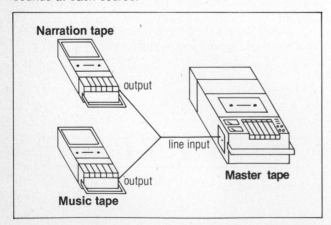

Narration tape

output

line input

Master tape

output

Music tape

You may have used a *mixer* while making the original recording in order to combine voices or add music or sound effects to the recording. The same procedure can be used during editing as sounds from separate tapes are blended

by controlling volumes and fading one in and the other out or under. Separate channels on the mixer are required for each sound component.

Through mixing, a number of tapes can be transferred onto a single track, or preferably different sounds can be dubbed to separate stereo or multitracks of a recorder. Copy the narration onto one track. Go back to the beginning, listen to the narration, and record music and sound effects on one or more other tracks in proper relationship to the narration. If mistakes are made, the music or effects can be redone without erasing the narration. Then the tracks can be played together or dubbed to another recorder as a single composite sound track.

When mixing sounds, be certain to make jumper cord connections properly (*outputs* from sound sources to *inputs* on mixer and from mixer *output* to recorder *input*). Balance playback levels carefully so sounds are heard in proper relationship (low level musical backgrounds under narration and realistic sound effect levels).

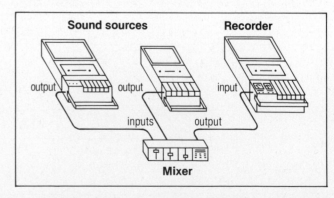

Sound sources Recorder

output output input

inputs output

Mixer

Consider passing a recording through an **equalizer,** which is a device with an arrangement of electronic filters. It permits you to selectively boost or subdue certain frequencies to modify the tonal quality of a recording. For example, with an equalizer, a 60 cycle background hum (caused by fluorescent lights) can be removed. Thus, by using an equalizer you can enhance the quality of a recording.

VARIABLE SPEED RECORDING

Normal speaking rate is 150 to 200 words per minute (wpm). An educated person reads printed material at 300 to 500 wpm. Therefore, the thought processes within the human brain can function much faster than the rate at which information is presented aurally to an individual. One way of closing this gap between speaking rate and potential degree of assimilation is to speed up the speech rate. The technique of increasing the word rate of recorded speech *without* distortion in vocal pitch (to avoid the "Donald Duck" speech effect) is called *time-compressed, rate-controlled,* or *accelerated speech.*

It also may be desirable to slow speech down for special listening purposes. Thus the general term *variable speed control* is applied to indicate the range of potential playback speeds that can be employed. A widely used recorder that has this capability (for accelerating speech) operates by deleting minute segments of sound and tying the remainder together, thus allowing an increased playback speed of speech with almost proper pitch and intelligibility. For slowing speech down, tiny fragments of sound are separated and the gaps filled with an extension of each previous sound.

compressed speech
Normal Sound

compressed speech
Deleted Portions

compressed speech
Remaining Segments

compressed speech
Reproduced Accelerated Sound at Near Normal Voice Pitch

To prepare an accelerated or slowed-down recording, copy the final master recording, played at normal speed, onto another tape placed in the variable speed recorder. Adjust the speed setting on this recorder to compress or to expand the sound to the desired degree. The resulting compressed or expanded tape can be used to make duplicates on regular equipment.

There is research evidence that speech compressed up to 50 percent (from 175 to 275 wpm) is understandable and can appreciably shorten the communication and learning time for a given message. But for many uses, especially those requiring understanding of technical material, compression of 10 to 30 percent is preferred by students (Foulke).

There are many potential uses for time-compressed speech. It can save time when an individual wants to scan

a recording rapidly (like leafing through pages of a book) to review a speech or lecture, to reduce boredom by keeping students alert and concentrating while studying recorded material, and most important, to shorten the time necessary when listening to a recording.

SYNCHRONIZING NARRATION WITH VISUALS

When sound is recorded on an audio track of a videotape, the picture and sound will be seen and heard in proper relationship. They are perfectly synchronized and always will remain so. But when pictures are separate from the recording, as with slide/tape or filmstrip/tape equipment, then another method must be used to synchronize the pictures and sound. This can be done by putting a signal on tape along with the recording to indicate when to advance to the next slide or filmstrip frame in accordance with the final script. An audible or an inaudible picture advance signal can be used.

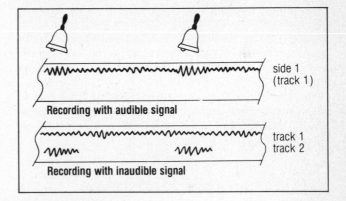

side 1 (track 1)

Recording with audible signal

track 1
track 2

Recording with inaudible signal

Audible Signals

An audible signal informs the projectionist or the individual user when to change to the next picture. Such a signal could be made with a musical instrument, a door chime, a bell, or a hand clicker. The best source for a controlled tone (preferably somewhere from 100 to 440 cycles per second —known as "hertz" or Hz) is an electronic audio "tone generator" keyed from a push button. The sound from any source should be of approximately one-half-second duration. It is mixed with the master recording onto the same track or a stereo track of the recorder.

Inaudible Signals

Because audible signals are often distracting and may be annoying to the viewer, an inaudible signal can be used to control picture changes. Tape recorders with built-in inaudible signal generators are available. The recorder is connected to the slide or filmstrip projector. The signal, generally 1000 Hz, is recorded at picture change spots on the *second track* of a monophonic audiocassette as you listen to the recorded narration on the first track. Upon playback each signal triggers a mechanism that activates the projector to change the picture in synchronization with the narration.

Some cassette recorders can also record an inaudible 150 Hz signal as a "stop" cue to pause the tape. This permits the user to reply to a question or do something else and then restart the tape.

Each signal on the tape requires sufficient time spacing from the previous signal to permit the projector to complete its projection cycle. Check the instructions with your equipment for the minimum allowable time (often 2 seconds) between signals.

Verbal Signals

If the resulting audio and visual materials are to be used by students studying independently, consider indicating slide or frame changes verbally along with some identification of the forthcoming picture, like—"turn to frame 12," or "in the next slide of the city hall. . . ."

When audible or inaudible control signals are used, it becomes possible for the student to inadvertently get the picture and sound out of synchronization and thus lose the meaning or continuity of the message being communicated. Furthermore, the signal method of controlling the advance of pictures eliminates the opportunity of going back to review either the narration or the picture, although some sound slide and filmstrip units, now on the market, can maintain synchronization in reverse (picture and sound) as well as in forward operation.

By verbally indicating changes and identifying the visuals, the student is free to go back to rehear or resee any material and then to easily move ahead with picture and sound synchronized.

DUPLICATING TAPE RECORDINGS

When your editing has been completed, with all sound and signals on a tape, this tape becomes the **master** recording. Label it as such. Make a copy of it onto an open reel or a cassette. Mark it as the *first-generation* copy. Punch out the "tabs" on the back edge of the cassette to prevent accidental erasure. File the master tape in a safe, cool, dry place.

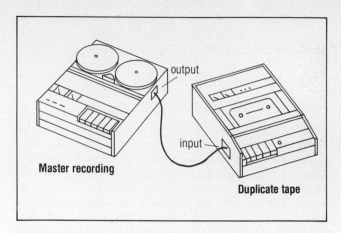

Master recording

output

input

Duplicate tape

Use the copy for making the required number of duplicates.

If the recording is made only on one side of the tape, without a 1000 Hz synchronization signal on the second track, you can repeat the same recording on the second side so that the user need not rewind the tape after each use.

High-speed equipment is available in many media centers for producing one or a number of copies, most often in audiocassette form. Commercial tape duplicating services are available also.

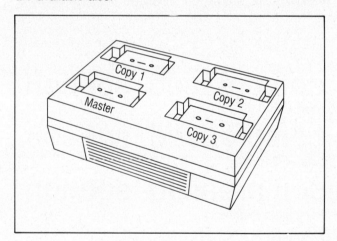

Copy 1

Copy 2

Master

Copy 3

Review What You Have Learned About Sound Recording Procedures:

1. What factors are important in selecting a narrator?
2. What are five important practices when preparing for a recording?
3. How would you make a recording to include music from a disc along with the narrator's voice?
4. If you find it necessary to make a correction on tape while recording, what procedure would you use?
5. What are two advantages for including music along with a narration? What can be a major drawback?
6. What is the difference between physical and electronic tape editing?
7. What equipment, materials, and procedures would you use for splicing tape?
8. What are advantages in using a compressed speech recording?
9. What method might you select to add an audible signal to a tape recording for indicating slide changes?
10. For use with a slide series, in what form might the duplication be put on tape if the recording requires one side only?
11. Do you have available a tape recorder that allows the addition of inaudible signals to tape? If so, how would you proceed to add signals to a tape?
12. What method do you have available for duplicating tape?

REFERENCES

Recording Sound

Alkin, Glyn. *Sound Recording and Reproduction*. New York: Focal, 1981.

Audio Principles Series. Slide/tape programs. Bloomington, IN: Indiana University, Audio-Visual Center—
X-115—*Basic Sound*
X-116—*Microphone Characteristics*
X-117—*Microphone Usage*

Nisbett, Alec. *Use of Microphones*. New York: Focal, 1979.

Oringel, Robert S. *Audio Control Handbook*. New York: Hastings House, 1983.

The Power of Sound in AV Presentations. Slide/tape presentation V10-39. Rochester, NY: Eastman Kodak Co., 1983.

Using Music

Kichi, John. "Music Rights and How to Obtain Them." *Educational and Industrial Television* 12 (December 1980):40–42.

Seidman, Steven A. "What Instructional Media Producers Should Know About Using Music." *Performance and Instruction Journal* 20 (March 1981):21–23.

Yungton, Al. "Scoring Shows with Music Libraries." *Audio Visual Directions* 5 (September 1983):43–46.

———. "There is No Perfect Music for the Job: What You Should Know About Copyright Laws and Music Libraries." *Audio Visual Directions* 3 (December 1981):35–37.

Variable Speed Recording

Foulke, Emerson. "Variable Speech Control in Audio-Tutorial, Training and Media Center Applications." *Third Louisville Conference on Rate-Controlled Speech*. New York: Foundation For the Blind, 15 West 16th Street, 1975.

Music Libraries

Emil Ascher, 630 5th Ave., New York, NY 10111

Brand Studios Projection Music Library, 10 Industrial Way, Brisbane, CA 94005

Capital Records Production Music, 1750 Vine St., Hollywood, CA 90028

DeWolfe Music Library, 25 W. 45th St., New York, NY 10036

Folkways, 43 W. 61st St., New York, NY 10023

Musicues, 1156 Avenue of the Americas, New York, NY 10036

Network Production Music, 4429 Morena Blvd., San Diego, CA 92117

NFL Music, 330 Fellowship Rd., Mt. Laurel, NJ 08054

Soper Sound, P.O. Box 498, Palo Alto, CA 94301

Valentino, Inc., 151 W. 46th St., New York, NY 10036

Sound Effects Libraries and Reference

BBC Sound Effects Library, c/o Films for the Humanities, P.O. Box 2053, Princeton, NJ 08540

The International Sound Effects Library, Emil Ascher, 630 5th Ave., New York, NY 10111

SFX Sound Effects, P.O. Box 401, Skokie, IL 60077

Turnbill, Robert B. *Radio and Television Sound Effects*. New York: Holt, Rinehart and Winston, 1951.

Part Four

PRODUCING INSTRUCTIONAL MEDIA

13. Printed Media

14. Display Media

15. Overhead Transparencies

16. Audiotape Recordings

17. Slide Series and Filmstrips

18. Multi-Image Presentations

19. Video and Film

20. Computer-Based Instruction

Chapter 13

PRINTED MEDIA

- Planning Checklist
- Organizing and Writing
- Editing the Text and Photographs
- Producing the Materials
- Application to Printed Media Groups
- Preparing to Use Printed Media

An important category of media, useful for instructional, informational, or motivational purposes, is comprised of those types prepared on paper. These include:

- **Learning aids,** such as guide sheets, job aids, and picture series
- **Training materials,** such as handout sheets, study guides, and instructors' manuals
- **Informational materials,** such as brochures, newsletters, and annual reports

Each of these groups is unique, serving a special communication purpose. See the definitions and descriptions of each group on page 165. There are elements in planning and preparation that are common to all of them. These we will examine first, and then consider specific requirements for each group.

PLANNING CHECKLIST

When you have decided to consider developing some form of printed media, start by giving attention to this checklist:

- Have you clearly expressed the **idea** to be served by the proposed materials (page 28)?
- Have you stated the **objectives** that are to be accomplished by using the materials (page 29)?
- Do you have a clear understanding of the nature of the intended **audience** or **student group** who will read or learn from the proposed materials (page 30)?
- Have you determined whether the type of **printed media selected is suitable** to achieve the objectives?
- Are you aware of **when** the proposed materials will be needed, in what **number of copies,** and **how** they will be used?

- Have you gathered the necessary **information** or **subject content** for the topics to be communicated, or analyzed the **task** that needs to be learned in terms of the objectives (page 32)?
- Have you considered whether there will be need for **assistants** and, if so, who might you ask to help or to provide professional or technical services (page 30)?
- Do you know what **funds** will be available for producing and duplicating the printed media?

Once these questions have been satisfactorily answered, and appropriate actions taken, you are ready to do the writing and to carry out the production.

ORGANIZING AND WRITING

Whether you will be preparing printed materials in the form of a simple checklist of items as a guide sheet, or as a compilation of articles for a newsletter, your ability to communicate effectively in writing is essential. While you cannot be taught how to become a competent writer by reading these few pages, here are some suggestions that you should apply to better insure that your materials will be effective:

- Procedures, facts, events, ideas, or whatever content is to be treated should be organized so there is a logical flow of information, providing for a continuity of thought. The reader must be able to easily follow each idea in turn, making sense out of what is presented.
- The arrangement of information and instructions should parallel the logical order of understanding and action the reader will need to take.
- Keep the potential reader in mind as you write. This will affect your selection of vocabulary, the number of exam-

ples you use, and other practices, all of which contribute to suitable reading and comprehension.

- Use simple words that are readily understood, unless technical terms are required, use them with moderation, after explaining and illustrating their meaning.
- Use proper grammar in sentence construction. In part, this refers to the placement and relationship of the elements of speech (subject, predicate, modifiers, and so forth) and the correct use of punctuation, especially commas and semicolons. (For ready reference, have at hand a good dictionary.)
- Set standards for consistency in capitalizing words and abbreviating expressions.
- Use short sentences rather than long, complex ones containing multiple clauses.
- Use short paragraphs because the appearance of lengthy ones on a page discourages careful reading. Each paragraph should treat one concept or aspect of a topic.
- Use headings and subheadings to identify and separate sections of the writing.
- As appropriate, periodically summarize the information presented. This repetition can increase the reader's understanding of the subject and also place the various elements of the topic in relative perspective.
- Consider the inclusion of a glossary containing definitions of important terms and possibly an index if the publication will be an extensive one.

One valuable test of effective writing is the ease with which readers can complete the reading with satisfactory understanding. Ease of reading is determined by applying a **readability formula.** Such a procedure employs factors like average number of syllables per 100 words and average number of sentences per 100 words. By comparing the averages of these two factors, within an article or a guide, a grade level of reading difficulty can be ascertained. Fry, a recognized reading authority, has designated a Readability Graph using these variables of syllables and sentences. Here are some examples from Fry's graph:

Syllables per 100 words	Sentences per 100 words	Grade level
124	7.5	4
138	5.0	7
148	4.3	9
154	4.3	11
162	4.0	12

EDITING THE TEXT AND PHOTOGRAPHS

You, or the person who does the original writing, will create the writing according to personal experience and knowledge. It is easy to make assumptions about what the reader may already know, and as a result one may fail to express oneself in ways that are clearly understandable. It is advisable that someone else review your writing and evaluate it in terms of organization, writing style, interest level, clarity,

and probably the treatment of technical content. Editing is the procedure of making necessary changes in the writing so that it says, to the reader, exactly what you mean.

The person (or persons) who will be responsible for editing should be selected carefully. Possibly there should be one individual competent in the subject so as to check the content for accuracy and another individual responsible for grammatical structure and literary expression. It is desirable for the latter person to have an academic background in English composition and experience in writing. Thus, the qualifications for editing should include strengths in these two areas—subject content and English composition.

Editing or rewriting may also be necessary for reasons beyond the content and grammatical purposes. Articles and study materials may have to be adapted to fit available space (known as "copyfitting"). Headings and subheads might need to be added to separate and identify sections. Introductions may have to be written in order to relate a series of articles so there is a smooth transition from one section or article to the next one.

Another phase of editing takes place when photographs are selected for inclusion with the text. Usually more than one picture is taken of a scene or a topic. Specific criteria must be used for choosing photographs—how well the subject is visualized, how well the picture is composed, how suitable is the technical quality, and so forth.

Picture editing may also include "cropping" photographs to improve composition. This requires putting marks at the edges of a print to indicate the visual area to be printed, thus eliminating unnecessary or distracting parts of the photograph.

PRODUCING THE MATERIALS

The production of printed media starts with editing. When adjusting the length of an article, there must be an awareness of space requirements and a plan for layout. To carry out production, a number of decisions and technical steps require attention.

Format and Layout

Some types of printed media require only a simple format; for example, a heading followed by a list of steps as for a procedure. Other types are more complex and may include a number of headings, printed text, art work and/or photographs, captions, and special items like a masthead or a logo. The manner in which these elements are combined becomes the layout. Each page of a publication may require a separate layout.

In designing a page, give attention to these matters:

- Placement of titles, headings, illustrations, and page numbers
- Number of columns
- Length of lines
- Width of margins
- Emphasis on specific content

The following sections of this chapter explain how many of these matters may be treated.

There can be three levels of layout preparation:

- **thumbnail**—simple, quick sketch to try out ideas
- **rough**—more detailed drawing to proper size, showing general placement of all elements
- **comprehensive** or final—showing each element to size, with color and accurate placement of all that will appear in final copy to insure correct spacing. This may be done with photocopies duplicated from final printed text, including headings and other elements. This layout is often called the **dummy**. It will be used as the guide for completing paste-up work.

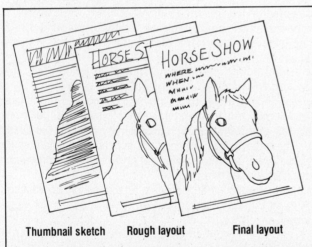

| Thumbnail sketch | Rough layout | Final layout |

In preparing a layout, many of the procedures described under the Visual Design and Layout section (page 107) of the Graphics chapter should be considered. Questions like the following may be important to consider:

- Does the design attract and hold attention?
- Should the layout be formal or informal?
- Is the continuity among all elements maintained so that the reader follows easily and understands?
- How many columns of text should be used on a page?
- Will this be a single color (generally black) product or will more than one color be used?
- Will the product be flat or folded, in a single sheet or multiple pages?

In seeking answers to the above questions, the following facts will prove helpful:

- The unusual, or what is novel, attracts attention and arouses interest.
- When first looking at a page, the eye usually starts at the upper left corner and moves across the page and down to the right. This is true of most printed pages unless the page has a special layout, designed to attract attention differently.
- White space is important on a page to separate elements and to create a feeling of openness.
- For reading ease, a column should be neither very wide nor very narrow.

Manuscript Text

The text may be prepared in final form on an electric typewriter, as the print-out from a high-quality word processor, or typeset by a printer. The decision of which method to use depends on these factors—desired level of professional appearance, money available, preparation time available.

Typesetting, while the highest quality and most expensive, may not be necessary for satisfactory results. A good typewriter, or a computer-controlled word processor that prepares clean, crisp copy can be used. Various type fonts (separate IBM typeface balls, for example) may be available and some machines will **justify** the right-hand margin. This latter procedure involves automatically changing the spacing between letters and words so that every line is of the same length.

> The format is teams of three to four faculty who meet once a week to create the syllabus and calendar; to choose readings, films and guest speakers; to plan field trips; to discuss the progress

Right Margin Ragged

> The format is teams of three to four faculty who meet once a week to create the syllabus and calendar; to choose readings, films and guest speakers; to plan field trips; to discuss the progress of

Right Margin Justified

If more than one typeface or type size is available, then decisions in terms of appearance and amount of coverage need to be made. Examine samples of prepared work that you receive in the mail or through your organization. Here are some helpful facts:

- For a typewriter, there are two type sizes (in various typeface styles)—*elite* and *pica*. Elite prints 12 characters per inch (letters, numbers, symbols, and punctuation marks), while pica prints 10 characters per inch. A typewriter makes six lines per vertical inch.

> This section of the report prese
> the solar water heater industry. The
> which achieved a reasonable degree of
> described. Next, the economic, perfor
> which produced the growth and subsequ

Elite-Size Type

> This section of the report
> overview of the solar water hea
> products and technologies which
> of market success are classifie
> economic, performance, and buil

Pica-Size Type

- Typesetting characters are measured in **points.** Textual lines commonly are printed in 10 or 12 point type. For appearance and ease of reading, "leading" or clear space between lines may be added. Thus, "10 point on 12" means 10 point lettering with two points of blank space between lines. Space between words is one-third the character point (4 points space for 12 point lettering).

> One possible source of power was heat.
> The English had been clearing the forest
> land for centuries. First, they burned the
> wood directly, then what was left had
> been nearly completely cleared by char-
> coal production.

10 point on 10

> One possible source of power was heat.
> The English had been clearing the forest
> land for centuries. First, they burned the
> wood directly, then what was left had
> been nearly completely cleared by char-
> coal production.

10 point on 12

- The length of a line is measured in **picas.** There are six picas per inch and 12 points equal one pica (thus, 72 points per inch).
- As you look at a page in this book, you will see that chapter titles are printed in 30 point, main section headings are printed in 11 point, text is printed as 9 point on 11.

If a number of items is to be listed, assist the reader to identify and study them by:

- Starting the first line of an item at the left margin, then indenting the first word of each following line
- Providing space between items for visual separation
- Numbering each item if the list is sequential
- Preceding the first word of an item with a large dot (called a "bullet") or other attention-getting symbol

To give special emphasis to a word, phrase, or sentence, consider these techniques:

- Enclose the expression in "quotation" marks.
- <u>Underline</u> the expression.
- Print words in CAPITAL LETTERS.
- Use a **bold-face** type style for words.
- Use an *italics* type style for words.
- Draw a box around the phrase.

Headings and subheads are bold and larger than the text. They are designed to attract attention and serve as the lead-in to what follows. There should be consistency in letter style between headings and text that follows. Headings can be prepared by typesetting, dry-transfer letters (page 125), or pressure-setting lettering (page 127), with excellent results. Give attention to the suggestions on page 122 when preparing lettering.

Proofreading

Once the text is typed or set, and headings are prepared, all writing should be proofread. A printer supplies **galley proofs** for this purpose before printing. Photocopies of typewritten originals should be made for proofing. Check the galleys or copies carefully against the original manuscript.

Errors in printing or typing should be marked and corrected. When correcting copy, refer to a large dictionary or proofreading reference for the basic symbols that are used to communicate to the printer. At this stage, changes made by the author on typeset material, may prove to be costly.

The galleys or typed copy will accurately indicate space to be used. Therefore, the final layout—the **dummy**—can be prepared at this time.

From the proofread materials, final copy for reproduction, with all corrections, is prepared. This is called the **repro copy.**

Art Work and Photographs

Illustrations, in the form of diagrams, photographs, cartoons, charts, or graphs, when prepared for printed materials, can serve various purposes. They may be decorative, designed to attract attention to the page and even to add to the pleasure realized from the reading. Of more importance, art work and photographs illustrate and clarify what is written, thus contributing to understanding. Furthermore, pictures may provide information beyond what can be stated. Finally, illustrations may contribute significantly to the retention of information.

Line illustrations are prepared as black-and-white drawings (page 135). Toning and texture patterns, consisting of closely spaced dots or lines, can be added to art work to

1. A *thrombus* forms in a blood vessel, when intact and ruptured platlets adhere together on a vessel wall. Factors which contribute to thrombus formation are:
A) decreased smoothness of vessel lining
B) decreased rate of blood flow
C) increased blood coagability

2. An *aneurysm* is a weakening or thin spot in an artery wall that may balloon out and finally rupture.

3. an infarction results there when is complete blockage of the blood supply in cerebral or coronary arteries. Prolonged blockage deprives tissue of nourishment this causing cell death and permanent tissue damage

Proofread Galley Proof

1. A *thrombus* forms in a blood vessel when intact and ruptured platelets adhere together on a vessel wall. Factors which contribute to thrombus formation are:
   a) decreased smoothness of vessel lining
   b) decreased rate of blood flow
   c) increased blood coagulability

2. An *aneurysm* is a weakening or thin spot in an artery wall that may balloon out and finally rupture.

3. An *infarction* results when there is complete blockage of the blood supply in cerebral or coronary arteries. Prolonged blockage deprives tissue of nourishment, thus causing cell death and permanent tissue damage.

Repro Copy

create the impression of shades of gray. Art work should be prepared either of a size to fit the area indicated on the layout dummy, or of a larger proportionate size which can be reduced. See suggestions for doing art work and sources for ready-made art on page 112.

Art work also may be required for special parts of a publication—main title, logo (an identifying symbol that may be used in future issues), and as a cover page.

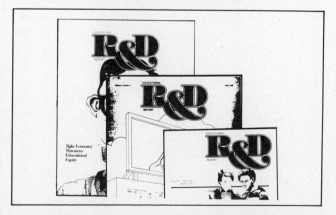

Photographs will be made as black-and-white prints unless four-color reproduction is possible or necessary. Black-and-white pictures should be printed to final size, or can be scaled to the available space for reduction by the printer. For the latter, prints on 5 ×7 inch paper are suitable. Photo-

graphs should be printed on single-weight, glossy finish paper. Full-color reproduction can best be made from a color slide. A size larger than 35mm ($2\frac{1}{4}$" × $2\frac{1}{4}$" or 4" × 5") is preferred for quality reproduction. Review the Photography chapter for detailed information on taking pictures and making prints.

Paste-up

Procedures have been developed through certain computer-assisted design (CAD) systems which allow the preparation of final, camera-ready copy, including headings, text, art work, and proper spacing for photographs. While this process bypasses the need for paste-up, it is not widely used because equipment, at this time, is very expensive.

The present, normal procedure is to gather all elements that will comprise each page and "paste" all parts on paper or cardboard, in accordance with the layout dummy. If a number of pages will have a similar format, use preprinted sheets (containing nonreproducible blue lines) showing columns for text areas and other elements to appear on each page. With these sheets, the paste-up process is simplified. When the paste-up is finished, the page is ready for duplication or reproduction.

See the explanation of the paste-up procedure on page 136. Use the temporary rubber cement mounting process (page 128) or a waxing unit (page 136). Note how both black-and-white line copy and continuous-tone photographs are handled. Also, if the final material requires more than a single color, see how separate, registered parts of a page should be prepared.

Color

For printed materials, the use of color, other than black, may be important to separate elements or call attention to something of particular importance on a page. On the other hand, color may be used just as an attractive feature.

The two ways to obtain color are to print on *colored paper* or to use *colored inks.* Whenever a second ink color is to be used, costs go up appreciably because a separate press run is required for each color. A careful choice of paper (weight, color, finish) can contribute appreciably to the appearance of any printed materials.

Reproduction

If you are working with a printer, a page-size litho (high-contrast) negative and then a metal printing plate are made from each paste-up sheet. As an alternative, for short-run reproduction, the printer may make a paper plate directly from the paste-up on a plate maker. Then, either a positive contact print from the litho negative or trial copies from the plate are run on a printing press to become **page proofs.** Examine the page proofs carefully to be certain all items are correctly placed (especially check headings and the match of captions with photographs). Then be sure everything is aligned accurately.

The final step in the production of printed media is to make the necessary number of copies. A few copies, prepared on an office copy machine, usually is called **duplication.** A long run (10 copies or more), with a printing press, is termed **reproduction.** See page 139 for explanations of the common duplication and reproduction processes.

Binding

Binding can be accomplished in various ways:

- Stapling the upper left corner or along the left edge to secure all sheets.
- Punching holes and putting all sheets in a two- or three-ring binder.
- Drilling holes and punching all pages together with plastic strips (Velo product).
- Punching holes and securing sheets with metal or plastic spirals.
- Stitching and/or glueing all sheets, then covering with binding cloth.

APPLICATION TO PRINTED MEDIA GROUPS

The planning, writing, editing, and production stages for making printed media have been described in terms of general procedures. Specific matters relative to each of the three printed media groups need special attention.

Learning Aids

These three forms of printed media—guide sheets, job aids, and picture series—are designed to provide step-by-step instructions for performing a task. The purpose may be to assemble and then operate equipment, troubleshoot for repair or maintenance, or to apply the steps of a procedure. The information and directions provided must be specific,

---

Folding Sheets of Metal
on the Bar Folder

Making a Single Hem

1. The steps that should be followed for making a single hem on the bar folder:

   a. Loosen the locking screw.

   b. Place the sheet against the fingers and move the gauge until the line marked on the sheet is slightly outside the folding blade.

   c. Tighten the locking screw.

   d. Insert the edge of the metal to be folded between the folding blade and the jaw.

   e. Hold the metal firmly aginst the gauge fingers with the left hand and place the right hand on the operating handle.

   f. To fold the edge, pull the operating handle with the right hand as far as it will go. Keep the left hand on the sheet until the sheet is held in place by the wing.

   g. Return the operating handle.

   h. Remove the sheet and place it with the fold facing upward on the beveled part of the blade as close as possible to the wing.

   i. Pull the operating handle with a swift motion to flatten the hem.

2. The hem can be flattened by placing the sheet with the fold facing upward on the beveled part of the blade as close as possible to the wing, and then pulling the operating handle with a swift motion.

Guide Sheet

THE PLANER

The planer, or surfacer, is used to bring material to uniform dimensions after any warp has been removed on the jointer. It is also used to smooth rough-sawn surfaces.

1. Measure each piece of faced stock to find greatest thickness.

2. Set depth-of-cut gauge to 1/16 inch less than thickest piece of stock.

3. Check direction of grain with hand. Know direction of rotation of cutter head.

4. Plane (surface) all pieces from which 1/16 inch will be removed at this depth-of-cut setting.

   Note: Stock normally remains straighter and warps less if an equal amount is removed from each side of the piece when planing.

5. Remove each piece as it emerges from the planer.

6. If necessary, reset depth of cut for thinner pieces.

7. Set depth-of-cut gauge to plane all pieces to final predetermined thickness.

Job Aid

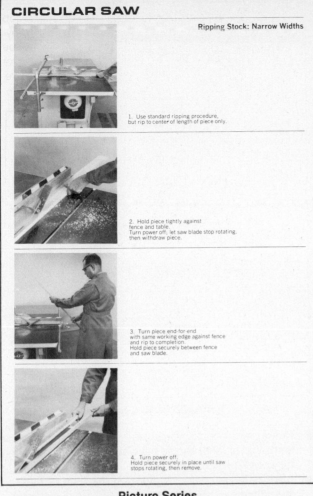

CIRCULAR SAW

Ripping Stock: Narrow Widths

1. Use standard ripping procedure, but rip to center of length of piece only.

2. Hold piece tightly against fence and table. Turn power off; let saw blade stop rotating, then withdraw piece.

3. Turn piece end-for-end with same working edge against fence and rip to completion. Hold piece securely between fence and saw blade.

4. Turn power off. Hold piece securely in place until saw stops rotating, then remove.

Picture Series

simple, and clear because most often the student, new employee, technician, or whoever uses the material, does so individually, without receiving any assistance.

Depending on the information or task to be learned, the material may consist entirely of words (guidesheet), some illustrations (job-aid), or many pictures (picture series).

When planning, writing, and producing these forms of printed media, give attention to the following:

- Analyze in detail the information, task, or skill to be taught and identify the arrangement of parts or the order of steps and sequences.
- If the task is complex, "storyboard" the various elements of a sequence (see page 48) so you will be certain there is a logical, smooth flow of information and instruction.
- Design the layout so the materials look appealing and are easy to follow. Use headings to guide the reader through the content.
- When writing directions, use simple English and explain or illustrate technical terms when each one is first introduced.
- Decide whether, and if so at what points, line drawings and/or photographs are needed to carry or clarify details. Remember that photographs are more expensive to prepare for printing than are line drawings.

- For special attention, highlight safety practices or other critical elements on the layout.
- If appropriate, provide a summary list of procedures and add questions for review to help the user relate and remember what was presented. For example: "Did you do _____ before doing _____?" "What tool do you use to _____?" "What should you do if _____ happens?"
- If photographs of a person performing a procedure are to be made, place the camera in a subjective position (page 92) and show the action in the scene with a close-up shot.
- Check diagrams and art work at a preliminary production stage to be certain of accuracy in depicting events and details.
- Have contact prints (page 102) of pictures made and select those for use before enlargements for printing are made.
- Early in planning, decide on the final form material will take—single copy for use at a work station, few copies for handing out to students on request, inexpensive multiple duplicates, or quantity reproduction for wide distribution.
- Another technique for analyzing a complex task is to develop a "flow chart" as used in computer-based instruction (page 249). The flow chart includes sequences of procedural steps and decision points for the student or trainee.

Training Materials

These forms of printed media—handout sheets, study guides, and instructor manuals—provide directions, information, and activities for various instructional purposes. While a handout briefly treats a limited portion of a topic, a study guide contains more details. They are both designed for student or trainee use. They must clearly communicate what is to be learned and how it should be studied. On the other hand, instructor manuals are to guide and assist those who conduct the instruction.

Although the content and its treatment will be different for materials to be used by students and those to be used by instructors, similar attention needs to be given to layout and graphic production techniques. Handout sheets and study guides should be carefully organized and simply written. Use outlines, lists, and brief statements. Employ logical page design, careful use of legible headings, typewritten or typeset text, and illustrations that motivate interest and communicate understanding. Depending on how extensive the development of a handout or a study guide, these matters of content should be considered:

- **Title**—a descriptive statement for topic, unit, session, or course.
- **Rationale** or **purpose**—reasons why the study to be undertaken is important, explaining where the topic fits into the overall program.
- **Prerequisites**—units, courses, or experiences the student should have successfully completed before starting.
- **Objectives**—list of learning objectives to be accomplished (see page 29), informing student what is required

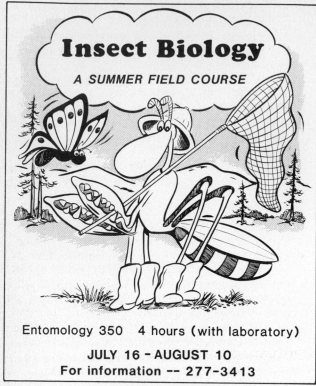

Insect Biology

A SUMMER FIELD COURSE

Entomology 350   4 hours (with laboratory)

JULY 16 - AUGUST 10
For information -- 277-3413

Handout Sheet

Name _____ IS Time _____ Date _____

Read Smart, "Physical Characteristics and Skills," pp. 345-378 and "Physical Growth, Health and Coordination," pp. 493-527.

Objective: To test for motor coordination differences among two different groups.

ACTIVITY 18: Motor Coordination and Growth   (2 hours)

Choose a child in the middle years to closely observe and put the sex and age here_____ Ask the child to do the following things for you and observe the ease with which they are accomplished. Supply a general description in Column 2.

(1) Task	(2) General Description
1. Stand on one foot.	
2. Touch nose.	
3. Close eyes and touch nose.	
4. Stand heel to toe.	
5. Catch a ball or other object.	
6. Throw.	
7. Wind a thread or yarn.	
8. Cut a circle.	
9. Jump, touch heels.	
10. Balance a rod vertically.	
11. Do a sit-up (with help)	
12. Do a sit-up (without help)	
13. Walk a straight line.	
14. Draw a circle.	
15. Dribble a ball.	

Find out as much as you can about the child's favorite active games and include in your description anything you think may have a bearing on the child's capabilities.

Now ask a teenager or adult to do the same tasks. Write the sex and age here _____. Summarize a general description below, following all instructions for the middle years child.

Study Guide Sheet

and criteria against which learning and performance will be evaluated.

● **Schedule**—calendar of activities and required deadlines for completing phases of program

Topic: Immobilization due to Fractures

Objective: To apply splints to five parts of the body

Time	Contents	Activities/Resources	Notes

8-10 min	Fractures
	upper arm
	forearm
	ankle
	knee
	back
		Introduce topic
		Show film clip #12A
		Discuss cause of
		  fractures at various
		  parts of the body	

10 min	Upper arm 8 steps	Show film clip #12B Demonstrate procedure Assign Workbook U-5	
15 min		Individual and team practice Checklist #5	See supply list #4
10 min	Forearm 7 steps	Show film clip #12C Demonstrate procedure Assign Workbook U-6	
15 min		Individual and team practice Checklist #6	Supply list #5
6 min 6 min		**Discussion** In what 3 places should you prevent movement to immobilize a fracture of the forearm?	Answer: broken bone end wrist elbow
		--- Break ---	
10 min	Ankle 5 steps	Show film clip #12D Demonstrate procedure Assign Workbook U-7	
15 min		Practice Checklist #7	Supply list #6
12 min	Back 6 steps	Show film clip #12E Demonstrate procedure Assign Workbook U-8	
20 min		Practice Checklist #8	Supply list #7
		Discussion If a person must be turned and you suspect a back fracture, what should you do?	Answer: Turn entire body as unit

Instructor's Manual Page

● **Resources**—books, supplies, equipment, and other items that student should have or which will be available for specific uses.

● **Information**—content material in the form of outlines, descriptions, reproduced diagrams or articles (duplicated with permission).

● **Activities**—learning experiences in which the student may participate (during class period, for self-paced learning, or for group work).

● **Exercises** or **worksheets**—questions to be answered during or following activities (be sure to provide sufficient space for writing answers).

● **Problems**—application situations, case studies, or similar practical activities that require students to make use of information, principles, or procedures that have been learned.

● **Answers**—acceptable answers to exercises, worksheet questions, and other participating activities.

● **Projects**—follow-up of major activities that require student to apply further what has been learned.

● **References**—lists of other resources (books, journal articles, media, persons, or places) that may be assigned for use or might be of future value.

● **Tests**—written exercises to measure learning; may be self-administered by student as self-check on learning, or as preparation for instructor's test.

● **Evaluation forms**—rating scales, checklists and other instruments to evaluate results of projects, participation in class or group activities, and so on, making the student aware of how learning will be evaluated.

● **Grading criteria**—description of how grading will be

made, indicating the percentage or point basis value of each activity, project, or other requirement.

This list of possible items to include in a study guide is extensive. Your own situation should indicate which of them are relevant for use. Because outlining, writing directions, and providing information in specific detail comprise the major portions of these materials, careful editing and proofreading are essential.

When planning an instructor's manual, although it is obvious, realize that the individuals who will use the manual probably have not been involved in planning the instructional program. While they may be highly qualified in the subject or skill area, the instructors have no familiarity with the manner in which the content is organized and treated. Furthermore, many newly appointed business and industrial trainers have limited, if any, teaching experience on which to base their classroom activities. Therefore, the manual must carefully lead instructors through the teaching requirements and provide sufficient information about resources and other necessary course logistics.

An instructor's manual may be divided into these parts:

- An **introduction** which explains the goal for the instructional program and specifies the major outcomes to be accomplished (often expressed as "terminal objectives"); may include information on how the manual is organized and should be used.
- Individual **units** or **topics** that comprise the course, each one treated as a separate section in the manual.
- A **supplement** or **appendix** containing test questions and/or answer keys, printed scripts for media, performance checklists, and other evaluation forms.

In developing each unit in the manual, consider the following:

- Specify objectives and requirements for students so the instructor will clearly understand what needs to be taught in the unit.
- Design pages in a split-page format, as is a two-column script.
- Along the right side, present directions, information, questions (possibly with suggested answers), group and individual activities for instructor and students, lists of references and resources, and so forth.
- Use the left side of the page to indicate suggestions for

timing of lesson components, for handling the content, for student assignments, and for obtaining or using media and other resources (may reference a specific resource by number or, in the case of visual materials, a small-size reproduction).

- Use an outline form in presenting directions and describing information to save reading time
- Employ cues or "visual signals" to direct attention; these may be boldface headings or graphic symbols that mark locations of certain items (questions, summary points, video or other media forms to be used).
- Space items on the page so the instructor can add notes relating to specific items.

Informational Materials

This category of printed media, consisting of brochures, newsletters, and annual reports, is the most complex to develop and requires production skills at a higher level than that needed for the other two groups. Most often, materials in this category are used to attract attention and to create a positive impression about an activity or an organization.

These attitudinal objectives require special attention during the planning phase and when applying the production techniques described earlier in this chapter. Let us review these matters as they relate to this category of printed media.

At the start of planning, consider the series of questions listed at the very beginning of the chapter. Of these, establishing clear objectives, recognizing the nature and interests of the potential readers, and being aware of financial constraints are probably most important. By responding to these questions, you can initiate planning that will likely lead to the successful development of your materials.

As you gather information for the publication, realize that in order to attract the reader you must create some excitement about the topic. This may mean attempting first to choose content that will arouse interest by being important to the reader, and then to treat the content in an unusual way—provocative, controversial, amusing—aimed at involving or otherwise stimulating the reader. Make certain the writing is simple, clear, to the point, and accurate in content.

Decide on an overall design in keeping with your treatment. Many of the usual rules of layout are sometimes violated as new ideas are explored; however, this may result in lively new arrangements of images and space that make a strong visual impact.

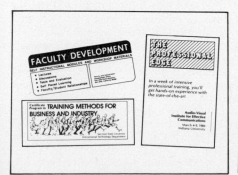

Brochures

Newsletters

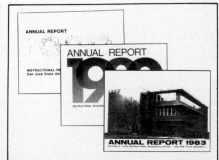

Annual Reports

The visual structure of a page is enhanced by the careful grouping of items; recognize the need for sufficient white space and establish a practical column width for the text or articles. Also, limiting the length of paragraphs will contribute to the readability of printed materials.

As you move to the production phase, here are some suggestions:

- Use a simple art style for preparing illustrations or select from available high-quality clipbook art (page 112).
- Use lines, tones, and texture patterns to emphasize certain parts of the layout or to create special effects.
- Select colors that arouse interest and create some enthusiasm for the subject.
- Develop tables and graphs in a form that makes their meaning explicit as well as interesting (page 114).
- Design an attractive, timely cover or first page, in good taste, and that has eye appeal. You want it to cause the reader to pick up the publication and start reading!

As stated previously, much of the information presented under the writing, editing, and production sections relate directly to this category of printed media. The above suggestions serve to reinforce some points already made.

PREPARING TO USE PRINTED MEDIA

Now the printed media are ready for distribution and use. Distribution may be accomplished by mailing, bulk delivery to departments within an organization, and by personally handing the printed media to individuals or having it available for pickup at a meeting or in an office.

Attempt to gather reactions from those using or reading printed media in order to evaluate its effectiveness. Consider giving attention to these factors:

- General appearance and appeal
- Organization of content
- Clarity and readability of content
- Value of illustrations
- Usefulness of content
- Response from users in the form of action relating to objective initially established

Review What You Have Learned About Printed Media:

1. What are the three categories of printed media treated in this chapter and the forms each category can take?
2. What *two* relationships are used to determine the readability level of written material?
3. What is the approximate readability level of this book as based on page 167 of this chapter?
4. What are three ways that special attention can be called to a word or a phrase on the printed page?
5. Define these terms—cropping, copyfitting, dummy, duplication, galley proof, justify, repro copy, reproduction.
6. What type of printed media might be selected for each purpose?
   a. To quickly review how to perform an operation that was learned some time ago.
   b. To give students a complete self-study package on a topic.
   c. To announce a new program to be available for members of an association.
   d. To provide a simple outline for all members of an audience attending a meeting.
   e. To keep members of an organization informed of events.
   f. To follow a sequence of steps in assembling a complex product.
   g. To help a new engineer who will teach a course in his speciality.

REFERENCES

Faulkner, Linda, and Hinds, Mary. "Planning and Perception: Principles for Designing the Printed Page." *Performance and Instruction Journal* 20 (March 1981):26–28.

Fry, Edward. "A Readability Formula That Saves Time." *Journal of Reading* 11 (April 1968):513–516, 575–578.

Hartley, James. *Designing Instructional Text.* New York: Nichols, 1978.

Hartley, James, and Burnhill, Peter. "Fifty Guidelines for Improving Instructional Text." *Programmed Learning and Instructional Technology* 4 (February 1977):65–73.

How to Produce a Newsletter. Washington, DC: Association for Educational Communications and Technology, 1979.

Jonassen, David H. *The Technology of Text.* Englewood Cliffs, NJ: Educational Technology Publications, 1982.

Levie, W. Howard, and Lentz, Richard. "Effects of Text Illustrations: A Review of Research." *Educational Communications and Technology Journal* 30 (Winter 1982):195–232.

Turnbull, Arthur T., and Baird, Russell N. *The Graphics of Communication: Typography, Layout, Design.* New York: Holt, Rinehart and Winston, 1975.

Wilson, Thomas C., et al. *The Design of Printed Instructional Materials: Research on Illustrations and Typography.* Syracuse, NY: ERIC Clearinghouse on Information Resources, Syracuse University, 1981.

Chapter 14
DISPLAY MEDIA

- Planning the Display
- The Chalkboard
- Display Easels and Flip Charts
- Cloth Boards
- Magnetic Chalkboards
- Bulletin Boards and Exhibits
- Planning to Use Display Media

When an instructor or speaker is to address a small class or an audience (up to 25 persons), the use of some type of nonprojected media may be appropriate. This category includes the **chalkboard, flipchart, cloth board,** and **magnetic board.** In each instance, the user displays materials, often for visual reinforcement, as the presentation proceeds. To this group of display media used by a speaker, we add **bulletin boards** and **exhibits.**

These media forms may be used in informal, yet dynamic ways. A speaker can focus and hold attention on an item being displayed, yet modify and adjust a presentation as it proceeds. Students or members of an audience can participate by handling and showing materials.

PLANNING THE DISPLAY

The selection and use of display media should be preceded by giving attention to these phases of planning:

- Have you expressed your **idea** or indicated the **topic** to be treated (page 28)?
- Have you stated one or more **objectives** to be served by the display media (page 29)?
- Have you considered the **audience** before which you will use the display media (page 30)?
- Have you enumerated the **subject content** to be included (page 32)?
- Have you decided which **type of display media is most suitable** for your use?
- Have you planned the layout of the content, possibly in the form of a **storyboard** (page 48)?

Many of the skills described in the Graphics chapter can be utilized when preparing materials for use on display surfaces. Particular reference should be made to these sections:

- Illustrating, including enlarging and reducing techniques (page 112).
- Procedures to ensure satisfactory legibility of lettering for nonprojected materials (page 121).
- Lettering techniques and aids suitable for preparing nonprojected materials (page 124).
- Dry mounting techniques to preserve and display pictorial materials (page 130).

THE CHALKBOARD

The chalkboard is probably the most common and most widely used type of display media. Everyone is familiar with the chalkboard from experiences during school days. Key words, outlines, lists, diagrams, graphs, and sketches are some of the visual and verbal things that can be written or drawn on the chalkboard.

For success in using the chalkboard, planning what will be displayed and employing simple preparation techniques are important considerations. Here are some suggestions:

- Plan to build explanations on the board, point by point, as the presentation proceeds.
- Prepare by writing or drawing lengthy materials on the board before the class or meeting is to start. Cover with a pull-down projection screen or paper held with tape; remove when ready to use.
- Place a few dots across the board in advance of use so that a line of writing will appear level.
- Use light chalk marks to place lines for guidance in drawing complex diagrams.

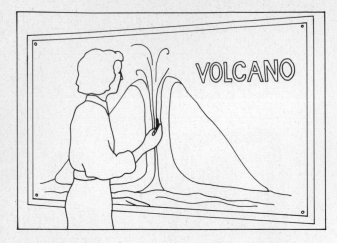

- Use templates made of plywood or heavy cardboard for tracing frequently used shapes.
- Use colored chalk, with a strong and intense color, to highlight parts of a diagram or important words.

Chalkboards, while most frequently black, brown, or green in color, may also have a white, nonglare, plastic surface. Writing on the white surface is accomplished with special water-based felt pens. The marks are removed with an eraser or a damp cloth. Such surfaces are often multipurpose since they can serve as a projection screen and, with a steel backing, can be used as a magnetic board.

DISPLAY EASELS AND FLIP CHARTS

When a chalkboard is not available, or you prefer to prepare verbal or simple diagrammatic materials prior to a meeting, consider using a display easel or flip chart. Individual visuals or a series, prepared on large sheets of cardboard, can be placed on a table-top easel or a floor stand. Each item may be displayed in turn.

A flip chart consists of a number of large sheets of paper fastened together at the top edge and set on an easel. Writing or drawing is accomplished with felt pens. Color can

be used for emphasis and to make the graphics more attractive. Each sheet can be turned over easily and the next one revealed during the course of the presentation. Leave a blank sheet of paper at the end of a sequence to provide a break before starting the next set of sheets

CLOTH BOARDS

Cloth boards are display surfaces that can be used to develop concepts or processes by the progressive addition of prepared parts. There are two kinds of cloth boards. One consists of a high-grade **felt** or **flannel** covering over plywood, masonite, or lightweight wall board. It is set on an easel (with a slight back tilt) for use. Pastel or light colors are advisable for the cloth covering. Prepare lettered materials in a bold size on cardboard. Diagrams and mounted pictures can also be used. Back the cardboard with strips of felt or sandpaper. This textured or rough backing will adhere to the flannel or felt surface.

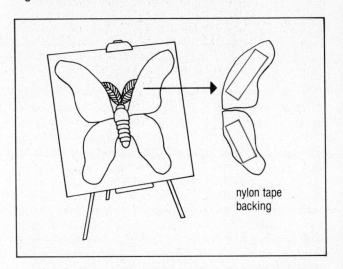

nylon tape backing

A preferred type of cloth board, the **hook-and-loop board,** safeguards against the possible slippage of flannel or felt covered surfaces. The surface material consists of a cloth containing countless tiny nylon loops, while display materials are backed with small strips of tape having numerous nylon hooks. In use, the nylon loops intertwine with the nylon hooks, joining the two surfaces securely. The

strength of this union is sufficient to hold not only paper and cardboard objects but also heavy three-dimensional objects. A one square inch of intertwined hooks and loops will support up to ten pounds of weight.

Commercial hook-and-loop boards are available. See sources on page 172. If you prefer to prepare your own board, you can purchase the nylon loop material as yardage from a major distributor of fabrics. The material should be adhered to plywood or other stiff backing with contact cement or white glue. Also, use a strong adhesive to attach small pieces of the tape to the back of objects.

MAGNETIC CHALKBOARDS

The magnetic chalkboard is more flexibly useful and permits more versatility than either of the cloth boards. The user can draw or write on the surface with chalk (or felt-tip markers on a white surface board) in addition to positioning prepared objects. The magnetized materials hold their position where placed and can be moved at will. The painted writing surface does not interfere with magnetic attraction.

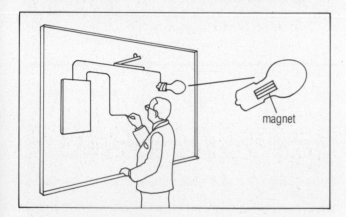

magnet

A magnetic board consists of sheet steel or a fine-mesh steel screen covered with a writing surface. Small magnets are cemented or taped to the back of objects or to graphic materials. Magnets may be of metal or of a magnetized rubber strip with adhesive backing. Plastic letters with magnetic backing are also available.

As previously noted, a white magnetic board can combine a projection surface with the display of objects.

BULLETIN BOARDS AND EXHIBITS

A bulletin board is generally a two-dimensional display prepared on a wall surface, while an exhibit is a three-dimensional display like a set of free-standing panels, a self-contained booth, or other functional display. While the previously described display media are selected and designed mainly for instructional purposes, the reasons for using a bulletin board or an exhibit can be much broader. They may be used for general information on a topic, or to publicize and stimulate an interest in an organization, a program, a product, or a service.

In addition to the planning steps listed at the beginning

of the chapter, decide on a theme or key idea to be treated in the display. Then plan the display on paper. Consider the design principles when arranging headings, visuals, and text or captions so as to attract attention, often that of persons passing by the display. Thus, communication must take place quickly and effectively. The most important point to keep in mind is *not to overload* the display with too many words or pictures.

Here are other suggestions for creating good bulletin boards and exhibits:

- Make sure the display will be located where there is a good flow of traffic.
- Use attention-getting devices like an unusual design, striking color combinations, a startling or unusual leading statement, or moving parts in the display (operated by a motor run by small batteries).
- Use pictures to illustrate the heading, thus reducing the time needed to read words.
- Project slides on a small rear-screen as an integral part of the exhibit
- Select three-dimensional objects or create a three-dimensional appearance from flat pictures by raising them from the surface with blocks of supporting wood.
- Include sound, if appropriate, by having the visitor push a button to operate an audio cassette player containing a continuous-loop tape.
- Involve the spectator with a question, a guessing game, a contest of skill, or by offering something to be picked up and taken away.
- Illuminate the display carefully with one or more floodlights, or highlight a key part which can create a feeling of excitement.
- Allow the display to remain in place for a suitable amount of time that attracts attention.

Finally, as you assemble the bulletin board or exhibit, ask yourself these questions:

- How will attention be captured so that interest is stimulated?
- How will attention be held until the message has been read completely?
- Can the message be easily understood in one reading?

PLANNING TO USE DISPLAY MEDIA

Once the preparation of materials is completed, make plans for their use. Make sure the bulletin board or exhibit will be available for viewing at times suitable to the anticipated audience. Refer to Chapter 8 for suggestions on using and evaluating your display materials.

Review What You Have Learned About Display Materials:

1. To which type of display media does each statement relate?
   a. Revealing, in turn, a sequence of related items on a portable display surface
   b. Flexible, in that both displaying objects and writing on the surface can take place
   c. Commonly found in most classrooms
   d. Including three-dimensional objects as part of a long-term display
   e. Displaying heavy, three-dimensional objects suspended on a board.
   f. Various techniques can be employed: using templates, preplanning with guide marks or lines, and writing in color
   g. A flat surface on which information can be displayed and viewed by persons over a period of time

2. To which items would you give attention when designing and preparing any display media?
   a. Plan to display each piece of material when it will support the verbal presentation, then remove or cover the item.
   b. At one time, show all the details required for the total presentation so the audience will see everything in relationship.
   c. By employing such eye-catching techniques as bright colors, unusual shapes, and by manipulating objects or writing on the surface the use of display materials can be enhanced.
   d. Give attention to the size of objects and to printed words so all can be seen by the audience.

SOURCES FOR DISPLAY EQUIPMENT AND MATERIALS

The Advance Products Co., Inc. 1101 East Central, P.O. Box 2178, Wichita, KS 67201

Bulletin Boards and Directory Products, 724 Broadway, New York, NY 10003

Eberhard Faber Inc., Crestwood, Wilkes-Barre, PA 18773

Expo Communications, Inc., P.O. Box 1699, Wilmington, NC 28402

Instructo Corp., 1635 North 55th Street, Paoli, PA 19301

Charles Mayer Studios, 168 East Market Street, Akron, OH 44308

Oravisual Company, Inc., 321 15th Ave., South, St. Petersburg, FL 33701

Weber-Costello Co., 1900 North Narragansett Ave., Chicago, IL 60639

Weyel International, Crimson Technical Sales, Inc., 325 Vassar Street, Cambridge, MA 02139

Chapter 15

OVERHEAD TRANSPARENCIES

- Features of Overhead Projection
- Dimensions of the Working Area
- Necessary Skills
- Preparing Transparencies
- Making Transparencies Directly on Acetate
- Making Transparencies of Prepared or Printed Illustrations—with No Size Change
- Making Transparencies of Printed Illustrations—with Size Change
- Generating Transparencies with a Computer
- Completing and Filing Transparencies
- Preparing to Use Your Transparencies

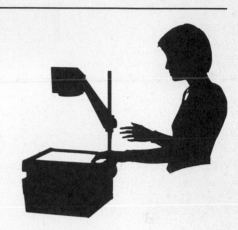

Transparencies are large slides for use with an overhead projector by a presenter positioned at the front of a lighted room. They project a large, brilliant picture.

Transparencies can visually present concepts, processes, facts, statistics, outlines, and summaries to small groups, to average-size classes, and to large groups.

A series of transparencies is like any other instructional media in that it requires systematic planning and preparation. Before you actually set about making your transparencies, therefore, always consider this planning checklist:

- What **objectives** will your transparencies serve (page 29)?
- What factors are important to consider about the **audience** that will see the transparencies (page 30)?
- Have you prepared an **outline of the content** to be included (page 32)?
- Are **transparencies an appropriate medium** to accomplish your purposes and to convey the content (page 42)? Might they be combined with other media for even greater effectiveness (page 174)?
- Have you **organized the content** and made **sketches** to show what is to be included in each transparency (page 48)?

FEATURES OF OVERHEAD PROJECTION

When showing visual materials with an overhead projector, you can make your presentation effective by using these techniques:

- You can show pictures and diagrams, using a pointer on

the transparency to direct attention to a detail. The silhouette of your pointer will show on the screen.

- You can use felt pen or a special pencil to add details or mark points on the transparency during projection.

● You can control the rate of presenting information by covering a transparency with paper or cardboard and then exposing the data when you are ready to discuss each point.

● You can superimpose additional transparent sheets as overlays on a base transparency so that you separate processes and complex ideas into elements and progressively present them.

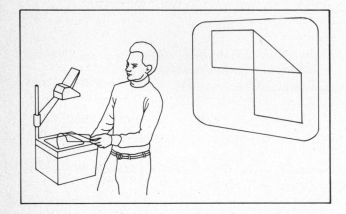

● You can move overlay sheets so as to rearrange elements of a diagram or a problem.

● You can duplicate inexpensively on paper the material to be presented as transparencies. Distributing copies to the class or audience will relieve them of the mechanics of copying complex diagrams and outlines.

● You can show three-dimensional objects from the stage of the projector—in silhouette if the object is opaque, or in color if an object is made of a transparent color plastic.

● You can simultaneously project other visual materials (slides or motion pictures) that illustrate or apply the generalizations shown on a transparency.

Principles from learning theory, as described in Chapter 2, and evidence from research studies reported in Chapter 3 indicate the importance of student participation during learning. Develop some transparencies that involve the learner by requiring the completion of diagrams, replies to questions, or solutions to problems. Or, provide students with paper copies of the content of transparencies and instructions for activities relating to your presentation.

DIMENSIONS OF THE WORKING AREA

The area of the stage (the horizontal glass surface) of most overhead projectors is 10×10 inches. The entire square can be used for the transparency, but it is better to avoid the extreme edges. Also, since a square is less attractive for most purposes than a rectangle, it is well to work within a rectangle having a height-to-width ratio of about 4 to 5. Thus a convenient transparency size, made on $8\frac{1}{2} \times 11$ inch film, has a $7\frac{1}{2} \times 9\frac{1}{2}$ inch opening (other formats are 8×10 inches and $7\frac{1}{2} \times 10$ inches). This is normally projected with the $9\frac{1}{2}$ inch dimension in horizontal position because the eyes have a greater field of vision horizontally.

It is difficult to view some parts of vertically oriented transparencies in rooms with low ceilings or with suspended lighting fixtures. Avoid mixing horizontal and vertical transparencies in a presentation because this can be annoying to an audience and bothersome to the presenter.

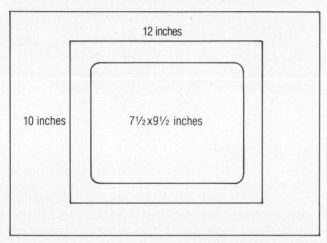

You can buy cardboard or plastic frames with the opening cut in them, or you can make your own from 6- to 10-ply cardboard (10×12 inches outside dimensions). An outline of the open area of a frame is printed inside the back cover of this book.

Whatever the size of the mask opening you plan to use, prepare all art work, pictures, and lettering to fit within this opening. An alternative is to use this proportion if size changes are to be made by photographic or other enlargement or reduction.

NECESSARY SKILLS

The production of transparencies requires knowledge and skills in many graphics areas described in Chapter 11. Also, abilities in photographic copy work are necessary.

Design and Art Work

Limit the content of a transparency to the presentation of a single concept or a limited topic. Do not try to cover too many points in a single transparency. A complex transparency may be confusing and unreadable to the viewer and thus ineffective. Design a series of transparencies rather than a crowded single transparency.

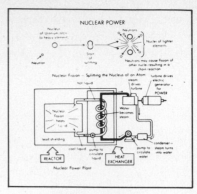

Original Subject

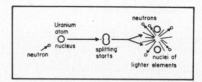

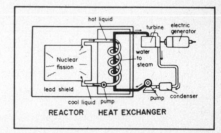

Subject Divided into Parts for Series of Transparencies

If you select diagrams and printed materials from books or magazines to convert to transparency form, be alert to certain limitations:

- The format may be vertical rather than horizontal as recommended above.
- The quantity of information included in printed materials may be more than can properly be presented in a single transparency.
- Materials printed in a book, to be read and studied closely at the reader's own pace, may be too dense and thus not suitable for projection which permits viewing for only a limited time.
- Finally, realize that copyright may impose limitations and it may be necessary to request permission for certain uses (page 57).

Therefore at first glance some printed materials may look suitable for use, but may in actuality be too small, too detailed, and even illegible as transparencies. It may be necessary to modify an illustration by:

- Eliminating unnecessary details like page numbers, printed text, and unwanted labels or subtitles.

- Modifying visual elements and replacing necessary lettering with larger size type.
- Simplifying a complex diagram by dividing it into segments for separate transparencies or by applying masking and overlay techniques to be described subsequently.

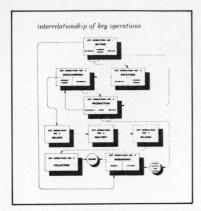

Original Subject

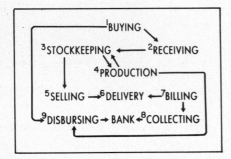

Redesigned for Transparency

As you plan diagrams and outlines or captions and labels for your materials, consider the applications of these graphic techniques:

- Planning the design and art work (page 106)
- Illustrating and coloring techniques (page 112)
- Legibility standards for lettering projected materials (page 121)
- Lettering materials and aids (page 124)

Adding Color

The addition of color to parts or areas of a black-line transparency can be used to separate details, to give emphasis to key elements, and to improve the appearance of the subject when projected.

Color must be transparent and can be applied in various ways:

- Use sharp-tipped felt pens directly on a transparency to write and draw lines (see page 178).
- Use broad-tipped felt pens to color areas on a transparency. Color may be applied as a solid area, although

overlapping strokes build up layers of color and these irregularities become visible. Felt pens can also be used to color by applying parallel lines, as in hatching, or by making dots, as in stippling.

- Use diazo or thermal films (page 181), which are available in a number of colors.
- Use transparent color adhesives and tapes, available in a wide range of colors and black-and-white shading patterns. Color adhesives can be applied to areas of any shape, and a number of colors can be used on a single transparency sheet.

When applying color adhesives, follow the instructions on page 119 and consider these additional suggestions:

- Adhere the color piece to the underside of the transparency so it cannot be damaged during use.
- Place the color piece in place over one line of the visual and ease it into place, smoothing by hand so no trapped air or wrinkles result.
- Use care so as not to cut through the acetate when trimming the color to match a line.
- Do not slip off the line being traced since the blade cut will be visible as a dark mark when projected.
- Apply two or more colors to an area for special effects.

Making Overlays

One of the most effective features of overhead projection is the **overlay technique.** As indicated on page 175, problems, processes, and other forms of information can be divided into logical elements, prepared separately as transparency sheets, and then shown progressively for effective communication.

In preparing a transparency in overlay form, first make a sketch of the total content. Decide which elements should be the **base** (*projected* first) and which elements, from the original sketch, will comprise each **overlay.** Make separate **masters** for the base and each overlay. Then prepare a transparency from each master, using one or more of the techniques described in the following pages.

When mounting the final transparent sheets, attach the base to the underside of the cardboard frame and the overlays to its face on the appropriate sides (see page 189 for details in mounting).

To insure proper alignment of all layers of the final transparency, carefully **register** the master drawings to the original sketch and also each piece of film to its master when printing. Do this by placing a guide mark (+) in each of two corners (outside the projected area) on the master exactly over the marks on the original sketch; or, preferably, use punched paper and film aligned on a register board (page 107).

See the facing page for an example of how an overhead transparency, consisting of a base and two overlays, is prepared. Notice the registration marks placed in corners of the original sketch and then on the master drawing, with them appearing on the transparent overlay sheets. Also see the arrangement for mounting the overlays on the cardboard frame.

176

Original Sketch

Master Drawings

Transparency Copies

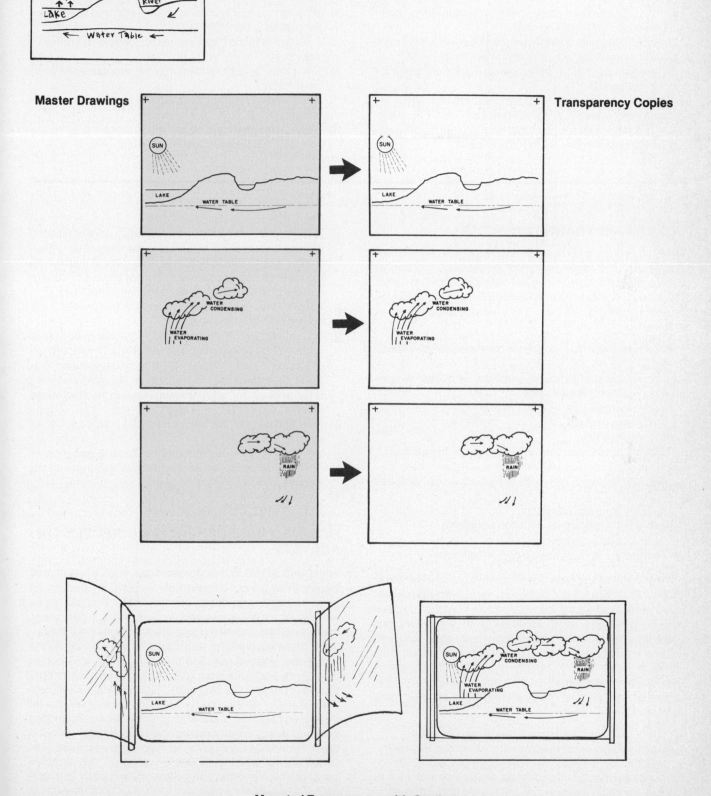

Mounted Transparency with Overlays

177

Review What You Have Learned About Planning Transparencies:

1. Of all the features of overhead projection, which ones seem of most value to you?
2. Should you plan transparencies to be viewed horizontally or vertically, or does it matter? Reasons?
3. What are the inside dimensions of a standard transparency mount?
4. How might you plan for student participation when developing a series of transparencies?
5. For what reasons might it *not* be advisable to use available printed pages as transparencies?

6. What things can you do to modify a printed illustration for use as a transparency?
7. For what reasons might you use coloring on a transparency?
8. What should be the minimum letter size used on transparencies?
9. What lettering aids might you use in preparing transparencies?
10. To insure proper alignment when preparing to make overlays, what procedure is followed?

PREPARING TRANSPARENCIES

Many processes have been developed for preparing transparencies. They range from very simple hand lettering or drawing to methods requiring special equipment and particular skills. Of these, the most practical and proven techniques are considered here. The methods are grouped as follows:

1. Making transparencies directly on clear acetate with felt pens
2. Making transparencies of prepared or printed illustrations—with no size change
   a. On thermal film
   b. On electrostatic film
   c. On diazo film
   d. As a picture transfer on pressure-sealing acetate
   e. As a picture transfer on heat-sealing acetate
3. Making transparencies of printed illustrations—with size change
   a. High-contrast subjects
   b. Halftone and continuous-tone subjects
   c. Full-color subjects
4. Generating transparencies with a computer

Which method or methods to use? First, consider those most appropriate to your purposes, the subject matter, and the planned use for the transparencies. Your final decision should be based upon accessibility of equipment and materials, on your skills and available time, costs, and, certainly not of least importance, on your standards for quality. Refer to Table 15-1, Summary of Methods for Preparing Transparencies, page 187.

Finally, ask yourself these questions as you start your preparations:

● Is the layout of the subject simple and clear?
● Have you checked the accuracy of content details?
● Will this be a single-sheet transparency or should you consider using overlays to separate elements of the subject—or might masking and uncovering of parts be effective?

● Will you plan to write some information on the finished transparency in addition to what is presented? (It is often more effective to add details to an outline or diagram as the transparency is used, or to involve the audience with activities relating to the transparency during use.)
● Is color important to the subject and treatment? If so, how should it be prepared?
● Is it satisfactory to do the lettering freehand, or should you consider using a lettering aid, like a boldface typewriter, dry-transfer letters, or a pressure-lettering machine (see page 124)? Remember to make lettering large enough for easy viewing—$\frac{1}{4}$ inch minimum size (Adams).
● Will duplicates of the transparency be needed for other users?
● Will paper copies of the content of the transparency, or related information, be needed for distribution to the audience?

MAKING TRANSPARENCIES DIRECTLY ON ACETATE

With these simple techniques transparencies are prepared quickly. They are not durable. For repeated use, neater and more permanent methods are advisable. But use these techniques for trying out your visuals; then, if necessary, make revisions before redoing them in permanent form.

On paper, outline the boundaries of the opening in the mount that you will use. Still on paper, make a sketch or position an illustration for tracing. Then, using the appropriate tools, put your drawing or tracing on acetate. Complete the transparency as directed later in this chapter. As indicated on page 118 under Adding Color, the fine-tipped felt pens are designed for drawing lines and writing, while the broad-tipped ones are useful for coloring areas. Select felt pens that are suitable for use on acetate. They produce transparent colors that can be either *permanent* or *washable.* The permanent-type inks require a plastic cleaner or other solvent (such as lighter fluid) for removal. Marks made

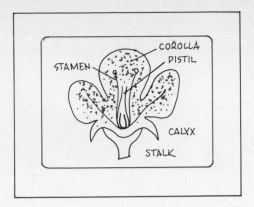

with water-base felt pens can be removed with a damp cloth.

To protect the surface of a transparency made with washable inks, cover it with a clear sheet of acetate.

Some inks tend to run slightly; therefore use pens lightly. Also, colors may not take to the acetate evenly; therefore consider using a stippling method (small dots of ink) to color large areas.

In addition to using felt pens on acetate, transparent tapes in various widths and colors can be used for lines and making graphs or diagrams. Dry-transfer letters (page 125) are available in transparent colors. They can be transferred to the acetate surface.

Review What You Have Learned About Transparencies Made Directly on Acetate:

1. Are all felt pens suitable for use on acetate?
2. How can you remove the markings of a permanent felt pen?

3. Do felt pens cover areas evenly?
4. How would you protect from smearing a transparency that has been made with water-base felt pens?

MAKING TRANSPARENCIES OF PREPARED OR PRINTED ILLUSTRATIONS—WITH NO SIZE CHANGE

To make transparencies as reproductions requires the preparation of one or more master drawings on appropriate paper and then the duplication of these drawings on transparent material. The methods of copying on thermal film and on electrostatic film, as described in this section also are suitable for making transparencies from printed illustrations (usually black-and-white high contrast). The final two methods, as picture transfers, are used with illustrations printed in color on clay-coated paper. Whichever method is being considered, replace small-size lettering to insure satisfactory legibility (page 121).

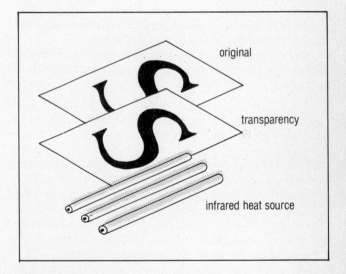

On Thermal Film

This is a rapid process. It is completely dry, and transparencies are ready for immediate use. Heat from an **infrared-light source** passes through the film to the original. The words and lines on the original must be prepared with "heat absorbing material," such as carbon-base ink or a soft lead pencil. These markings absorb heat, and the resulting increase in temperature affects the film, forming an image on it within a few seconds.

No special paper is required for the preparation of original printing or master diagrams. White bond paper is preferable and graph paper with light blue lines can be used. The lines do not reproduce.

The film to make thermal transparencies is of two types. (Check specific instructions with the product you use):

- Single sheet film
- Film with a disposable transfer sheet

Several brands of thermal copy machines are available. In each, the principle of thermal reproduction is the same. Most carry the original and film on a belt or between rollers around the infrared light source. The basic steps in the use of these machines are:

1. Set the control dial as directed (usually at the white indicator). Turn on the machine if necessary.
2. Place the thermal film (with the notch in the upper right corner when the film is held vertically) on the original material.
3. With the film on top, feed the two into the machine.

179

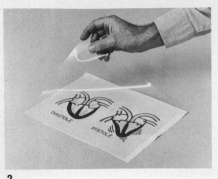

2

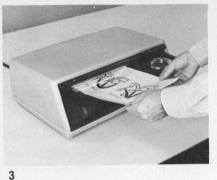

3

4

4. When the two emerge, separate the film from the original. Discard the transfer sheet if necessary.

The transparency image may be *too light,* or *too faint,* for satisfactory projection; if so, print it again with a fresh sheet of film, *increasing* the exposure time (that is, with the machine running at *slower* speed). A faint thermal transparency has been *underexposed.*

If, on the contrary, the transparency is *too dense,* or if there is *unwanted background tone,* print the transparency again with a fresh sheet of film, shortening the exposure time (that is, with the machine running at *faster* speed). A thermal transparency that is too dense has been *overexposed.*

Clean the belt or glass surface of the machine frequently to ensure transparencies without dirt spots.

The thermal process requires attention to a number of details and permits the use of a number of effective techniques:

- Original materials for thermal reproduction should be prepared with black India ink, soft lead pencil, or a typewriter having large type and a carbon ribbon. Electrostatic (Xerox) paper copies have carbon-based images and make excellent originals for thermal reproduction. Black printing inks are satisfactory, but inks of other colors, regular ball-point pens, and spirit duplicated copies (purple) are not suitable for thermal reproduction.
- If using India ink, make sure the ink is completely dry before the master is run in the copy machine.
- If you are uncertain about correct exposure or about the reproduction quality of a diagram, use test strips of the thermal film before exposing a whole sheet. Cut four to six vertical strips from one piece of film, being sure to clip

a corner of each for proper placement on the diagram.
- In addition to black-image film, various colored-image, tinted (color background for black image), and negative films (black background with clear or colored image) are available. While some presenters believe tinted-background film reduces eye fatigue, there is evidence that better visual acuity is achieved with clear-background transparencies (Snowberg).
- Add color to areas on a transparency with felt pens (page 178) or use color adhesive material (page 119).
- A thermal transparency, because of its opaque image, can serve as a master for the diazo process to be described later in this chapter.

On Electrostatic Film

With appropriate film (sources, page 190), most plain paper copiers can be used to prepare transparencies. See page 141 for a description of the electrostatic process. Be sure to match the film type to the machine. After fanning a few sheets, they are loaded onto the paper feed tray. The process proceeds as with paper duplication.

Black line electrostatic film is available with either a clear or colored background. In addition, color electrostatic copy machines can reproduce multicolor originals in overhead transparency format. Some machines can reduce and enlarge.

Here are some suggestions to consider when planning to make electrostatic transparencies:

- Any original, either drawn, typewritten, or printed, can be used to prepare a transparency.
- Commercial copy machines vary in their ability to duplicate original material into a quality transparency.
- A transparency of uniform density and maximum clarity

Overexposed

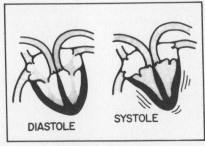

Correctly Exposed

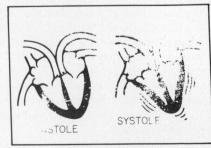

Underexposed

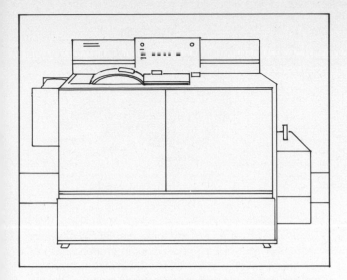

will be made from an original having a good black line image on white paper.

- By covering sections of an original, separate transparency masters can be made when overlays are to be prepared. Light lines that result at the edge of a mask can be erased.
- Colors and gray areas result in a transparency with varying degrees of density and clarity.
- Overall, this method does not result in as sharp and intense a black image transparency as do other methods.

On Diazo Film

Diazo films have been designed especially for the preparation of brilliantly colored, crisp-line transparencies. The term *diazo* refers to the organic chemicals, **diazo salts,** that, along with **color couplers,** are coated on acetate. If the coating on the film is exposed to **ultraviolet light,** it is chemically changed so that no image will appear. But if the coating is *not exposed* to ultraviolet light and is *developed* in an **alkaline medium,** like fumes of commercial aqueous ammonia (ammonium hydroxide), the diazo salts combine with the color coupler in the film to form a colored image. By using various color couplers during manufacturing, any one of about 10 colors can be coated on acetate.

To prepare a diazo transparency, put lettering or drawing on a translucent or transparent sheet and expose it in contact with the diazo film to ultraviolet light; then develop the film in ammonia vapor. The opaque marks on the master diagram prevent the ultraviolet light from affecting the film next to them, hence color appears in these areas when the film is developed in the ammonia.

1. From a sketch, prepare a master drawing on translucent paper using black inks or other materials that make opaque marks.
2. Cover the drawing with a sheet of diazo film. (When the film is held vertically, the cut corner should be in the upper right position, with the sensitive side of the film facing you. Place this side against the master.)
3. Place the master drawing and film in the ultraviolet-light

1

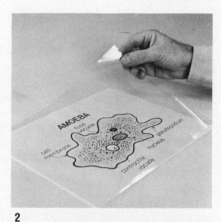

2

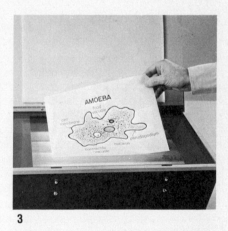

3

4

5

6

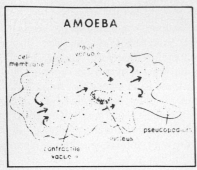

Overexposed

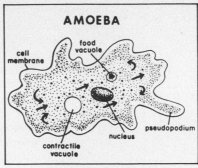

Correctly Exposed

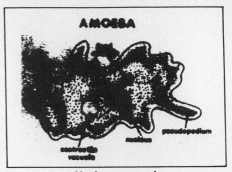

Underexposed

printer. (The master should be between the light and the diazo film.)

4. Set the timer or speed control for proper exposure.
5. After the exposure time has elapsed, transfer the exposed film to the container of ammonia vapor or insert the film in the diazo developing section of the machine.
6. In a short time the image will appear. Keep the film in the ammonia until the color appears fully. (Overdevelopment is not possible.)

The transparency image may be too *light,* or *faint,* for projection; if so, print it again with a new sheet of film, *reducing* the exposure time. A *faint* diazo transparency has been *overexposed.*

If, on the contrary, the transparency shows some *unwanted tones* in the background, print it again with a new sheet of film, *increasing* the exposure time. A *muddy* diazo transparency with unwanted background has been *underexposed.*

Of key importance to the success of the diazo process is the paper used in making the master. Ultraviolet light must pass through the paper in those areas that are *not* to result in color on the film, while opaque marks are *necessary* on the master to block the ultraviolet light from reaching the film in order for color to appear. Thus for the master a transparent film or translucent paper is essential. Tracing paper, commonly used for engineering and drafting work, is highly translucent and is recommended. Select a grade with a fine fiber texture (often called *vellum*) and use the same kind for *all* masters. If you change your master paper you may also change the correct exposure and thus waste film until the correct exposure is redetermined.

Another factor for success in the diazo process is the quality of the opaque marks made on the tracing paper. Black india ink makes good opaque lines. Regular pencils and typewriter ribbons are often not suitable. (Pencils can be used for shading effects.)

The diazo process requires attention to a number of details and permits the use of a number of effective techniques both in preparing the master and in modifying the transparency. Among these are:

● Exposure time in the printer is critical, but development time is not. Film must be developed long enough to obtain maximum color and can be removed from the ammonia any time thereafter.

● If areas of a transparency are to be in solid color, attach construction or other opaque paper in proper place on the translucent master. It will entirely block the ultraviolet light and result in a rich, even colored area.

● If more than a single color is needed on one sheet of diazo film, areas can be colored by applying one or more pieces of color adhesive material (see page 119).

● A thermal transparency can be used as the master for color diazo work.

● Diazo films do not have a long shelf life. Keep unopened packages under refrigeration until ready for use. Allow the film to reach room temperature before use.

● If an ink spot or other imperfection must be removed from the translucent master, the paper can be cut and the spot removed. The resulting hole in the paper has no effect on exposure or on the final transparency.

As a Picture Transfer on Pressure-Sealing Acetate

In this process the inks of printed pictures (color or black-and-white, on **clay-coated paper**) adhere to specially prepared acetate to make a transparency of the picture. The picture and the acetate are sealed together with heat and/or pressure, then submerged in water to dissolve the clay coating and soak the paper free from the inks. The inks remain on the acetate. A plastic coating applied after the acetate is dry transparentizes and protects the picture side.

A similar method that employs dry mounting equipment is explained after this. With either process, magazine pictures to be "lifted" must be printed on the aforementioned clay-coated paper. When selecting a picture always test it for clay as described below.

1. Test the selected picture for clay coating by rubbing a moistened finger over a white area of the page, outside the picture. A white deposit on the finger indicates the presence of clay.
2. Separate the adhesive-backed (Con-Tact shelf paper) acetate from its backing sheet.
3. Carefully adhere the acetate to the *face* of the picture.
4. Use a roller or other tool to apply pressure over the entire surface. Roll or rub in all directions.
5. Submerge the adhered picture and acetate in a pan of cool water.
6. After 2 to 3 minutes, separate the paper from the ace-

1

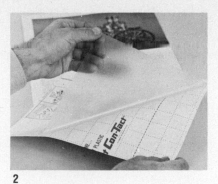
2

3

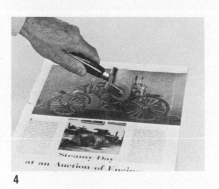
4

5

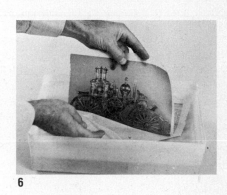
6

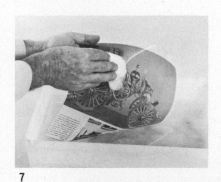
7

8

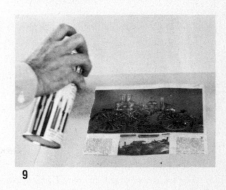
9

8. Blot the transparency between sheets of paper toweling or hang it to dry.
9. When the transparency is dry, coat the picture side with plastic spray to transparentize and protect the image by flooding the surface lightly and evenly.
10. When the spray coating has dried, mount the transparency for use.

Suggestions for avoiding problems when making a picture transfer include:

- Select bright pictures with sharp color contrasts for best-looking results.
- Keep the surface of the picture to be used as clean as possible because dirt, dust, and oil from fingers can prevent the ink from adhering to the film.
- Do not trim the picture before starting. Allow excess paper around the picture for film overlap.
- Cut the acetate so none will extend beyond the edge of

Picture Transfer Transparency

tate. The inks of the picture and some clay adhere to the acetate.
7. With wet cotton, slowly and carefully rub the image side (ink side) of the acetate to remove the clay. Be sure to loosen all the clay. Then rinse the transparency in clear water.

183

the paper. If the film laps beyond the page it will stick to the table.

- Applying pressure to the acetate and picture is important to insure a good seal. Use a roller on both sides and in all directions. If a roller is not available, any blunt tool, like the back of a comb wrapped in a handkerchief, can be used.
- If a transparency does not fit the standard cardboard frame, cut a piece of cardboard with the necessary opening.

As a Picture Transfer on Heat-Sealing Acetate

This method is, in principle, the same as the preceding pressure-sealing one. In this method, heat is used, and a dry-mount press is required to provide the heat and pressure as was used in **laminating** (page 133). In comparison with the pressure-sealing method, the results with this technique can be superior for two reasons. First, the adhesive material on the Seal-lamin film is evenly distributed. Second, the pressure exerted in the press is sufficient and even over the entire picture surface, while the rubbing used with the Con-Tact film can result in uneven pressure. For both reasons, with the method described below, inks should transfer fully from the page to the laminating film.

1. Set the dry-mount press temperature recommended for the heat-sealing acetate (275°F for Seal-lamin).
2. Test the picture for clay (as in the preceding process).
3. Refer to page 133. Follow steps 2 to 7 for the laminating process.
4. Follow steps 5 to 10 of the previous picture transfer process on page 183.

Note the suggestions on page 183 for avoiding problems when making a picture transfer. In addition to those listed, for this process allow the picture and film to curl naturally when removed from the dry-mount press. They will straighten out when cooled or in water.

The heat lamination unit described on page 134 can also be used to make high-quality picture transfer transparencies. With this machine, if you have a page with useful pictures on both sides, apply the laminating film to both sides. Then carefully split the paper so that each picture, with its adhering film, is separated. Follow the regular procedure described above to make a transparency from each picture.

These picture-transfer methods can also be used to prepare large-size, full-color materials for display in an exhibit. By setting a transparency in front of a light source, the picture will be illuminated.

Review What You Have Learned About Transparencies of Diagrams and Printed Illustrations:

1. What method can be used to make a transparency consisting of any writing material on any kind of paper?
2. What is the main thing to check when preparing or selecting a diagram to be made into a thermal transparency?
3. Is the kind of paper for the master important in the thermal process?
4. If lines on a thermal transparency are very weak and thin, is the film *over-* or *underexposed*? The next time, must you *slow down* or *speed up* the machine?
5. How might small areas be colored on a thermal transparency?
6. Explain in your own words the principle of the diazo process.
7. What material is used as a master in the diazo process?
8. How would you prepare a diazo master requiring a solid color to cover a large area?
9. What is the order in which materials are placed in the ultraviolet-light printer?
10. Which is critical for time—ultraviolet exposure or ammonia development?
11. If the lines on a diazo transparency are very weak, does this condition indicate *over-* or *under*exposure? Would you expose for a *shorter* or *longer* time?
12. Which process might be used to reproduce a printed diagram consisting of three colors?
13. What is the first step in preparing to make a picture transfer from a picture?
14. How is pressure applied in both picture-transfer methods described?
15. Why must all the clay be removed from the transferred picture?
16. To what other process is making a picture transfer with a dry-mount press related?

MAKING TRANSPARENCIES OF PRINTED ILLUSTRATIONS—WITH SIZE CHANGE

Three techniques may be used. Two are black-and-white photomechanical processes that are employed in graphic arts when copy is prepared for offset platemaking and printing. The third technique produces full-color transparencies.

Before you proceed, review the following:

- Legibility standards for projected materials when printed items are considered for reproduction (page 121)
- Permission to reproduce copyrighted materials (page 57)
- Use of your camera, especially with reference to:

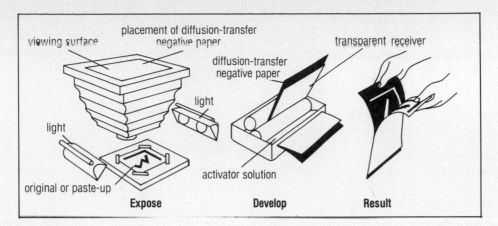

viewing surface — placement of diffusion-transfer negative paper — transparent receiver

diffusion-transfer negative paper

light

light

original or paste-up

activator solution

Expose   **Develop**   **Result**

- Sheet-film cameras, if available (page 74)
- Camera settings (page 77)
- Correct exposure (page 84)
- Close-up and copy work (page 96)
- Processing film (page 99)
- Making prints (page 101)

High-Contrast Subjects

One widely applied method is **diffusion transfer** (PMT) which requires a large process camera (page 138). A transparency is made directly when, after exposure in the camera, an 8×10 inch or 8½×11 inch paper negative is processed and the image is transferred to a film positive. (See the illustration above and the Eastman Kodak and Le Tissier references for further details of this process as it applies to preparing transparencies.)

The other technique is to use a **high-contrast film** like Kodalith Ortho Type 3. This is preferable to using regular black-and-white sheet film for transparencies as it is easier to work with, has a clear base for projection, can produce more saturated black lines, and dries quicker by reason of being thinner.

The high-contrast process for preparing paper prints from line copy is described on page 138. A similar procedure of film exposure (on 4×5 inch film is preferable) and negative processing is applied for preparing high-contrast film transparencies.

Once the negative is prepared, it is printed in an enlarger onto an 8×10 inch or larger sheet of the *same* high-contrast (Kodalith) film. Place a mounting frame for an over-head transparency on the easel under the enlarger and fill it with the image from the negative. Expose the film. Process the film in the same developer and other chemicals as were used to develop the negative. (See illustration of the process below.) Dry and mount the transparency for use.

Here are other suggestions for preparing transparencies, when working with high-contrast materials:

- Prepare a negative rather than a positive transparency by: (1) using a reversal negative in the diffusion-transfer process, or (2) making a contact print of the Kodalith high-contrast-camera negative onto a piece of high contrast film and then using this positive in the enlarger to make a negative transparency.
- Separate a subject into its components for preparing overlays by: (1) blocking out areas on the original art when filming, or (2) making a number of negatives of the subject equal to the number of overlays needed and then mask out all but necessary areas on each negative by opaquing. Print each negative, in the enlarger, onto a separate sheet of high-contrast film. Be sure to maintain registration during this process.
- Add color to areas on a transparency with colored adhesive. Apply it to the base (shiny) side of the transparency. (See page 119.)
- Because high-contrast transparencies made by either of the above processes will scratch easily, use the transparency as a master for reproduction by the diazo process to make additional copies, often in color. (See page 181.)

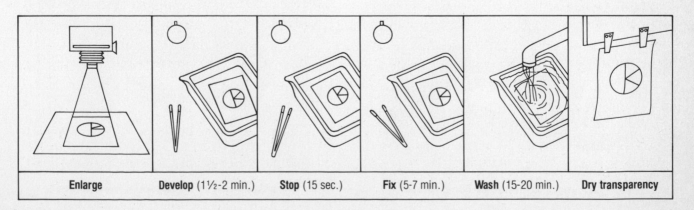

Enlarge   **Develop** (1½-2 min.)   **Stop** (15 sec.)   **Fix** (5-7 min.)   **Wash** (15-20 min.)   **Dry transparency**

185

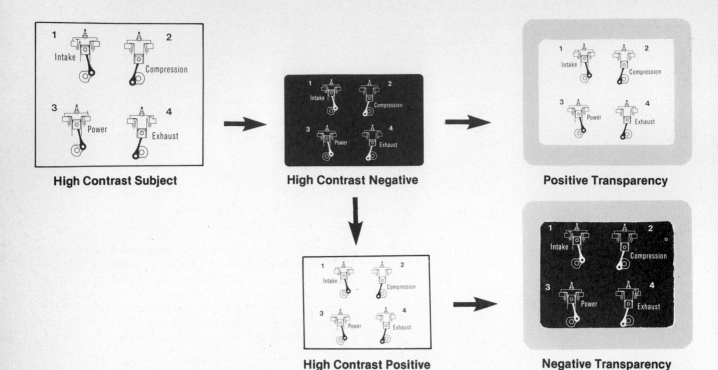

High Contrast Subject

High Contrast Negative

Positive Transparency

High Contrast Positive

Negative Transparency

Halftone and Continuous-Tone Subjects

In this category of subjects are halftone illustrations printed in books and magazines, photographs, and original works of art which contain shades of gray varying from white to black. Be sure to recognize the difference between such subjects and high-contrast subjects, which consist only of black marks on white paper.

Although the high-contrast process is used primarily for reproducing line subjects, it can be adapted to prepare continuous-tone transparencies from negatives of halftone and continuous-tone subjects. With the diffusion-transfer process, halftone negative paper is used, or exposure is made on regular negative paper through a halftone screen. With the Kodalith method a regular black-and-white, continuous-tone negative is enlarged onto a sheet of high contrast film. Then the film is processed in *paper* developer (diluted 1:12 with water). The use of stop bath, fixer, and wash are the same as in the regular high-contrast procedure.

Most of the transparency making techniques described in this chapter, except for the picture-transfer methods, start with masters that result in high-contrast transparencies (black or colored lines on a clear background or the reverse as black or colored backgrounds with clear or colored lettering and lines). A continuous-tone transparency made by these methods is of fair to poor quality, but the results of the diffusion-transfer and high-contrast methods, just described, are highly acceptable.

Full-Color Subjects

Full-color transparencies can be prepared from 35mm

A Halftone (or Continuous-Tone) Subject

Transparency of Continuous-Tone Subject

slides, color negatives, color photographs, and flat art. There are various commercial products and commercial services available:

- Eastman Kodak Ektagraphic Overhead Transparency material which requires an Ektaflex Printmaker and Ektaflex PCT color printmaking products. With this equipment and materials, standard photographic techniques can be employed to prepare full-color transparencies.
- The 3M/840 Full Color Transparency maker reproduces slides and all forms of paper originals into full-color transparencies. This service is available on a per unit charge through 3M offices.
- The Polaroid Colorgraph 8×10 inch overhead transparency material, requires a Polaprinter for preparing full-color transparencies from 35mm slides, opaque originals and transparency art.

GENERATING TRANSPARENCIES WITH A COMPUTER

Transparencies, in full color, can be created by applying graphics methods with a computer. Starting on page 115, advantages for using computer graphics systems and the required equipment are described, followed by explanations of the procedures for preparing graphic materials.

With a microcomputer and a suitable graphics program, many kinds of visual information can be designed to appear on the display terminal. The outlines, diagrams, charts, and graphs can then be transferred to transparency form on

acetate or film by using a plotter or film recorder. This can be done in black-and-white or in multiple colors.

When designing transparencies with a computer, give attention to legibility matters presented on page 121. Be especially alert to the following:

- Do not crowd too much information onto a single transparency.
- Make characters (words and numbers) of a suitable size (minimum $\frac{1}{4}$ inch in height).
- Provide satisfactory space between lines of text.
- Make lines thick enough and of suitable resolution for satisfactory reading.
- For both characters and areas, select colors and their intensity so that strong images will be projected.

Table 15-1 **Summary of Methods for Preparing Transparencies**

	METHOD	EQUIPMENT	COST OF MATERIALS (APPROXIMATE)	TIME FOR PREPARATION	EVALUATION
DIRECTLY ON ACETATE	With felt pens (page 178)	None	$0.20–$0.30	Moderate	Suitable for quick preparation and temporary use; lacks professional appearance
OF ILLUSTRATIONS —WITH NO SIZE CHANGE	1. On thermal film (page 179)	Thermal copier	$0.40–$0.75	Very brief	Good method for rapid preparation of one-color transparencies from single sheets
	2. On electrostatic film (page 180)	electrostatic copy machine	$0.50–$1.00	Brief	Requires an expensive machine, but readily available, and a special film; quality not as good as other methods
	3. On diazo film (page 181)	Diazo printer and developer	$0.50	Moderate	Excellent method for preparing color transparencies; requires translucent originals; a variety of applications of the process are possible

187

Table 15-1 (**continued**)

	METHOD	EQUIPMENT	COST OF MATERIALS (APPROXIMATE)	TIME FOR PREPARATION	EVALUATION
	4. As picture transfer on pressure-sealing acetate (page 182)	None or cold roller laminator	$0.40	Short	Converts any magazine picture printed on clay-coated paper to a transparency; a simple process to apply; results are very effective if suitable original pictures are available
	5. As picture transfer on heat-sealing acetate (page 184)	Dry-mount press or heated roller laminator	$0.60	Moderate	Similar to above process, but takes a few minutes longer; equipment often available in graphic production center
OF ILLUSTRATIONS —WITH SIZE CHANGE	1. On PMT film positive (high contrast or continuous-tone copy) (page 185)	Process camera and developing unit	$1.00	Short	Requires large graphic arts camera and processing unit; simple, quick method for making quality transparencies with size change
	2. On high-contrast film (high contrast or continuous-tone copy) (page 185)	Camera and darkroom	$1.50	Long	Most complex process in terms of skills, equipment, facilities, and time; but useful when original materials must be changed in size; results in high-quality transparencies
	3. On full-color film (page 186)	Camera and processor or commercial service	$ 2.50–$10.00	Short	Requires special equipment or commercial service; result is high quality although expensive
GENERATING TRANSPARENCIES WITH A COMPUTER	On acetate or full color film (page 187)	Computer, program and plotter or film recorder	& 2.00–$20.00	Moderate	Requires training of operator; design ideas can be tried out and best selected; complex graphics easily prepared in full color with suitable equipment

Review What You Have Learned About Photographic and Computer Methods to Prepare Transparencies:

1. What one piece of equipment is essential for converting an illustration onto a transparency when a size change is required?
2. What two methods can be used to prepare black-and-white transparencies from small illustrations?
3. What type of film is used to prepare negatives of line copy illustrations?
4. How do you differentiate between a high-contrast and a halftone or continuous-tone subject for a transparency?
5. In what ways does the enlarging process differ from that of making the negative?
6. In what ways does the preparation of a continuous-tone transparency differ from the preparation of a high-contrast one?
7. How can you use the photographic process to make a transparency involving overlays?
8. Which of the following can be used as an original to make a full-color transparency?
   a. color photograph
   b. 35mm color slide
   c. color negative
   d. art in color on cardboard
9. How can a graphics image on a video screen, generated by a computer, be converted into a full-color transparency?

COMPLETING AND FILING TRANSPARENCIES

Mounting

Mounting adds durability and ease to the handling of a transparency, but it is not always necessary to mount transparencies in cardboard frames. Those prepared on polyester film will lie flat. Transparencies to be used only once can be left unmounted since the cost of frames and the time to attach them may not be justified. Some people prefer to keep transparencies unmounted for ease of filing in notebooks, but the standard-size cardboard or plastic frame easily fits into letter-size filing cabinets. If overlays are to be used, mounting is essential.

Tape a single-sheet transparency to the *underside* of the frame. Use masking or plastic tape rather than cellophane tape for binding.

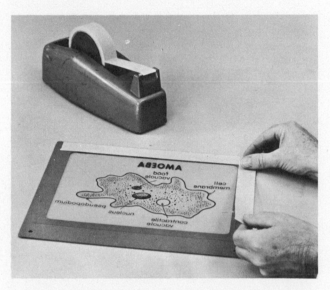

If the transparency consists of a base and overlays, tape the base to the underside of the mount as usual, and the overlays to the face. Be sure the overlays register with the base and with each other (page 176). Then fasten each overlay with a tape or plastic hinge along one edge of the cardboard frame.

Overlays for successive or cumulative use can be mounted on the left or right sides of the cardboard frame, also if necessary on the bottom and top (the top edge should be the last one used). Trim any excess acetate from the edges of overlays so opposite or adjacent ones fit easily into place.

After mounting overlays, fold and attach small tabs of masking tape or adhesive-back labels on the loose upper corner of each overlay. Number them to indicate the order of use. These tabs are easy to grasp when overlays are to be set in place over the base transparency.

Masking

To control a presentation and focus attention on specific elements of a transparency, use a paper or cardboard mask, as was mentioned on page 174. The mask may be

a separate unit, mounted to move vertically or horizontally, or a number of hinged opaque overlays.

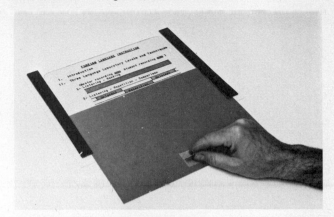

Adding Notes

Write brief notes along the margin of the cardboard mount for reference during projection.

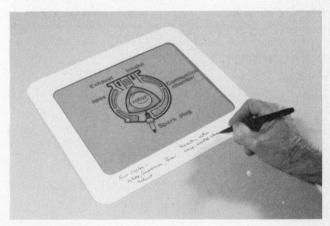

Filing

If your transparencies are in mounts 10×12 inches or smaller, they will fit in the drawer of a standard filing cabinet. File them under appropriate subject, unit, or topic headings.

PREPARING TO USE YOUR TRANSPARENCIES

Remember that the success of your transparencies will depend not only on their content and quality, but also on the manner in which you use them before an audience. Look over the transparencies and arrange them in order for use during your presentation or instruction. It may be advisable to number them in sequence. Write brief notes along the margin of the cardboard mount for reference during projection. Then follow the suggestions in Chapter 8 as you make the room arrangements for using the transparencies.

Review What You Have Learned About Completing and Filing Transparencies:

1. Do you correctly mount a single-sheet transparency on the underside or the face of the cardboard frame?
2. How is a transparency with overlays mounted?

3. What are two ways of masking transparencies?
4. What filing system for transparencies might you use?

REFERENCES

Adams, Sarah, et al. "Readable Letter Size and Visibility for Overhead Projection Transparencies." *AV Communication Review* 13 (Winter 1965):412–417.

Clark, Bill, and Creager, Agatha. "Communicating with Computer-Generated Transparencies." *Audio Visual Directions* 3 (October/November 1981):69-73

Green, Lee, and Dengerink, Don. *501 Ways to Use the Overhead Projector.* Littleton, CO: Libraries Unlimited, 1982.

LeTissier, David. *Instant Graphic Techniques with Instant Art and Agfa-Gavaert.* Maidstone, Kent, England: The Graphics Communications Centre, Ltd., 1981.

Magee, Rima. "Overhead Projection Transparencies in a Hurry.

Using Your Computer." *Photomethods* 26 (August 1983):39–41.

Making Black-and-White or Colored Transparencies for Overhead Projection. Publication S-7. Rochester, NY: Eastman Kodak Co., 1980.

Smith, Richard E., and Pearson, Jerry D. *The Overhead System: Production, Implementation, and Utilization.* 2nd ed. Austin, TX: University of Texas Film Library, General Libraries, Box W, 1979.

Snowberg, R. L. "Bases for the Selection of Background Colors for Transparencies." *AV Communication Review* 21 (Summer 1973):191–207.

SOURCES FOR EQUIPMENT AND MATERIALS

Arkwright, Main Street, Fiskeville, RI 02823

AVCOM Systems, Inc., Box 977, Cutchogue, NY 11935

Eastman Kodak Company, Motion Picture and Audiovisual Markets Division, 343 State Street, Rochester, NY 14650

Folex, Inc., 6 Daniel Road East, Fairfield, NJ 07006

G.A.F. Corp., P.O. Box 14954, St. Louis, MO 73178

Labelon Corporation, 10 Chapin Street, Canandaigua, NY 14424

Scott Graphics, Holyoke, MA 01040

J. S. Staedtler, Inc., P.O. Box 787, Chatsworth, CA 91311

Tersch Products, Inc., Industrial Boulevard, Rogers, MN 55374

3M Audiovisual Division, 3M Center, St. Paul, MN 35144

Visual Products Supply, 3044 Payne Ave., Cleveland, OH 44114

Chapter 16

AUDIOTAPE RECORDINGS

- Recordings to Motivate or Inform
- Recordings to Instruct

Besides being used to play music, audiotape recordings can provide motivation, convey information, provide drill and practice, or teach a skill. With worksheets and other materials, tape recordings can serve as the study guide and information source for various instruction and training needs. Recordings also can be used in conjunction with projected visual materials. The latter use is described with specific instructional media in other chapters of this book. This chapter considers tape recordings, by themselves or as the study guide along with other materials, for instructional purposes.

RECORDINGS TO MOTIVATE OR INFORM

If the objective of your recording is primarily to provide an introduction, an orientation or general information about a topic, then the preparation of the recording may require any of a number of techniques, from simply documenting a speech on tape to a sophisticated dramatic presentation or other multiple voice and sound combination.

Planning

Give attention to the following:

- Consider the **audience** that will use the recording so that you will be assured of meeting their interests and needs (page 30).
- Identify the broad **purpose** or specific **objectives** that will be served by the recording (page 29).
- Prepare an **outline of the content** to be treated, if appropriate (page 32).
- Develop a **script** in sequential, narrative, or dramatic form (page 50). This will indicate the role for each voice and the use of music or other effects.

- Detailed **notes** of the content may be all that is required for an informational narration.

Techniques

Because an audiotape recording is designed to appeal to only the sense of hearing, it is necessary to make an extra effort to hold a listener's attention. To do this, pace the message briskly and deliver it in a succinct, clear, personal, and informal manner. Then through careful editing of the recording, a product, superior to the initial recording, can be created.

As the listener interprets the verbal message, mental impressions are formed. To make these impressions as effective and memorable as possible, various techniques may be employed:

- Keep the recording short. While the complete recording may be 30 minutes in length, subtopics should be limited to only 5 to 8 minutes.
- Voice tone should be conversational, friendly, and direct.
- Condense speeches and other unscripted verbal presentations by eliminating extraneous material. Avoid long, complex sentences. Use a narrator's voice to make logical bridges from one section to the next, and to reinforce key points of content.
- Include music. It can provide a break between sections and continuity throughout the presentation.
- A recording that is dynamic and helps to maintain a high interest level may include a variety of features—two or more voices, music, sound effects, indications of being at different locations, and a summary of what has been presented.
- Allow for rapid review of material by using the compressed speech technique (page 153).

EXAMPLE SCRIPT OF TAPE
RECORDING TO INFORM

The purpose of this recording is to explore some of the more promising alternative energy sources in view of the critical energy crisis period we are now entering. The causes for this crisis are threefold:

First, costs for fossil fuels are increasing. Second, the reserve of fossil fuels is limited and rapidly diminishing. And third, pollution of the atmosphere is an unfortunate result of both coal and oil uses.

What options are available to us for moving away from these fossil fuels and providing suitable amounts of clean, dependable energy at suitable prices within our technological know-how for our *own* personal or family uses and on a national level?

Most exploratory work and actual development of alternative energy production are related to nuclear, solar, geothermal, and wind sources. These represent the major hopes for both the short and the long term when supplies of oil and natural gas become too expensive to use and reach depletion. Let's examine each one in detail. [and so on] . . .

The most common use of a tape recorder is to document a speech or other verbal event. In addition, for motivation and information purposes, a recording might be prepared to do the following:

- Report on highlights of a meeting.
- Serve as a magazine or journal containing a report on new developments, programs, or personalities in a subject area or an organization.
- Interview notable persons.
- Provide for a group discussion or a question-and-answer presentation of a topic.
- Introduce new personnel to a program, to a facility, to a procedure, or to the requirements for a task.
- Document role-playing situations, as performed by members of a group, for followup analysis.
- Present or explain problem situations for groups or individuals to solve.
- Serve as a major source of information and entertainment for blind persons.

Production

Preparing the tape will require attention to the various matters considered in Chapter 12. Follow those steps that are important in your situation in order to ensure a quality recording. If you plan to use more than a single voice, such as a speaker as well as someone to introduce, interpret, or summarize parts of the presentation, then you may be required to mix voices and edit the tape.

Packaging for Use

After the master recording has been completed and duplicate cassettes have been prepared, package the program for use. This should include attention to:

- Using an "indexing" method for locating sections of the recording, as described later in this chapter.

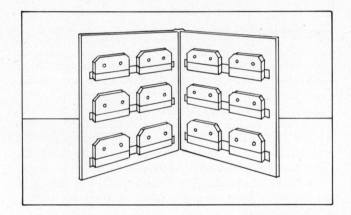

- Considering the advisability of copyrighting the recording (page 67).
- Making and applying labels to identify the program.
- Preparing information that outlines the content and suggests uses for the tape as related to listener job responsibilities in an organization. This material may be placed on the inside of the container cover or on separate paper.
- Obtaining containers for individual cassettes or albums for sets of cassettes.
- Providing for availability of cassette recorders to be used with cassettes.
- Planning distribution through a media service, an organization's office, or by mail.

RECORDINGS TO INSTRUCT

Although instructional recordings might be used with a group as all members listen and participate in the learning activities, it is more common to provide them for individual self-paced learning. Differences in learning rates, the desire to replay a section of a recording, and the need for time to think about something just presented all indicate that for effective learning to take place each person should be allowed to use a recording at his or her own pace.

A key feature of successful instruction with tape recordings is the opportunity for an individual to interact with the material being presented. This requires answering questions, solving problems, using the information, or applying the concepts and principles presented. The individual then receives feedback on answers, either on tape or on accompanying paper.

Tape recordings have essential roles in two instructional formats. One is the **audio-notebook** and the other is the **audio-tutorial** system (AT).

An audio-notebook consists of a study guide and workbook combination or separate worksheets that accompany the tape recording (Langdon). The tape may introduce the topic, explain or describe content, and periodically direct the user to activities in the workbook. The workbook may

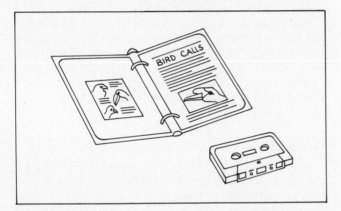

include questions, exercises, and problems that are to be completed. Answers and followup discussions are provided on the tape or provided on other pages. The study guide may contain supplementary reading materials, may direct the student to other references, and may provide self-check tests for the student.

In the audio-tutorial system a tape recording serves as the central part of a study unit (Postlethwait, Russell). The tape may provide information and also direct the student to various learning activities—reading pages in a textbook,

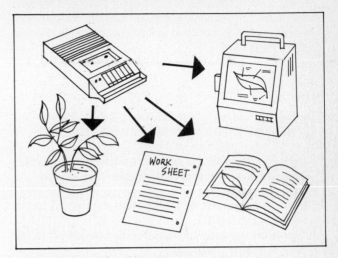

examining materials, making observations, performing experiments, completing worksheets, and so on. Feedback is provided on the tape as answers or for discussion. A major difference between the audio-notebook and the AT system is that with the latter, the resources and activities correlated

EXAMPLE SCRIPT OF TAPE
RECORDING TO INSTRUCT

The subject of this program is an examination of alternative energy sources. Please look over the objectives you are to satisfy on page 1 of the study guide. Then read the brief article by C. P. Nelson starting on page 2. It establishes a good base for the energy crisis that we now face. Identify the three main reasons for the crisis. Turn off the tape. Start again with me when you are ready. (Pause)

By reading the Nelson article, you should have identified these reasons for the critical energy crisis: (1) costs of fossil fuels are increasing; (2) the reserve of fossil fuels is limited and rapidly diminishing; (3) pollution of the atmosphere is a result of both coal and oil uses.

Now let's turn to the options. Most exploratory work and actual development of alternative energy production are related to nuclear, solar, geothermal, and wind sources. These represent the major hopes for both the short and long term when supplies of oil and natural gas become too expensive to use or become depleted.

We'll examine each one—first, nuclear power. The potential energy locked in the nucleus of an atom can be released by two processes *fission* and *fusion*. Look at the diagrams starting on page 5 of your guide as I discuss these two forms of nuclear energy. You should be able to differentiate between the two forms.

[and so on] . . .

with the audiotape generally go beyond what is immediately on paper for the student to use.

Another application of the audio recording may be in computer-use training. An audiotape can be used in conjunction with a new piece of equipment or software program. As the user listens to a recording, he or she is introduced to the equipment and to procedure details ("Raise the door on the right side . . . insert the disc . . ."). Images on the video screen can be pointed out and explained. Reference can be made to other essential information that will make understanding the computer clear and learning more rapid. Furthermore, computer memory can be saved when the audiotape is used instead of displaying information on the screen.

Planning

With either the audio-notebook or audio-tutorial method the following planning steps should receive attention:

- List **objectives** for the topic to be treated (page 29).
- Consider the **group** or **individuals** who will use the materials (page 30).
- Develop an **outline of the content** that relates to the objectives (page 32).
- Decide on the **activities** and **resources** relating to the content that will be used.
- Prepare the **script** or detailed notes for the recording (page 50).
- Write accompanying **worksheets, guides, activity sheets,** and so forth (page 166).

The tape portions of the lesson should be structured to contain the following:

- A motivational portion that introduces the lesson, lists the objectives, and indicates any special preparation required (unless these are provided in the worksheet)
- Explicit instructions for participation work
- Indications of correct responses for immediate feedback
- Summary and/or instructions leading to other materials or activities
- Pauses of sufficient length (or instructions to turn off the recorder or 4 to 5 seconds of music) to allow for completion of responses or performance of activities

Instead of initially preparing the script on paper, consider adlibbing the necessary information as a recording on tape, then type out the material, edit the script, and rerecord in final form.

Producing the Tape

A special feature of this type of instructional media is that the recording should convey a warm, personal feeling to the student using it. This means that your voice should be conversational and informal as if you were talking directly to a student. If you speak too slowly, the pace will drag and students will become disinterested. Consider using the variable speech control technique described on page 153 to

pep up a slow voice. Also, many of the suggestions offered under the previous Techniques section on page 191, should be applied here.

When students are first introduced to this method, they should be encouraged to stop the tape at any point and relisten to any portion. Remember that you are providing timely information, definitions, parenthetical explanations, and elaborations on other materials with which the student is directed to study or work. The tone of your voice places emphasis on important points and expresses authority not served through the written word.

The recording for instruction should not be a straight-through lecture. Listening to only a voice for 5 to 6 minutes may be acceptable, but when you approach 10 minutes of continual listening, student interest drops. A procedure such as this may be used: (1) information briefly presented on tape; (2) student participation through an activity or through replying to questions about the information; (3) a review of the results of the activity or the correct answers; (4) on to further explanation or new information.

Plan for variety in pacing, include various kinds of activities requiring differing lengths of time. All these techniques can contribute to maintaining student interest and alertness.

When you are ready to make the recording, refer to the preparation and recording procedures in Chapter 13.

Indexing the Tape

Instead of requiring that a person listen to the complete recording from start to finish, you may wish to provide flexibility for the student to select parts of a recording for listening and omit other sections. This may relate to material for certain objectives that a student needs, or chooses, to work with while electing to bypass other material. As a result of an answer to a question or problem, you may wish to "branch" the student to review or advanced material. Or the student may wish to go over again only certain sections of a recording.

Normally, in order to satisfy each of these needs, the student haphazardly moves ahead in the tape by using the fast forward control on the recorder, hoping to stop about where the listening should start. With the **Zimdex** method, index numbers are placed on the *second* track of a monophonic cassette tape that the student will use in order to locate accurately any section of the recording (Rahmlow et al).

The method works in the following manner:

1. On side 1 record the scripted information.
2. On side 2, starting at the beginning, record a series of numbers—1, 2, 3, 4, 5 . . .—for the full length of the tape. Leave a five-second pause after each number.
3. Return to side 1 and listen to the material recorded. Prepare a list of the sections of content that you wish to index for location.
4. Play side 1 again. Before the beginning of each section turn the tape over and listen for the number that corresponds to the starting point for that section. Note the index number. Your index may be:

SUBJECT	INDEX NUMBER
Advantages of solar energy	385
Collection process	354
Solar collectors	324
Scientific principles used	281
Storing energy for space heating	259
Heating water	231

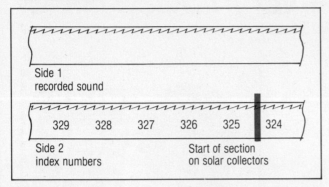

Side 1
recorded sound

| 329 | 328 | 327 | 326 | 325 | 324 |

Side 2
index numbers

Start of section
on solar collectors

A person wishing to locate a section of the recording starts on side 2 of the tape, moving the tape fast forward. Periodically the tape is stopped and played as the user listens for an index number. Right after the appropriate number is heard, the cassette should be turned to side 1. The beginning of the desired material is ready to be heard. If you have an extensive music collection on audiocassettes, the index numbers can help you to locate selections.

Preparing the Recording and Accompanying Materials for Use

When the master recording is completed, it will be necessary to:

- Make duplicate cassette copies of the tape (page 154).
- Prepare accompanying printed materials in final form and duplicate (page 139). It is often desirable to provide students with a printed copy of the narration. Some students prefer to read the words as they are spoken or use the script for review.
- Collect other materials that may be required as part of the unit.
- Consider the advisability of copyrighting the recording and/or complete package (page 67).
- Label individual items so that each is properly marked for easy identification.
- Package each unit or set of materials.
- Plan for distribution and use.

Review What You Have Learned About Making Audiotape Recordings:

1. How does a recording that is designed to inform differ from one designed to instruct?
2. What are *five* uses for which an informational audiotape might be prepared?
3. What are some techniques that may be used when preparing a motivational or informational recording?
4. What two formats might instructional tape recordings take? What is the difference?
5. What planning steps should be followed in preparing an instructional tape?
6. What are examples of the activities that can be correlated with the use of recordings for instruction?
7. What type of mood should the tape for an AT unit have?
8. What might be the limit of time for required listening to a section of an instructional tape without other activity?
9. Of the procedures described in Chapter 12, which ones will you refer to for help in making your recording?
10. Describe how you might prepare an index for locating sections on a recording?

REFERENCES

Langdon, Danny G. *The Audio-Workbook.* Vol 3 in *The Instructional Design Library.* Englewood Cliffs, NJ: Educational Technology Publications, 1978.

Postlethwait, S. N., Novak, J., and Murray, H. *The Audio-Tutorial Approach to Learning.* Minneapolis: Burgess, 1972.

Rahmlow, Harold F., Langdon, Danny G., and Lewis, William C. "Audio Indexing for Individualization." *Audiovisual Instruction* 18 (April 1973):14–15.

Russell, James D. *The Audio-Tutorial System.* Vol. 3 in *The Instructional Design Library.* Englewood Cliffs, NJ: Educational Technology Press, 1978.

Chapter 17

SLIDE SERIES AND FILMSTRIPS

- Preparing and Taking Pictures
- Processing Film
- Mounting Slides
- Editing Slides
- Duplicating Slides
- Recording and Synchronizing Narration and Other Sounds
- Filing Slides
- Preparing to Use Your Slide Series
- Filmstrip Format
- Making a Filmstrip from Slides
- Duplicating a Filmstrip
- Correlating a Filmstrip with a Soundtrack
- Preparing to Use Your Filmstrip

While a slide series and a filmstrip differ in physical appearance, they are planned in the same way and may be used for similar purposes. Both can convey information, teach a skill or affect an attitude through individual study or group viewing. See page 38 for a consideration of the similarities and differences between slides and a filmstrip.

Before making slides or a filmstrip, always consider this planning checklist:

- Have you clearly expressed **your idea** and limited the topic (page 28)?
- Will your program be for **motivational, informational,** or **instructional purposes** (page 28)?
- Have you stated the **objectives** your slide series or filmstrip should serve (page 29)?
- Have you considered the **audience** which will use the materials and its characteristics (page 30)?
- Have you prepared a **content outline** (page 32)?
- Have you written a **treatment** to help organize the content and then sketched a **storyboard** to assist in your visualization of the content (page 48)?
- Have you decided that a **slide series or a filmstrip is an appropriate medium** for accomplishing the purposes (page 42)?
- Have you prepared a **scene-by-scene script** as a guide for your picture taking and sound recording (page 50)?
- Have you considered the **specifications** necessary for your materials (page 53)?
- Have you, if necessary, **selected other people to assist** you with the preparation of visual and audio materials (page 30)?

In practice, many more slide programs are produced than

are filmstrips. Also, slides usually are the starting point for making a filmstrip. Therefore, we will first consider the necessary details for producing slides, then give attention to filmstrip production.

The production of a slide series involves nine major activities: (1) taking pictures, (2) preparing title slides, (3) processing film, (4) mounting slides, (5) editing slides, (6) duplicating slides, (7) recording and synchronizing sound, (8) filing slides, and (9) preparing to use the slide series.

The production of a filmstrip requires attention to five topics: (1) filmstrip format, (2) making a filmstrip from slides, (3) duplicating a filmstrip, (4) correlating a filmstrip with sound, and (5) preparing to use the filmstrip.

The information in Part III, Chapters 10 and 11, on photography and on graphic techniques is basic to the successful preparation of a slide series or a filmstrip. As necessary, refer to the page references indicated with the following topics.

PREPARING AND TAKING PICTURES

Your Camera

Although other format cameras can be used, the majority of slide series are made with 35mm and 126 cameras. Those cameras with adjustable lens settings (*f*/numbers), shutter speeds, and attachments for focusing are especially useful since their flexibility enables you to record various subjects under almost any light and action conditions.

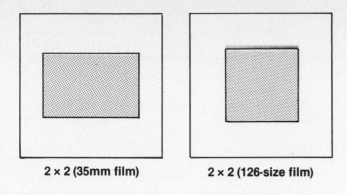

2 x 2 (35mm film)   2 x 2 (126-size film)

There are two major types of 35mm cameras:

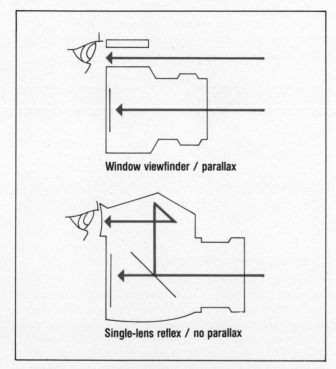

Window viewfinder / parallax

Single-lens reflex / no parallax

● One with a **window viewfinder** through which you see a picture slightly different from the one that the camera will record. This difference becomes greater as the camera gets closer to the subject. (Study the parallax problem described on page 96).

● The other (single-lens reflex camera) with a **reflecting mirror** and a **prism** which permit you to accurately view the same picture that the lens transmits to the film regardless of the distance from camera to subject.

The single-lens reflex camera is preferable for picture taking in which framing is critical, as in close-up and copy work. On the other hand, the window viewfinder camera has the advantage of small size and compactness.

Carefully study the three settings that are made on adjustable cameras—lens diaphragm, shutter speed, and focus. Understand the purposes of each, the relationship of one to another and to depth-of-field, and determine how each setting is made on your camera. These matters have been discussed and explained starting on page 77.

Accessories

You may find need for:

● A photographic light meter to determine exposure accurately (page 84)
● A tripod to steady the camera (When filming at shutter speeds slower than $\frac{1}{30}$ second *always* use a tripod.)
● An electronic flash unit or photoflood lights for indoor scenes (page 87)
● A close-up attachment to photograph subjects at close range and to do copy work (page 97)
● A cable release to eliminate any possibility of jarring the camera during long exposures

Film

Select a reversal film to prepare slides when only one or a few copies will be needed. If many copies will be required, use a color negative film and have inexpensive positive slides made by contact printing the negative. If you wish to see the slides you shoot in less than five minutes, consider using Polaroid's *Autoprocess 35mm system*. It consists of color and black-and-white films for use with any 35mm camera. The film is developed in a manually operated processor for immediate viewing.

In addition to considering the factors about film explained on page 81, select film on the basis of:

● The main light source that will strike the subject (day-

Table 17-1 **Commonly Used 35mm Color Films**

		EXPOSURE INDEX(ASA)		
FILM	TYPE	DAYLIGHT[a]	PHOTOFLOOD[a] (3400°K.)	USE
KODACOLOR VR	—	100	25 (80B)	Negative for prints and slides
KODACHROME 40	Type A	25 (85)	40	Slow speed; good color resolution for copy work
EKTACHROME 64	Daylight	64	20 (80B)	Moderate speed; general purpose
EKTACHROME 200	Daylight	200	64 (80B)	High speed for subjects under low light level outdoors, or indoors under fluorescent light

[a]Numbers in parentheses are filters recommended for converting film to use with other than recommended light sources.

light, flash, or photoflood—3400° K or 3200° K; see the note about fluorescent lights on page 86)

- The anticipated light level (low, moderate, high)
- The desired reproduction of colors
- The expected number of pictures to be taken (based on 35mm 20- or 36-exposure cassettes or rolls)
- The manner of film processing (by film laboratory or by yourself)

The data about selected films given in Table 17-1 are correct as of the time of writing; but changes and new developments can be anticipated. Carefully check the data sheet packaged with your film for the latest assigned exposure index and other details.

Exposure

Correct exposure is based on proper camera settings for the film used and for the light conditions under which pictures are to be taken. Your camera may have a built-in exposure meter and semi- or completely automatic features for determining exposure.

Film information sheets provide general exposure data for average conditions (page 84). For proper exposure, reversal color films permit only a narrow range of camera settings, limited to from $\frac{1}{2}$ to 1 f/stop on either side of the correct setting.

Lighting

As indicated in Table 17-1, color films are designed for use with specific light sources. Select your film accordingly, although with a proper light-balancing filter a film can be used under other than the recommended light conditions. When such a filter is used, the exposure index of the film is reduced (example: Ektachrome 64 with exposure index 64 when used with photofloods requires a No. 80B filter and the exposure index is reduced to 20). Refer to the film information sheet packaged with each roll for detailed information about light-balancing filters.

The conditions under which a subject is to be filmed may require the use of artificial lighting. To boost the light level, flashcubes or electronic flash units can be used. They are handy and easy to operate, especially when small areas must be illuminated. Make sure that your camera is synchronized (at the recommended shutter speed) with the flash. Then apply the information and formula on page 87.

For more carefully controlled lighting use photoflood lamps. They are available in various sizes and are used with separate metal reflectors or have reflectors built into them. In place of regular photoflood lamps, consider using highly efficient **sealed quartz halogen lamps** of approximately 1000 watts. Avoid flat lighting created by placing lights beside the camera only. Instead, establish a lighting pattern involving a key light, fill lights, and supplementary background and accent lights. Study the purposes and placement of these lights and methods for determining exposure with them as described on page 88.

Composition

Composition must take place in the viewfinder of your camera when you film each scene; therefore study the general suggestions for good composition on page 94. As you select subjects, keep in mind the proportions of the slides you are preparing.

If you can, prepare all slides with a uniform format—preferably horizontal.

Using Masks

A wide variety of paper, cardboard, or plastic masks is available or can be designed to be combined with slides when mounted. They can block out unwanted areas of film, change proportions of a slide, create unusual outlines of projected images, or allow more than a single image to be put on one slide.

The possible use of specific masks should be considered when a slide program is being planned and when individual subjects are being composed in the camera.

 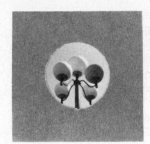

Close-up and Copy Work

Close-up and copy techniques are often very useful when preparing color slides. Your script may call for close-ups of objects, for details in a process, or for copies of titles, maps, pictures, and diagrams. For these purposes, as has been mentioned, the single-lens reflex camera is necessary by reason of its accuracy in viewing. Refer to page 96 for guidelines; these deal with viewfinding, parallax, focusing, lens diaphragm openings, exposure timing, and equipment or attachments that you may need for this special kind of photography. Also refer to suggested procedures for copying flat materials.

A reminder—always remember to obtain a release when preparing to use copyrighted materials.

Illustrations

For preparing illustrations and diagrams:

1. Plan the art work in terms of the slide proportions (page 106). An outline of the open area of a 35mm frame is printed inside the front cover of this book.
2. Select suitable backgrounds (page 119).
3. Use appropriate illustrating, drawing, and coloring techniques (page 112).
4. Use suitable copy techniques to photograph each illustration as a slide (page 96).

Titles

Titles should serve the purposes noted on page 61. Be sure to take account of the legibility standards for projected materials (page 121) as you select lettering sizes for titles, captions, and labels. Then:

1. Word each title so it is brief *and* communicative.
2. Select materials or aids for appropriate lettering (page 124).
3. Prepare the lettering and art work (page 112), keeping in mind the correct proportions of your slides (page 107 and inside front cover), using simple yet effective design features (page 107), and selecting appropriate backgrounds (page 119).
4. Use close-up copy techniques to photograph each completed title on color film (page 96).

High-contrast photography and related techniques can be used to make a variety of title slides. Each method starts with a high-contrast, black-and-white negative prepared on Kodalith Ortho Type 3 or Kodak Reproduction Film 2566 (a substitute for Kodalith that can be processed in Dektol paper developer, resulting in high contrast images). These films are used mainly in graphic arts printing as described on page 139. Kodalith can also be used to prepare overhead transparencies (page 185) and the same processing steps are followed with 35mm Kodalith film.

The accompanying diagram (page 200) illustrates the types of title slides that can be prepared by starting with a 35mm Kodalith negative. A description of each process, as numbered in the diagram, follows:

1. **Clear letters on black background**—Use the negative directly as a slide.
2. **Colored letters on black background**—Color the clear lettering by dipping the negative in transparent watercolor dye, by rubbing the emulsion (dull) side of the negative with a colored felt pen, or by adhering a strip of color adhesive material to the negative (page 119).
3. **Clear letters on colored background**—Place the negative in contact with a piece of diazo film; expose to ultraviolet light and develop the diazo film in ammonia vapors (page 181).
4. **Clear letters over subject**—Copy the negative and the 35mm subject scene on the same frame of film by double exposure. Be sure the letters will be over a dark area of the scene. By underexposing the subject one-half stop it will appear slightly darker so the lettering can stand out.
5. **Black letters on clear background**—Contact print the negative onto another piece of Kodalith film to make a positive (page 139).
6. **Black letters on colored background**—Prepare as in (5) above, and then dip the film in concentrated watercolor dye.
7. **Black letters over subject**—Prepare as in (5) above, and then seal the positive film and the 35mm subject scene together in the same frame. Because the two pieces of film will absorb heat, possibly separate and

lose focus when projected, copy the slide onto a single frame of film.
8. **Colored letters on clear background**—Place the positive film in contact with a piece of diazo film and treat as in (3) above.
9. **Colored letters over subject**—Prepare as in (8) above, and then seal the resulting film bearing the color image and the 35mm subject scene together in the same frame. See the suggestion in (7).
10. **Colored letters on colored background**—In the same frame, mount together (3) and (8) above. See the suggestion in (7) above.

The above ten examples are only some of the creative possibilities for combining colors and using positive and negative images, along with other techniques, for making title slides. As a further example, black-and-white original titles and art work, filmed on Kodak Vericolor Slide Film 5072 (ISO 25) at various exposure settings, and with or without various filters, will result in very attractive multi-color slides (C-41 process). See the references by Balint and Coyne for using this film and other special methods for preparing multi-colored title slides.

A variation of the **progressive disclosure** technique, used with overhead transparencies (page 176), can be applied to slides. Use this method if a series of titles or a list is to be shown, one item at a time, but in cumulative order.

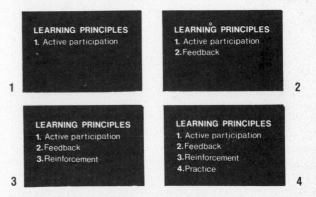

Prepare the total list on white paper, then use the *negative slide* method for preparation as described previously. Frame the entire list, then cover all items but the first one with a sheet of white paper. Film the first item. For the second picture, uncover the second item and film the two that are exposed. Repeat with three items uncovered, and so forth. The result will be a series of slides, each one revealing an additional title or item on the list.

After the slides are completed, use felt pen or colored adhesive (page 119) to color the words comprising the list. Use a different color for each new item introduced on a slide. In this way the new item will stand out and be separated from the prior items on the list.

Using the Computer

Title and graphics slides in full color can be created on the color video screen of a computer terminal. To develop such materials a computer graphics system is required. This includes a microcomputer or time-share computer terminal,

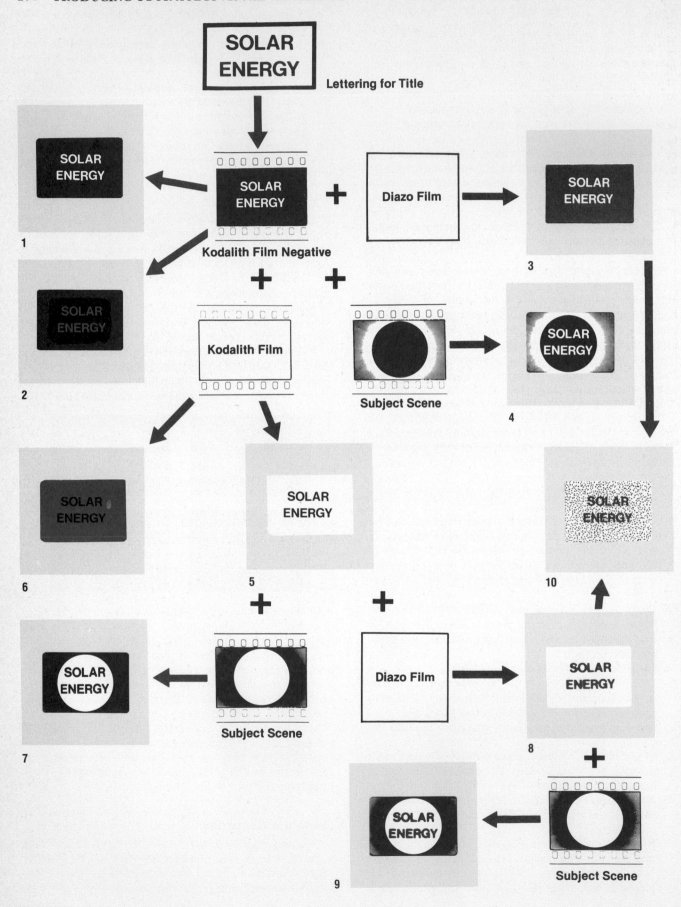

a keyboard, a graphics tablet (optional), a video display screen, and a film recorder. Starting on page 115, the advantages, equipment details, and procedures for preparing graphic materials with a computer are described. The range of full-color materials that can become slides, in addition to titles, includes charts, graphs, diagrams, and numerous art forms.

It is possible to photograph an image on the video screen with a regular 35mm SLR camera equipped with a close-up lens. Mask the lens so as to eliminate reflections of shiny camera parts from the screen. Eastman Kodak distributes a "print-imaging" outfit to make this method more suitable. There are limitations in terms of image sharpness, color

reproduction, and distortion due to curvature of the cathode ray tube. It is therefore preferable to use a special camera —the film recorder—that overcomes these disadvantages.

Computer-generated slides can be produced if your organization has all necessary equipment. An alternative is to have the terminal equipment and transmit electronic images of slides via telephone to a company which has the film recorder. There are also production companies which will handle the complete job, preparing slides according to your requirements.

The choice of which service to use will depend on:

- The number of slides needed over a period of time
- How quickly slides will be needed
- The level of slide quality (resolution, color selection) that would be acceptable
- Any confidential nature of slide content
- Financial considerations relating to equipment purchase or lease

Scheduling and Record Keeping

As you plan the slides, make a list of scenes that can conveniently be filmed together. Then schedule each group. Organizing the work thus will save time and facilitate your picture making. See the example on page 55.

Then, as you prepare to make slides, consider the suggestions on page 55. Keep a record of the scenes filmed, the number of times each is taken, the camera settings used, and any special observations. Develop a form similar to the sample log sheet on page 56.

Remember to obtain a release from the persons appear-ing in your pictures. See the sample release form on page 57.

For some purposes you may use a **documentary** approach for preparing a slide series. This means that you shoot slides treating a topic or event *without* knowing the specific content of scenes. Upon examining the resulting slides you *then* develop your script from them. See further discussion of the documentary method on page 233.

PROCESSING FILM

One advantage and convenience in using reversal color film is that after exposure, a roll can be sent to a film-processing laboratory (through your local photo dealer). The slides are returned mounted in cardboard or plastic frames ready for projection. But, if desired, many color films (Ektachrome, and color negatives) can be processed with kits of prepared chemicals. Time can thus be saved between filming and seeing the completed slides; moreover, if a number of rolls is ready at about the same time, money can also be saved. The requirements and some of the cautions that must be observed when processing both reversal color films (E-6 process) and color negative films (C-41 process) are outlined on page 100. Maintaining the required temperature with little variation is most important.

There is a variety of equipment available for processing color reversal film (E-6 process). The simplest way is that of hand processing film on a reel in a tank as illustrated on page 100. This procedure is suitable for handling a few rolls of film (generally no longer than 36 exposures per roll) during a processing period. It requires constant attention to maintaining proper temperature and controlling other variables.

If large quantities of film or longer lengths, as for film-strips, will be processed frequently, then continuous processing equipment should be used (see page 100). Film, when fed into such a machine, passes through a series of deep tanks containing the required chemicals. Temperature and strength of the solutions are automatically maintained. With a continuous processor, consistent slide quality will be better assured than when hand processing. On the other side, initial equipment expense is high, continual maintenance is necessary, and greater technical knowledge is required for successful operation of the equipment.

MOUNTING SLIDES

Slides can be mounted in commercial cardboard frames, in plastic mounts, or between glass plates. Each type has certain advantages and also disadvantages. Make your choice of the type to use based on the following:

- **Cardboard**—least expensive, frame easy to mark with felt pen; slide held firmly if properly sealed; requires heat to seal; bends and frays easily, leading to possible jam-ming in the projector (use rounded-corner type); film may buckle during prolonged projection and lose focus in some projectors.
- **Plastic**—rigid; edges do not fray; easy and quick-to-

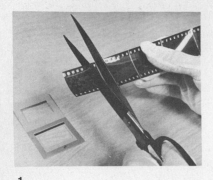

1

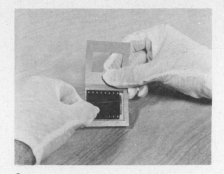

2

3

mount film; somewhat more expensive than cardboard; more difficult to mark frame; film may shift position and buckle in less expensive types; better quality plastic mounts include pins that engage the film sprocket holes, locking the film in place (known as **pin registration**) for precise placement of an image on the screen during projection.

- **Glass**—offers maximum protection for film; avoids buckle and loss of focus; heavy; most expensive; may include pin registration; requires more time to mount film; may not fit 140 slide tray; may break if dropped.

Slides may not need the protection of glass. In modern projectors they are removed from trays and returned to them mechanically during projection, and are touched by the hands only when being filed or rearranged in the tray. Also, some projectors automatically adjust focus as film heats and curls.

Mounting in Cardboard Frames

Use these tools and materials: cotton gloves, hand iron, cardboard mounts, gummed-back thumbspots, scissors—and the film to be mounted.

1. Cut the film along the frame line between the pictures.
2. Align the film in the mount.
3. Using an electric iron (set at "low"), seal all four sides.

If a large number of slides is to be mounted in cardboard, a faster method than using the hand iron may be preferred.

A relatively inexpensive slide-mounting press consists of 2×2 inch heated pressure plates. The folded mount, holding a 35mm film frame, is placed between the pressure plates, which are then tightly closed. After a few seconds the plates are opened and the sealed slide drops out. Semi- and fully automatic slide mounting units also are available commercially.

Mounting in Plastic Frames

A variety of plastic slide frames is available for mounting slides. With one-piece frames, the film is either placed in one section, the other part folded over and secured or the two halves are gently pried apart and the film slipped between. Other types consist of separate parts that are snapped in place to hold the film.

Check your photo supplier and other commercial sources (page 211) for specific kinds of plastic slide mounts and for automatic mounting equipment.

Fold-Over Frame   **Snap-Together Frame**
PLASTIC MOUNTS

A Hand-Operated Slide-Mounting Press

A Semi-Automatic Plastic-Slide Mounting Machine

Mounting in Glass

At one time slides were sealed in glass with slide-binding tape. This preparation required much hand work; eventually, also, the tape might loosen or become sticky. Moreover, some projector trays do not accommodate tapebound slides. More recently, with the availability of plastic frames into which the film and glass slip easily, mounting slides in glass can be accomplished quickly and with little effort.

Once slides are mounted, each one should be marked to indicate correct position for viewing—subject left to right and correct side up. Do this by placing a dot with a felt pen (permanent type on plastic or glass) in the *lower left* corner of the frame when the slide is viewed correctly. When placed in a tray for projection, the mark will be in the upper right corner.

EDITING SLIDES

Selections must be made from among all the slides—some are in addition to those called for in the script or are substitutes; others are multiple takes of the same scene but differ in exposure and composition.

Place all slides on a light box or other illuminated area for ease of inspection. Discard those so indicated on the log sheet prepared while filming. Examine the slides; eliminate the poorer ones until the remaining selection is limited to those of highest quality that fit or supplement the prepared script. Now revise the script as necessary and, if spoken or recorded commentary is to accompany the slides, refine it. Refer to the suggestions on page 58.

With the editing finished, your slide series is nearing completion. It may be advisable at this time to show the series and to read the narration to other interested and qualified persons. For suggestions for developing a questionnaire to gather reactions which may help you to improve your slide series, see page 66.

DUPLICATING SLIDES

If the number of duplicate slides that will be needed is known before photographing begins, then all duplicates can be made as high-quality originals when the original subjects or materials are photographed. Should additional sets be

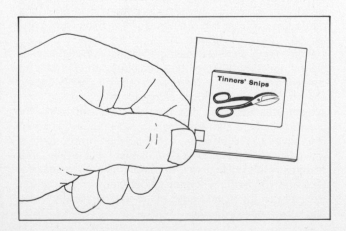

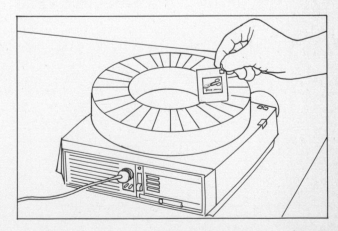

required after photography has been completed, a film-processing laboratory can make duplicates of the original slides. There are also methods you can use to prepare your own duplicate slides.

The two major problems that should be recognized when duplicating slides are an increase in contrast (loss of detail in highlights and in shadows) and a shift in colors from those in the original slide (or the need to improve or correct colors in an original slide). To control contrast, use Ektachrome Slide Duplicating film 5071 (balanced for 3200°K) or Ektachrome Slide Duplicating film SO-366 (balanced for electronic flash). With either of these films, the resulting duplicate would show only a minimum gain in contrast.

In order to balance colors as close to those in the original, or to correct colors (like to replace a blue cast with a warmer orange tone), place one or more special filters (CC filters for *color correction*) between the slide and the light source (or over the projector lens for the first method described below). It takes testing and experience to choose filters that will properly correct colors and result in duplicates that best reproduce original or desired colors in subjects.

Projector-Screen Method

Project the slide onto a matte-surface screen. Make certain that the projector is placed at a right angle to the screen in order to eliminate distortion of the image. A three-foot-size picture is suitable. Set the camera on a tripod as close to the projector as possible in order to avoid distortion when filming and to ensure that an even amount of reflected light is received from all parts of the screen. Adjust the image size or camera position until the image fills the viewfinder. It may be necessary to use a telephoto lens on the camera.

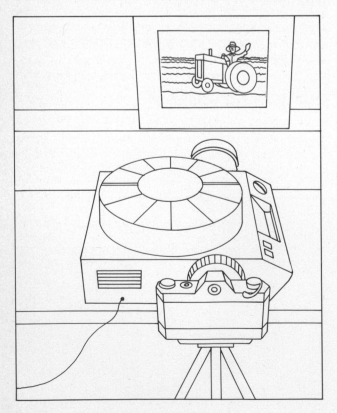

Use a film balanced for artificial light. Determine exposure with the meter built into the camera or use a reflected light meter. Bracket your exposure (take additional pictures at f/stops on both sides of the recommended exposure). With this method, it is difficult to avoid making a duplicate slide that will have more undesirable contrast than the original slide, and some loss of original color.

Available-Light Method

A slide can be copied directly by placing it against a window and using a bellows or extension tubes on the camera for 1:1 size focusing (page 97). A macro lens with the *flat field* feature is advisable for use in this and the following close-up methods. Set a piece of translucent glass behind the original slide to diffuse the light or use a holder for the slide with a built-in diffuser.

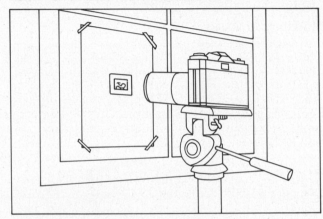

For this method, select a fine-grain film such as Koda-chrome (Daylight) to ensure as sharp a picture as possible. Use a 35mm single-lens reflex camera—preferably one that has a behind-the-lens exposure meter.

Slide-Duplicator Method

The most successful method of copying slides is with specially designed slide duplication equipment. It has a base with a translucent glass plate that holds the original slide to be copied. Beneath the glass plate is an electronic flash unit. A vertical bar holds the camera and the extension bellows. Some units include a camera; with others you attach your own 35mm single-lens reflex camera. A built-in light meter allows for accurate exposure determinations. Use a low contrast film like Ektachrome Slide Duplicating film 5071.

Professional slide-duplicating equipment, available in table-top and floor-stand models, includes cameras with accurate viewfinders that allow you to record *exactly* what you see on film (regular 35mm cameras usually record more subject area than you see through the viewfinder). The precision camera also includes film registration pins so the film moves to *precisely* the same position at the aperture with each advance (or reverse movement) of the film. Other features of this equipment may include bulk film magazines (100 and 400 feet), motorized camera movement, multiple exposure on the same frame, masking portions of a frame, and other special effects.

Many precision-camera systems are microprocessor controlled. The microprocessor controls camera functions and stores sequences and commands for future reference. These camera systems are used, not only for making duplicates and adding reference numbers in the corner of slides, but also to prepare title slides, special effect slides for multi-image productions and filmstrips.

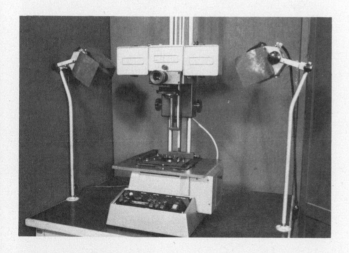

Videotape Method

For some uses, it may be desirable to transfer the slides, along with the sound track, to videotape. This may be a convenient way for showing slide/tape programs because the synchronization of picture and sound is automatic and only a videotape player and television receiver are necessary for viewing.

Aim the video camera onto a matte projection screen or a plain white wall. Use a three-foot image. Frame the image in the camera viewfinder with some clear or "bleed" space (video safe area) around the image to allow for possible loss of picture area on some television receivers. Connect the audio output from the tape playback unit to an audio input on the videotape recorder. Project the slides as the tape is played and record both the images and the sound track simultaneously on videotape.

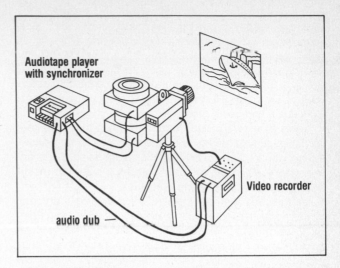

With this method, be alert to two potential problems. First, if at any time a clear white image is projected to the screen because there is no slide in the projection position, damage can result to the video tube. Second, because of the automatic exposure feature of many video cameras, it is difficult to create a black screen or a fade-out when a dissolve control unit is used with slide projectors. (The lens diaphragm setting of the video camera lens is automatically changed to overcome what seems to be an apparent underexposure situation.)

RECORDING AND SYNCHRONIZING NARRATION AND OTHER SOUNDS

Narration can be used with a slide series in the following ways:

- As informal comments while slides are projected
- As formal reading of narration as slides are projected
- As recorded narration and other sounds with an audible signal to indicate slide changes
- As recorded narration and other sounds with an inaudible signal which electronically controls slide changes

If a tape-recorded narration is to be prepared, refer to earlier suggestions concerning the selection and duties of personnel, recording facilities and equipment, recording procedures, sound mixing procedures, and tape editing (Chapter 12).

Synchronizing slides with tape requires the use of a **pro-**

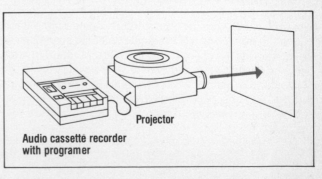

graming unit, either connected between the tape recorder and the slide projector, or built into the tape recorder, which is directly connected to the projector. A signal generated by the programmer (usually 1,000 Hz) is recorded on tape. This signal, when the tape is played back, closes a relay that advances the next slide into projection position. See page 153 for further information about recording signals and using programers.

Some recorders include control devices that on a programmed signal (150 Hz) will cause the tape to stop, allowing an observation to be made or a question to be answered. When ready, the user presses a button to restart the tape for the program to continue.

Another unit that can be useful for making a smooth, professional presentation is a dissolver, which is a compact box of electronics that increases or decreases electrical resistance in the lamp circuits of two projectors. It thus increases the intensity of light in one projector while it decreases it in the other. The two projectors are carefully aimed at the same screen so their images coincide. The images superimpose and change on command from the programer, with one image fading out and the next one on the screen fading in, thus creating an effective impression of gradual change and transformation as images blend one from another.

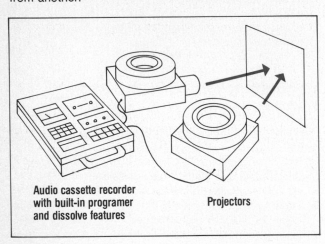

Audio cassette recorder with built-in programer and dissolve features

Projectors

When dissolves are used, the screen is always illuminated during slide changes, which helps to hold audience attention. Employing this technique of dissolving from one slide to the next creates a smooth visual flow as compared with the usual single projector slide change procedure that causes a sudden black screen between adjacent slides.

The simplest dissolve units have a few fixed dissolve rates for controlling two projectors. Other units allow for many precise rates of light control. Some sophisticated units can be programed to create numerous screen effects by controlling the on–off rates of projectors' lamps, of slide-change mechanisms, and even of shutters placed over projector lenses. Such controls can cause flashing, superimpositions, pop-ons, pop-offs, slide changes (forward, reverse, or hold), and simulated animation effects, all from static slides.

Programmers and dissolve units are widely used in multi-image slide programs. See Chapter 18 for further details.

FILING SLIDES

Initially most slides can be stored in the small 20- and 36-exposure boxes in which they are received from the processing laboratory. As quantities of slides are accumulated, some type of filing system becomes advisable so that individual slides can be located easily. Develop a numerical or color-coded filing system and consider the illustrated filing methods.

Plastic Sheet

Projector Tray

Slide Box

Storage Display Cabinet

For visual reference purposes, numbered groups of slides can be placed on a light box, photographed on black-and-white film, and the negative printed to 8×10 inches. This sheet, with its numbered slide pictures, can be referred to when slides are being selected for a presentation or other use. Then, using the numbers, the slides are located in the storage file. Each topic in a file can be treated in this way and the slide picture sheets filed for reference in a notebook. This method is preferable to having to take each slide from a file and look at it when deciding on a selection.

Slides on Light Box

Photosheets of Slides

Another way to locate specific slides that have been filed in a numerical system is by using a computer program. Most microcomputer systems have one or more inexpensive and

```
50 REM :SLIDE SUBJECT SEARCH
100 PRINT "SELECT SUBJECT CATEGORY"
110 PRINT "1-BUILDINGS 2-PEOPLE 3-ACTIVITIES"
120 PRINT "CATEGORY NUMBER ";:INPUT A
130 IF A=2 THEN 2000
140 IF A=3 THEN 3000
1000 PRINT "SUBJECT: BUILDINGS=SLIDE FILE B2."
1010 PRINT "SLIDE NUMBERS 125 TO 250"
1100 END
2000 PRINT "SUBJECT: PEOPLE=SLIDE FILE P1."
2010 PRINT "SLIDE NUMBERS 300,310,312,313"
2130 END
3000 PRINT "SUBJECT: ACTIVITIES=SLIDE FILE A2."
3010 PRINT "SLIDE NUMBERS 500 TO 570."
3020 END
```

easy-to-use filing (data base) systems which can maintain information on a slide file. Depending on how you set up the system, it can rapidly sort slides by any variable you desire (subject, location, photographer, film type, where used, and so forth). Many can also be used to generate listings of slides, or small printed labels which can be adhered to sets of slides or to individual slides.

By expanding a simple computer program like that above (developed in BASIC language on a microcomputer), you can identify slides under appropriate subject categories and subgroups.

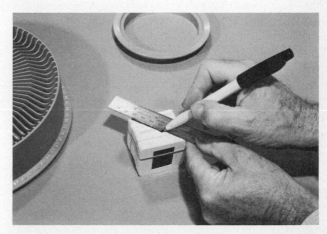

PREPARING TO USE YOUR SLIDE SERIES

When your slides have been completed, give attention to a number of final details as you prepare for their use.

The slides may have been designed for motivational, informational, or instructional purposes (page 28). You may now need printed materials—information or summary sheets, guide sheets, or worksheets. Prepare and duplicate them in sufficient copies for the audience or for individual users.

Arrange for duplication of the master tape recording, if one is to accompany use of the slides (page 154). Labeling and packaging the slides and tape recording may be necessary.

Projectors and viewers require that slides be placed in a tray for use. A simple technique can ensure that the slides are in proper order and in correct position (upside-down) for projection. Once you are certain all slides are positioned correctly, draw a diagonal line across the top cardboard edge of all slides. Then should you at any time find a break in the continuity of the line, look for a missing or improperly positioned slide.

A useful technique that allows you to transfer the attention of the audience or an individual from one topic to the next, or to the other matters (answering questions, participation activities, discussion, or whatever) by interrupting the projection, can be worked in at this stage. Do this by inserting a black slide or 2×2 inch pieces of cardboard in the tray at places where the screen should be dark. Then when a blank falls into projection position the light rays from the lamp are interrupted and the screen goes dark. When the

audience or individual is ready for the next slide, projection can smoothly continue—the blank is automatically returned to its position in the tray and the next regular slide is shown. With some model projectors, an empty slot in the tray will accomplish the same purpose of darkening the screen.

Finally, arrange for the projectors, recorder, and other necessary equipment. Check the physical facilities where the slides and tape will be shown and heard. For a successful showing, give careful attention to the details enumerated on page 66. If the materials will be for individual use, then rear-screen viewers should be available. Arrange to instruct those who will use the slides and tape in the operation of the equipment. It may be advisable to post operating instructions at each station.

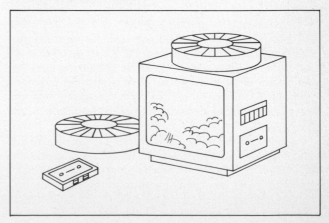

Review What You Have Learned About Preparing Slide Series:

1. Is your camera a window-viewfinder or a single-lens reflex type? What are its advantages or limitations?
2. What film or films would you select for use? How did you arrive at your choice(s)?
3. What purposes can be served by masking a slide?
4. What method of making slide titles would you use in each situation?
   a. A general main title having black lettering on a colored background
   b. "Overprinting" a word as clear lettering on a prepared slide
   c. A series of three titles to be shown progressively as negative slides with colored words
5. What equipment is required to prepare a high-quality color slide when a diagram is created by a computer?
6. How does the production of a *documentary* slide program differ from the nondocumentary method?
7. What is the designation for the reversal color film processing procedure you can carry out yourself?
8. What factors do you consider in making choices during editing of slides?
9. What are the three types of slide mounts and advantages of each kind?
10. How would you proceed to prepare duplicates of some original slides?
11. When preparing to duplicate slides, what two problems often occur? How do you handle each one?
12. What equipment is available for your use in synchronizing slides and a tape recording?
13. Why can a dissolve unit be used to present a smooth, uninterrupted slide presentation?
14. What method of filing your slides might you use?

FILMSTRIP FORMAT

A filmstrip consists of a series of illustrations and photographs in sequence, most often with an accompanying audio recording. Filmstrips are most frequently prepared on 35mm film, although certain viewers are designed to use 110, 16mm, or Super 8mm film sizes. The most widely used filmstrip format is the 35mm *half-frame* type (each filmstrip frame is equal in area to *one-half* that of a regular slide). The camera that is required to prepare such a filmstrip is termed a *half-frame* camera.

The half-frame filmstrip consists of pictures about 18×24 mm positioned on the film so that the long dimension runs across the film. The proportion of this half-frame picture area is three units high to four units wide. For composing titles and art work, an outline of the open area of a half-frame filmstrip frame is printed inside the front cover of this book.

MAKING A FILMSTRIP FROM SLIDES

The easiest way for making a filmstrip is to prepare slides and copy them with a half-frame camera onto reversal 35mm color film. Make a regular 2×2 inch color-slide series, including titles and captions. Be certain the selection and quality of pictures are acceptable and the sequence order is firm. Changes cannot be made easily once the filmstrip is prepared. Consider a try-out in slide form before conversion to filmstrip.

If a half-frame camera is not available or you require that a high-quality, professional-level filmstrip be prepared, the services of a commercial film laboratory can be used. Such a company will use a half-frame (often a 35mm motion-picture) camera which allows for precise film movement to accurately frame each image as the film is advanced.

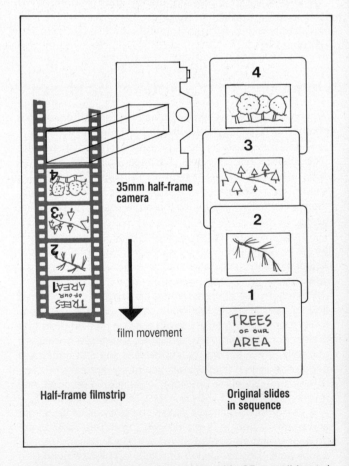

35mm half-frame camera

film movement

Half-frame filmstrip

Original slides in sequence

Most filmstrip preparation starts with 35mm slides, although some original materials are filmed on either 4×5 inch Ektachrome sheet film or as $2\frac{1}{4} \times 2\frac{1}{4}$ inch slides. These larger areas permit sharper reproduction; simplicity in addition of captions, labels, and other markings; easier control of color (through color and contrast masking at the laboratory); and generally better overall quality compared with starting from the smaller 35mm slide.

Remember! All the slides must be in horizontal format. While vertical slides are not recommended except for split frame scenes, they can be used by masking at the sides. The image from them will appear smaller on the filmstrip frame.

The format of a 35mm slide (2 to 3) differs from that of a half-frame filmstrip frame (3 to 4). Therefore in copying slides, *cropping,* or loss of part of the slide area, must be recognized. To be prepared for this loss, reproduce the cropping guide shown onto a piece of clear acetate. Place the guide under a slide. Within the rectangle is the approximate area (depending on the lens and aperture alignment of your camera) of the slide that will be converted to the filmstrip frame. By examining a few slides you will realize that the subject for each slide must be composed so that important objects or action are not close to the side edges or they may be lost during conversion to filmstrip.

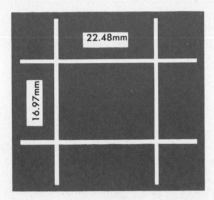

If you start with 126-size slides (square format), cropping is necessary at the top and bottom for adjusting to the 3 to 4 ratio.

Attach your half-frame, 35mm, single-lens reflex camera equipped with a proper close-up attachment to a copystand. The slides to be copied are placed over a simple light box or other back lighted surface. Align the camera perpendicular to the slide. A small bubble level is helpful in doing this. Each slide must be positioned correctly under the camera in relation to the direction of film movement in the camera. Refer to the illustration on the facing page to be certain you are positioning your slides properly.

Use a low contrast film like Ektachrome Slide Duplicating film 5071. Determine correct exposure from slides with the meter built into the camera. You may want to check the exposure reading with a reflected type light meter (page 84) held directly over the illuminated slide. Shoot a roll of test film, exposing for various camera lens settings, to be certain the correct exposure for your setup is determined.

The length of the filmstrip you can make is limited by the length of the film available for your camera. A normal 36-exposure roll of 35mm film can result in a strip of 60 single frames, leaving a few frames for leader and trailer. Longer filmstrips require special camera attachments or the services of a film laboratory.

Finally, if you send the film to a laboratory for processing,

be sure to indicate that it is to remain as a strip and is *not* to be mounted as slides.

Frequently it is desirable to print a small number in the lower corner of each frame for reference purposes. This is particularly important when a filmstrip is to be used for self-paced study. Frame numbers are added when the original filmstrip is made from slides or when duplicating the original. The numbers, appearing white on the frame, are prepared by double-exposure—first the scene and then a high-contrast negative containing the number of proper frame location (usually lower-right corner). While double-exposing is possible with many 35mm cameras, the need to do it properly for each frame as it is shot, and the care for placement of the number, makes this a difficult technique to accomplish outside of a film laboratory unless specialized filmstrip production equipment is available.

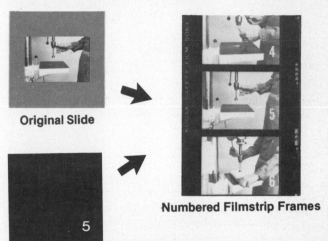

Original Slide

Numbered Filmstrip Frames

High Contrast Negative with Number

Whenever slides are to be converted to a filmstrip be sure to number the edge of each one in proper sequence to avoid an incorrect order or errors by the photographer. Mark original material with crop marks (felt pen lines in corners outside the visual area) to indicate the exact area of the visual to be filmed.

The American National Standards Institute (ANSI) has established specifications for identifying and marking the various parts of a filmstrip—start, focus frame, title, and end. A visual reproduction of the standards can be found in a DuKane Corporation publication. See the reference by Lord and Jewison on page 211.

DUPLICATING A FILMSTRIP

Preparing more than a few copies of a filmstrip by shooting each one from the original photographs or slides is impractical. It is advisable to have duplicates made from the master filmstrip by a film laboratory providing such service. Two methods are employed in making duplicates. Good quality copies can be made directly from an original filmstrip. Such a service is charged on either a per frame basis ($0.20), or on a per foot basis ($0.50 in quantity). There are 16 frames per foot on a 35mm half-frame filmstrip. This method is more economical when 12 or fewer duplicates are to be made.

The second duplication method is recommended when large number of high-quality strips are needed. An internegative ($3.00 per frame) is made from the original filmstrip and positive copies ($0.25 per foot) are run from it. A procedure similar to that used with motion pictures is followed. From the internegative one or more positive "answer" prints are made. Upon approval, "release" prints in the quantity requested are made.

See the list of film laboratories providing the services described here (page 212). Inquire about special services to fill your needs.

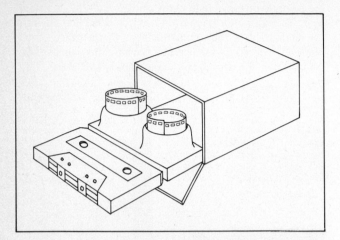

Upon completion, roll each filmstrip with the emulsion (dull) side on the inside to insure proper handling when projected. Store it in either a filmstrip can (source on page 212) or use a plastic box to hold the filmstrip and an audiocassette and label with the title.

CORRELATING A FILMSTRIP WITH A SOUNDTRACK

If narration, with or without music and sound effects, is to accompany the filmstrip, refine it to fit the pictures *before* the master filmstrip is prepared. This can avoid costly changes in visuals that may be required by any revision of the narration.

Refer to suggestions in Chapter 12 for the selection of a narrator and duties of recording personnel, recording facilities and equipment, recording procedures, tape editing, and sound mixing procedures.

Audible or **inaudible** frame advance signals can be added to the master recording. See page 153 for a discussion of how an audible signal can be generated and put on the tape. The choice of an inaudible signal depends on the type of filmstrip projector or viewer to be used. Most manufacturers require a 50-Hz (Hz-cycles per second) tone *superimposed* on the same track as the narration and music. Some equipment responds to a 1000 Hz cueing signal similar to that used for sound/slide recorders (page 205), placed on the *second track* of the audiocassette. With the 50-Hz system both tracks of the tape can be used for the program, often with the narration and the inaudible change signal on one track and the same narration with an audible signal on the second track. Or the program with the inaudi-

ble signal can be repeated on the second track so the tape does not have to be rewound after each use.

The 1000-Hz system uses both tracks so only one program can be presented on the cassette. One advantage of this method is that either the signals or program material can be changed without erasing the other one. With the 50-Hz procedure complete rerecording or dubbing from a multitrack master tape is necessary to make changes in either the signal or the narration.

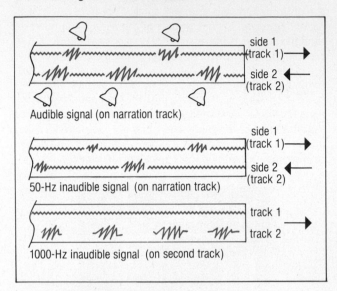

Check the filmstrip equipment that you will be using. Not only do different projectors/recorders require attention to the frequency of control signal, but they may require a different signal intensity level to trigger the advance mechanism as well as a different minimum time span between pulses. Many film laboratories that offer filmstrip service also provide complete sound services.

PREPARING TO USE YOUR FILMSTRIP

With the completion of the filmstrip and of your preparations for using it, consider the advisability of developing an instruction guide as described on page 166. Consider also correlating the filmstrip with other materials for use in an instructional program.

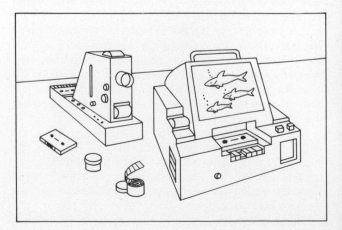

Review What You Have Learned About Producing Filmstrips:

1. How does the proportion of a filmstrip frame differ from that of a 2×2 inch slide?
2. Can a regular 35mm camera be used to prepare a filmstrip?
3. What are the two most common methods for making filmstrips?
4. What is the single most important matter that needs attention when selecting subjects, composing scenes, or choosing slides for a filmstrip?
5. If at least 25 copies of a filmstrip master are to be made, what duplication method is preferred?
6. What may be advantages and limitations for using audible and inaudible control signals?
7. If you are preparing a filmstrip, for the equipment you have available, what type of sound-picture synchronization would you use?

REFERENCES

Preparing Slides

Aneshansley, James, and Smith, Todd. *The Oxberry Slide Handbook.* Carlstadt, NJ: Oxberry, 180 Broad St., 1981.
Calderone, Glenn. "Computer-Generated Color Graphic Slides." *Technical Photography* 15 (January 1983):21–22.
"Computer Generated Slides Offer Wide Range of Capabilities and Prices." *Audiovisual Notes from Kodak.* Publication T-91-2-1, Rochester, NY: Eastman Kodak Co., 1982.
Effective Lecture Slides. Publication S-22. Rochester, NY: Eastman Kodak Co., 1980.
Gabera, Don. "Polaroid's Quick Access 35mm Slides." *Technical Photography* 14 (October 1982):8, 48.
Iserman, Ted. "Slide Effects for the 80's: A Guide to Precision Camera Systems." *Audiovisual Directions* 3 (October/November 1981):22–35.
Preparing 2×2-inch Slides for the Kodak Ektagraphic AudioViewer and AudioViewer/Projector. Publications S-15-88-AP. Rochester, NY: Eastman Kodak Co., 1980.
Special Effect Techniques for Slides and Filmstrips. Austin; TX: Stokes Slide Services, P.O. Box 14277.
Stecker, Elinor H. *How to Create and Use High Contrast Images.* Tucson, AZ: HP Books, 1982.

Art Work and Titles

Balint, Brian J. "Vericolor-ful Title Slides." *Functional Photography* 17 (July/August 1982):28–31.
Basic Art Techniques for Slide Production. Slide/tape presentation V10-15. Rochester, NY: Eastman Kodak Co., 1982.
Coyne, Martin. "Producing Multi-Colored Slides." *Technical Photography* 14 (October 1982):22–35.
Optical Titling. Slide/tape program V10-64. Rochester, NY: Eastman Kodak Co., 1982.
Reverse-Text Slides. Publication S-26. Rochester, NY: Eastman Kodak Co., 1983.

Producing Slide/Tape Programs

Creating Slide/Tape Programs. Filmstrip/tape program. Washington, DC: Association for Educational Communications and Technology, 1980.
Planning and Producing Slide Programs. Publication S-30L. Rochester, NY: Eastman Kodak Co., 1980.
Slide Shows Made Easy. Slide Series (6 parts). Rochester, NY: Visual Horizons, 180 Metro Park.

Sunier, John. *Slide/Sound and Filmstrip Production.* New York: Focal, 1981.

Producing Filmstrips

Lord, John, and Jewison, C. B. *Handbook for the Production of 35mm Sound Filmstrips.* St. Charles, IL: DuKane Corp., 1976.
Sunier, John. *Slide/Sound and Filmstrip Production.* New York: Focal, 1981.

SOURCES FOR EQUIPMENT, MATERIALS, AND SERVICES

Slide Masks

Eastman Kodak Co., Motion Pictures and Audiovisual Markets Division, 343 State Street, Rochester, NY 14650
GEPE, HP Marketing, 216 Little Falls Road, Cedar Grove, NJ 07009
Heindl & Son, P.O. Box 150, Hancock, VT 05748
Stokes Slide Services, 9000 Cameron Road, Austin, TX 78752
Visual Horizons, 180 Metro Park, Rochester, NY 14623
Wess Plastics, Inc., 50 Schmitt Blvd., Farmingdale, NY 11735
WTI, Inc., 22951 Alcalde Drive, Laguna Hills, CA 92653

Slide Mounts and Mounting Equipment

Byers Photo Equipment, 6955 West Sandburg Street, Portland, OR 97223
Kaiser Corporation, 3555 North Prospect Street, Colorado Springs, CO 80907
Pako Corporation, 6300 Olson Memorial Highway, Minneapolis, MN 55440
Pic-Mount Corporation, 40-20 22nd St., Long Island City, NY 11101
Seary Manufacturing, 19 Nebraska Avenue, Endicott, NY 13760
Wess Plastics Inc. (see above)

Slide Duplicating Equipment

Charles Beseler Co., 8 Fernwood Road, Florham Park, NJ 07932
Forox Corporation, 393 West Avenue, Stamford, CT 06902
Mangum Sickles Industries, Inc., 1200 Sickles Drive, Tempe, AZ 85281
Marron-Carrel, Inc., 2640 West 10th Place, Tempe, AZ 85281
Oxberry Division, Richmark Camera Service, Inc., 180 Broad Street, Carlstadt, NJ 07072
Radmar, Inc., 1282 Old Skokie Road, Highland Park, IL 60035

SlideMagic System, Maximilian Kerr Associates, Inc., 2040 State Highway 35, Wall, NJ 07719

Filmstrip Services

Berkey K+L, 222 East 44th Street, New York, NY 10017

Berry & Homer, Inc., 1210 Race Street, Philadelphia, PA 19105

Evergreen Film Service, 1416 West 7th Street, Eugene, OR 97402

Filmstrip & Slide Laboratory, Inc., 300 Park Avenue South, New York, NY 10010

Frank Holmes Labs, Inc., 1947 First Street, San Fernando, CA 91340

International Color Image Labs Inc., 2301 North San Fernando Boulevard, Burbank, CA 91504

Promocraft Productions Ltd., P.O. Box 14729, Cincinnati, OH 45214

PSI Film Laboratory, 3011 Diamond Park Drive, Dallas, TX 75237

Radmar, Inc., 1282 Old Skokie Road, Highland Park, IL 60035

Rolf Slide Laboratories, 21 Shoen Place, Pittsford, NY 14534

Slide Strip Laboratory Inc., 432 West 45th Street, New York, NY 10036

Stokes Slide Services, 7000 Cameron Road, Austin, TX 78752

Visual House, 23 East 39th Street, New York, NY 10016

Chapter 18

MULTI-IMAGE PRESENTATIONS

A multi-image presentation refers to the **simultaneous projection** of two or more pictures, on one or more screens, that may include sweeping panoramas, rapidly changing sequences, simulated or actual motion, or various combinations of these effects. The program may be designed to motivate, inform, or instruct, while creating a variety of moods in the audience. Most often the images are slides, but motion pictures and overhead transparencies may also be used in conjunction with slides.

The term **multi-image** should not be confused with **multimedia.** The latter refers to the *sequential* use of two or more different types of media in a presentation or as part of a learning package for self-paced study.

INTRODUCTION

In all conventionally projected visual forms—slide series, filmstrips, motion pictures, or video recordings—the images are single and presented sequentially. The only exceptions are when split-screen or some other special technique is applied in order to present a simultaneous comparison or relationship. The sequential continuity of separate images conveys meanings to the viewer in a linear manner. When two or more images are viewed simultaneously, their immediate interaction can be more dynamic for the viewer, capturing and holding attention. The nonlinear nature of a multi-image presentation makes it a powerful instrument of communication.

There is research evidence to indicate that the use of multiple images is an effective media form. In a study of multi-image projection, Perrin concluded that "the immediacy of this kind of communication allows the viewer to process larger amounts of information in a very short time.

Thus information density is effectively increased and certain kinds of information are more efficiently learned."

Besides being a motivating, exciting experience for the viewer, this efficiency of communication should be a major reason for using the multi-image presentation technique. The variety inherent in such presentations can be designed for audiences of any size.

PURPOSES THAT CAN BE SERVED

Many informational and instructional purposes can be served by multi-image presentations. Among these are:

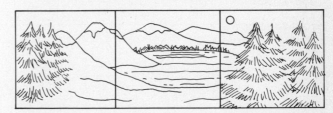

- Panoramic or wide view of a subject across two or more screens

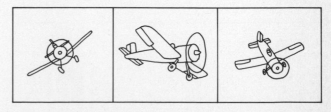

- Showing a subject from different camera angles or distances

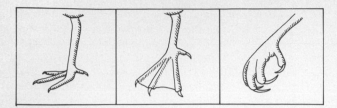

● Comparing or contrasting objects and events

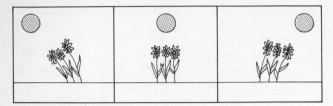

● Presenting sequential time segments relating to a single event

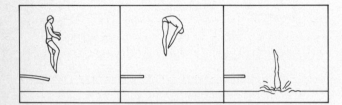

● Simulating motion of a still subject across multiple screens

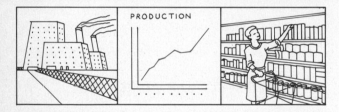

● Giving meaning to an abstract idea with several supporting visuals

● Emphasizing a fact or concept by repeating identical images

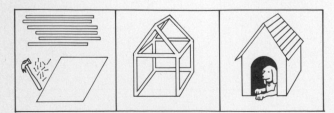

● Illustrating steps in a process

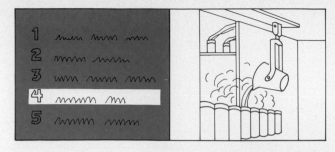

● Illustrating relationships, such as form to function, model or diagram to actual object

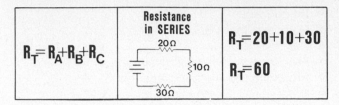

● Relating answers to questions or solutions to problems

● Relating parts of a subject to the whole

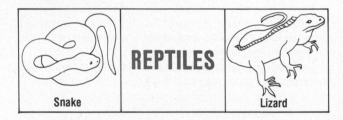

● Adding titles or captions to identify visuals

● Combining motion and still pictures

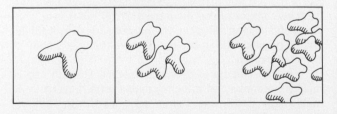

● Developing concepts aesthetically, like growth, change, or interrelationships

Any number of these techniques might be incorporated in a multi-image presentation. In addition, you probably will discover other purposes for which this instructional media format can be utilized effectively.

PLANNING

Because a multi-image presentation is complex, involving many images, it is necessary to effectively blend the images together to form understandable sequences. Therefore, careful, detailed planning is essential for successful results. Consider this checklist:

- Have you clearly expressed your **idea** and limited the topic (page 28)?
- Have you stated the **objectives** your multi-image presentation should serve (page 29)?
- Will the presentation be for **informational** or **instructional** purposes (page 28)?
- Have you considered the **audience** which will see the presentation (page 30)?
- Have you prepared a **content outline** (page 32)?
- Have you written a **treatment** to help organize the content (page 48)?
- Have you decided that a **multi-image presentation** is **a valid method** for accomplishing the objectives (page 42)?
- Have you made a **storyboard** or prepared a **script** for the images on each screen, including accompanying sound (page 48)?
- Have you considered the **specifications** necessary for your presentation (page 53)?
- Have you, if necessary, selected **other people** to assist you with locating, preparing, and eventually using materials, and to help with other technical matters relating to multi-image presentation (page 30)?

Because of using two or more screens and possibly multi-track sound, special attention should be given to preparing the storyboard or script. A form can be used that has separate columns for describing the images that will appear on each screen, columns for each sound element—narration, music, sound effects—and a column in which to specify the effect (slow dissolve, out, flash, etc.) for bringing the visual to the screen and the length of time for each component to be seen or heard. This form can also serve as a programing or cue sheet when readying the presentation for use.

PREPARING THE VISUAL COMPONENT

Because slides usually are the main medium used for a multi-image presentation, it is recommended that you carefully study the fundamental slide-making techniques described in Chapter 17. Many of them are essential for producing a high-quality multi-image presentation.

A multi-image presentation requires attention to many details. Too often those planning such a program become fascinated with the gimmickry of using multiple screens, multiple-track sound, and sophisticated control equipment. Visuals may be unorganized while the pictures and sound may have little relationship or continuity. The message may be overpowered by the techniques and prove to be an uncoordinated collection of visual and audio misimpressions.

If this is your first experience with multi-images, you may want to start with two images on a single screen (as described on page 206). Then use two screens, requiring four or more slide projectors. With this background move to the three-screen level for suitable subjects. Experience has shown that the ideal number of projected images for most communication functions is *three*. But you need not limit yourself to three screens if additional ones are necessary.

Suggestions

As you consider the number of images to use, realize that the visual attention of the viewer shifts as the slides change. Thus, the involvement of the viewer and the impact of the message can be controlled by the careful placement of images as they are shown separately or simultaneously.

The effectiveness of a multi-image presentation can better be assured by giving careful attention to these factors:

Multi-image Script/Cue Sheet						
Visuals Number/Description				Sound		
Left	Center	Right	Effect/Time (min/sec)	Narration	Music	Sound Effects

- Treat ideas and concepts one at a time. More than one message, either visual or verbal, divides the viewer's attention, and both messages may lose effectiveness. But do realize that a continuity of meaning must be created as the presentation proceeds.
- Use the screen purposefully to communicate your message by not projecting images on all screens at all times unless pertinent to the message. By limiting the number of images on the screens at any one time, you can give emphasis and draw attention to particularly important visuals.
- Select relevant pictures that directly treat the subject being developed and which are free of extraneous or distracting details.
- Think of pictures on two or more screens much like a multiple-page magazine layout consisting of a series of related images.
- Face people and objects on side screens inward toward the center screen if possible, unless a direction to simulate movement is necessary.
- Compose picture elements and organize picture relationships with greater care than for conventional presentations, taking into consideration that the center screen normally carries the major message and side screens show related pictures. Use this rule flexibly by purposefully changing the center of attention from one screen to another as the subject development may dictate or for reasons of variation.
- Juxtapose images at times in varying sizes and shapes by masking (page 198); a limited amount of mixing different-format slides (square, horizontal, and vertical) can provide visual variety.
- Allow a picture to remain on the screen long enough for the viewer to grasp its impressions or comprehend its message, but not for so long as to become boring.
- Establish a rhythm, but also plan to vary the pace of image changes which can contribute to interest and hold the viewer's attention.
- Select slides that are of good quality, both technically and aesthetically.
- Balance the appearance of simultaneously projected slides for aesthetic purposes by maintaining the same degree of brightness, contrast, and color intensity (unless a different contrast or other elements are part of the impression to be communicated).
- Prepare panoramic scenes across two or three screens by first securing the camera to a tripod, taking two or three pictures of sections of the subject by turning the camera on the *axis of its lens,* and matching the vertical element along the edge of one shot with the same element in the adjacent slide.
- To avoid visible frame edges between adjacent slides in a multi-image panorama, use soft-edge or seamless masks. These are 35mm black-and-white film frames that have been exposed to give precise tones from clear through gray to black shaded edges. A soft-edge mask is mounted along with each of the three panorama slides, shot as described above. The images on the screen will have soft edges and when aligned properly will blend, giving the impression of a single, continuous image across the screens.

Original Slides

Image on Screen Resulting From Soft-Edge Mask

- Use cut pieces of 2×2 inch cardboard as slides to block light to a screen when no image is to be projected from a projector (with some model projectors, leaving the projection slot empty will result in a dark screen); or for pleasant effect, use a solid color slide (for example, made from colored acetate or a piece of unexposed and fully developed color diazo film, as described on page 181, placed in a slide mount) instead of a cardboard blank.
- Select and arrange slides for a sequence by studying them while on a light box or light table (page 203), but ultimately judge their quality and relationships while being projected onto a screen.
- Mark all slides with numbers and screen designations (L, C, R) according to the script or cue sheet. Then, when slides are placed in a tray, draw a felt pen line across the top edge, starting with slide number 1 and progressing diagonally to the last slide in a tray. (See the illustration on page 207.) If, at a later time, you notice a break in the line, one or more slides are either out of correct order, upside down, or backwards.
- If the presentation extends beyond a single tray with any projector, plan to shift to the next tray when a slide, placed in the "0" slot of a full tray, is in the projection position.

Motion on the Screen

Although a multi-image presentation is primarily composed of slides, there are ways to create motion in the program. The most obvious technique is to include a brief 16mm film segment. When the film fills a logically blended interval between slide sequences, it can make a significant impact on the message.

By moving images across screens or from one part of a screen to another part, an effect of motion can be created. The illusion is strengthened by giving careful attention to the time interval rate of slide changes and to dissolve rates.

Animation, which is the illusion of movement created by a series of still images (see page 241), can be accomplished with slides. Careful slide-by-slide storyboard planning (in terms of the number of projectors and images on

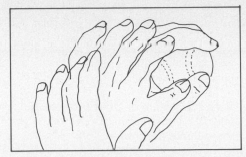

the screens), accurate registration of slides, and sophisticated programing methods can result in a smooth and believable flow of action. Examples of such simulated motion could be an arm or foot motion, a zoom from an overall long shot to close-up detail of an object, and special techniques like flashes, spins, wipes, and freezes.

In creating motion with slides, attention to these factors is of utmost importance:

● Each slide should be shot in a pin-registered camera and placed in pin-registered mounts (see page 202).
● Change of slides, through careful programing in synchronization with the sound track, takes place with split-second timing. (It usually takes a minimum of two seconds to recycle a projector for showing the next slide. However, multiple projectors, used with dissolve units, will allow you to change images more rapidly.)
● Prior to use, projector images must be aligned on each screen so that slide changes will perfectly match.

While these matters need attention in any multi-image program, they are particularly critical when a sophisticated animation technique is to be applied. During your own experiences while developing such programs, you will discover many more useful practices. Also see suggestions in references at the end of the chapter.

PREPARING THE AUDIO COMPONENT

Just as careful attention is given to the audio portion of a regular sound-slide series, similar considerations should be given in developing the more intricate multi-image presentation. Here are some particularly important suggestions:

● Select music to set a mood and a rhythm for the presentation but do not allow it to interfere with or override the narration.
● Visuals may indicate the type of music needed; on the other hand, a piece of music selected as consistent with the topic being treated, may provide the framework for the choice of visuals.
● Use of authentic sound effects gives realism to the presentation.
● Do not overnarrate; let the visuals carry much of the information or message.
● For some purposes, through the use of more than one voice, the presentation can be more dynamic.

(Other suggestions regarding selecting and working with a narrator, sources for music, recording techniques, and mixing sounds onto a single audio track can be found in Chapter 12.)

If images are to be controlled manually by the projectionist, while following the script during the presentation, then a monaural tape recorder or high-output cassette recorder may be used for recording the narration, music, and sound effects.

A preferred method, which allows more flexibility when mixing sounds and is essential when images are to be changed automatically, requires use of one or two *stereo* or *multitrack* reel-to-reel or stereo cassette tape recorders.

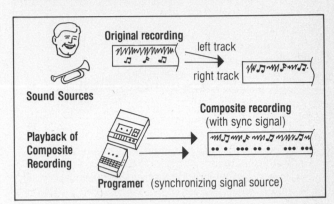

These stereo recorders must be the type which allows recordings to be made separately on each of the two tracks. Record or dub the narration on the first track of one stereo recorder; then music and other sound effects can be transferred from their original recordings, in proper relation to the narration, on the second track. If any errors are made while recording on either track, the recording can be erased and a rerecording made without damage to the other track.

Then the two tracks are mixed onto the first track of the other stereo recorder by connecting the output of the playback stereo recorder to the input of the first track of the second recorder. The second track of the resulting tape remains blank for program cueing signals.

PROGRAMING THE PRESENTATION

In a single-screen presentation, verbal comments and audible or inaudible cue signals are used to pace the pictures correctly with the narration. Review the information on synchronizing slide/tape programs and using dissolve units on page 205. Carefully programing all elements is of even greater importance in the complexities of a multi-image presentation.

The starting point for synchronizing pictures and sound is the storyboard or script sheet. Once it is finalized, it becomes the cue sheet for programing the presentation. There are several practical ways to do this. Here are some suggestions, from simple methods to those that require complex and costly equipment.

Manual Control

If you do not have programing equipment available, or if you want to try out a program before going through the electronic programing procedure, you can make slide changes by hand. Obviously, this is a primitive method in terms of equipment now available. It is limited as to the

217

sophistication level you can attain (only one dissolve rate and no special effects) and is as reliable as is your ability to push slide–change buttons with no errors.

The simplest method of manual control is to hold the slide change controls in your hand (attached directly to the projectors or through dissolve units). Control the image changes manually by pushing the control button for each projector according to the narration or music cues on the script.

A preferred method is to listen to the narration recorded on the first track of a stereo tape recorder, and to verbally record, at proper times, the slide changes onto the second track (call out "left . . . right . . . center . . . right . . . left . . ."). During the presentation the operator wears a headset, listening to the slide change cues on the second track and pushing the appropriate projector control button. The audience only hears the music and narration sound track.

Tone-Control Programer

A programer is used to synchronize the visual and audio portions of a multi-image presentation. In principle, the equipment is similar to that needed for synchronizing a single-projector slide/tape program or a two-projector program with dissolve. But now the programer should be able to handle many more functions. Essentially, it must turn projector lamps on and off at any rate—slow or fast—while advancing or backing up a slide tray, or holding a slide in projection position.

The effects that can be created include:

- **Fade-in/fade-out**—lamp on one projector gradually going off or coming on to either have an image appear on a dark screen or to disappear, leaving the screen blank.
- **Dissolve**—lamp gradually going off on one projector while the lamp comes up on a second projector; thus one image gradually fades out as a second, superimposed one, fades in.
- **Cut**—instantaneous reduction of light from one projector while the light comes fully on from a second one; thus there is an immediate change in images on the screen.
- **Flash**—repeated ON–OFF changes of a lamp while a slide remains in the projection position; this technique is effective for calling attention to the image.
- **Freeze**—projecting two images together on the same screen as in overprinting or superimposing.
- **Wipe**—removal or addition of one or more images or placement of images across two or more screens by

fading or dissolving; thus giving the impression of "wiping off" or "wiping on" the image or images across the screens.
- **Animation**—flashing a series of images in sequential order and possibly recycling them to create the illusion of motion.

By combining a number of these techniques a range of creative effects can be accomplished with a programer. With a **tone-control programer** from six to ten different tones could be placed on an audiotape, in synchronization with the program sound track. Each tone would control a projector or dissolve function. By pairing tones you could cause a slow (using one tone) or a fast (using another tone) dissolve to take place. Programing is done in **real time**—the actual clock time required to run the program (12 minutes, 18 minutes, or whatever time).

The more expensive and complex units enable you to program and control a greater number of projectors and a wide range of special effects. In addition, some programers can be directed to control other on–off switches in a meeting room or spot lights, a curtain, the projection screen, and so forth.

Electronic Microprocessor Programer

With the application of computer technology and the design of microprocessors, very reliable, easy-to-use multi-image programers can encode and then direct projectors and other devices during playback to carry out countless operations. Commands are stored in the microprocessor's memory as digital "bits." Errors are easily corrected by erasing a signal and replacing it with a new one. You can program various combinations of visual effects, see them on the screens, and then select the best one for transferring to the synchronization track of the audiotape containing all sounds.

Popular brands of microprocessors can be programed to generate the command functions for a multi-image presentation. There are also "dedicated" units which are microprocessors designed solely for multi-image use. Their keyboards consist of buttons, each of which serves a specific multi-image function (dissolve, fade, flash, hold, etc.). When a key is depressed, the program cue is stored on audiotape.

Programing can take place in real time, as described with the previous tone-control units, or in **leisure time**. The latter means that you can press the keys for an order and the necessary speed factor (for example—a *dissolve* taking place for *5 seconds*), sit back and think about it or even try it with the slides. Then when you are ready, work on the next command or sequence of commands. The memory of the microprocessor will take this information and properly store it in relation to the prior cues. Then, on playback, the visual effect will take place accurately. It follows the previous cue in real time (for example—the five-second dissolve and then the next command in proper time relation). This way of being able to program at your own leisure pace reduces much of the pressure that is inherent in real-time programing. You will find that it takes about 1 hour to program 1 minute of screen time using the leisure-time method.

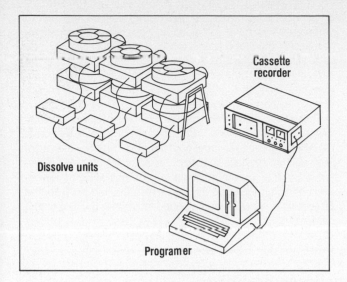

Some microprocessor units can be connected to a video monitor for visual display of the programed cues or to printers for obtaining a paper copy of the commands. With this visual facility, any errors in a program can be located easily and immediately corrected.

PREPARING TO USE THE PRESENTATION

As with any presentation involving instructional media, attention to numerous details must be handled prior to the actual program use. Some of these are enumerated in Chapter 8. With a multi-image presentation the likelihood of problems arising is greatly increased due to the complexity of the electronic equipment and number of projectors required. Give yourself plenty of time for setting up and checking all equipment. Therefore, special attention should be given to the following:

- For the presentation, use a single, wide screen, or sufficiently large screens, all of the same type (matte, lenticular, or beaded) so the images have visual impact for the audience and can contribute to physical and psychological factors relating to realism and personal involvement.
- Make certain to use good projection practice—set projectors 90 degrees to the screen to avoid the keystone effect, align so images will be at the same height, and adjust images either as close as possible side by side without overlapping or with narrow well-defined space (like the black edge of the screens) between them.
- When using slides, obtain similar model projectors and lenses so the brightness of images will be the same.
- If a three-screen, multi-image presentation is to be made, then at least six slide projectors and a single multi-channel dissolver unit or separate dissolvers for each set of projectors on a screen are needed, all activated through a programer with sufficient control capabilities. In addition, a good quality reel or heavy duty audiocassette recorder is required.
- Check the electrical power requirements of all the equip-

ment and be certain that the circuit for the electrical outlets can handle the load. Bring extra extension cords and multiple outlet boxes.
- Make your setup so that either the projectors aim down a center aisle or, preferably, are set on high tables to project over the heads of the audience.
- The biggest fear in a multi-image presentation is the possibility of a projector missing a cue or changing a slide with a random signal, thus losing synchronization. With modern, solid-state equipment this is becoming more unlikely, but you should have a script with you. Recognize that it is very difficult to recue a complex program after it starts without returning to the beginning.
- When you plug equipment into an electrical source, make certain that the programer and projector control units (dissolvers) are hooked into the *same* electrical outlet so that any current fluctuations are the same for all equipment. This can avoid one major cause of miscues.
- Be sure to check yourself and any assistants on the operation of all equipment and rehearse the presentation before the first showing.

TRANSFER TO VIDEO OR FILM

Most multi-image programs are designed for a single, major showing or are used repeatedly in one location. The setup of a multiple projector show is costly and time consuming, thus making the production available for use at a number of different locations is difficult. If there are requests and advantages for using the program with many audiences, it may be advisable to convert it to a single, synchronized medium—videotape or 16mm film.

The disadvantages for doing this are the reduced image size and lower image quality. Also, the 3 to 4 ratio of the film and the television screen is so different from the three-screen format that there will be white or black space at top and bottom of the frame. These can be serious drawbacks if comparison is made with the sharply projected images on large screens. The portability and ease of showing the program must outweigh these limitations.

Transfer Procedure

Set up a matte-finish projection screen or use a flat, white wall surface. The total projected image area of the slides can be about 4 feet wide. Mount the film or video camera on a sturdy tripod, directly behind the projectors, perpendicular to the screen and parallel to the axis of the projectors. Run a test to adjust lens setting and focus. Record color bars and an audio tone (see page 234) on the videotape to set standards for playback.

Then prepare to transfer the sound to the videotape or to the sound track of the film. Connect from output of the tape recorder to input on the video recorder or sound system of the motion picture camera. Check recording level.

Run and record the program. Check the resulting video recording for acceptability. However, film must be sent to a laboratory for processing before it can be viewed. Two po-

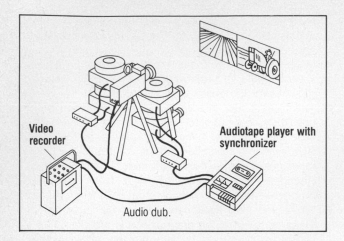

Video recorder

Audiotape player with synchronizer

Audio dub.

tential problems when transferring the slide images to videotape may have to be faced. One relates to potential damage of the video tube if a bright white light is projected to a screen instead of a slide. A second problem is the difficulty in recording a fade-out because the camera's automatic exposure feature attempts to correct what is considered to be an underexposure.

The above described method for transferring a multi-image program to videotape may result in a satisfactory recording. If a higher quality product is desired, there are professional services that offer "aerial image" transfer systems. Such a system consists of a combination of mirrors and prisms against and through which the images from the projectors pass while being aimed into either a video or film camera.

Review What You Have Learned About Multi-Image Presentations:

1. What are two general reasons why a multi-image presentation might be selected for use with a group?
2. List six instructional purposes that such a presentation can serve.
3. Which of these statements are true according to information in this chapter?
   a. Project images on each one of three screens continuously during a presentation.
   b. In planning, relate a multi-image presentation to a multipage magazine layout.
   c. Each separate screen can be used to treat its own idea or concept.
   d. In composition, generally face people on side screens toward the center one.
   e. Major messages most often should be on the center screen.
   f. For a special effect, music can be used to dominate the narration.
   g. Sound effects and two voices are useful audio techniques in a multi-image presentation.
4. If you have a stereo recorder available, what simple method could you use to synchronize a two-screen presentation?
5. Do you have a programer available? If so, what type is it?
6. What unit accepts signals from a programer and controls the illumination changes on a projector?
7. What is the difference between real-time and leisure-time programing?
8. What is the name of a special effect associated with each description?
   a. Turning the projector lamp on and off rapidly while showing the same image.
   b. Projecting two slides on the same screen for a period of time.
   c. Allowing the screen to go black as light intensity in a projector is slowly reduced.
   d. Showing an image move across the screens by dissolving a series of slides
9. When preparing to show a multi-image program, what are five matters that require attention to ensure a successful presentation?

REFERENCES

The Art of Multi-image. Washington, DC: Association for Educational Communications and Technology, 1978.

Fradkin, Bernard. A Review of Multiple Image Presentation Research. Stanford, CA: ERIC Clearinghouse on Informational Resources, 1976.

Goldstein, E. Bruce. "The Perception of Multiple Images." AV Communication Review 23, Spring 1975, pp. 34–68.

Gordon, Roger L. The Art of Multi-Image. Abington, PA: The Association for Multi-Image, 1978.

Images: An Overview of Multi-image Production. Slide/tape presentation V10-12. Rochester, NY: Eastman Kodak Co., 1980.

Images, Images, Images—The Book of Programmed Multi-image Production. Publication S-12. Rochester, NY: Eastman Kodak Co., 1981.

Lewell, John. Multivision. New York: Focal/Hastings House, 1980.

Perrin, Donald G. "A Theory of Multi-image Communication." AV Communication Review 17, Winter 1969, pp. 368–382.

SLIDES TO VIDEO/FILM TRANSFER SERVICE

Aerial Image Transfer Service, 875 Avenue of the Americas, New York, NY 10001

California Communications, 6900 Santa Monica Blvd., Los Angeles, CA 90038

Panorama Productions, 2353 De la Cruz Blvd., Santa Clara, CA 95050

Slide Transfer Service, 74 First Avenue, Atlantic Highlands, NJ 07716

Chapter 19
VIDEO AND FILM

Video recordings and motion picture films share a close relationship in that each one can portray a subject in motion, along with natural or appropriate sounds. Motion and sounds together give the medium its interest, its pacing, and its strongest feature—its sense of continuity or logical progression. In addition, both video recordings and films can present information, describe a process, clarify complex concepts, teach a skill, condense and expand time, and affect an attitude. Although this relationship exists between video and film, each one has certain advantages and limitations in terms of equipment, production, and uses.

ADVANTAGES AND LIMITATIONS

Advantages of Video

- Unlike film, videotape is not sensitive to light, thus loading and unloading tape in the recorder is easier.
- Videotape does not have to be developed, while film must be sent to a film laboratory for processing. Therefore, tape may be viewed right after recording, allowing for immediate evaluation of results.
- Videotape is lower in cost than is film and it can be reused many times.
- Sound and picture are more easily recorded together (in synchronization) on videotape than on motion picture film.
- When editing is completed, a video program, including all pictures and sound, may be ready for immediate use or duplication.
- Duplicate copies of a video recording are convenient to make and are lower in cost than are film copies.

- Convenience of packaging and handling for use are distinct advantages

Advantages of Film

- Film has a higher resolution, better color fidelity, wider exposure latitude, and a greater contrast range than does videotape. Thus film is capable of producing a superior projected picture, especially when greatly enlarged for group viewing.
- The 16mm film format and projectors are standardized, well-established, and universally available, while a videotape prepared in one format cannot be used on a player of a different format.
- Outside of frequent cleaning and occasional adjustments, motion picture cameras and projectors require less maintenance than do the electronic and mechanical features of video cameras and recorders.
- As of today, it is easier and less expensive to accomplish the special techniques of animation and time lapse on film than on videotape.

From the above analysis, it is evident that there are advantages and limitations in both video and film. Each media form might be used for a complete production, or intermixed, depending on production requirements and conditions. As technology continues to improve the electronics of the video medium, and as greater standardization is realized, it is evident that video will increasingly become the major production format. Therefore, in the information that follows, primary attention will be given to the production of video recordings. Reference will be made *only* to motion picture methods for special procedures in production starting on page 240.

221

PLANNING

Because making a video recording seems to be so easy, there may be a tendency to choose a topic, pick up the camera and recorder, and start shooting. This procedure will rarely result in a satisfactory product. Before video recording or filming starts, planning is necessary—whether for treating a limited topic or an extensive subject. Organize the content and plan the method of visualizing the subject to be recorded or filmed. Consider this planning checklist:

- Have you expressed your **ideas** clearly and limited the topic (page 28)?
- Have you stated the **objectives** to be served by your production (page 29)?
- Will your materials be for **motivational, informational,** or **instructional** purposes (page 28)?
- Have you considered the **audience** that will use the material and its characteristics (page 30)?
- Have you prepared a **content outline** (page 32)?
- Have you considered whether a **videotape recording, or film, is an appropriate medium** for accomplishing the objectives and handling the content (page 42)?
- Have you written a **treatment** to help organize the handling of the content (page 48)?
- Have you sketched a **storyboard** to assist with your visualization of the content (page 48)?
- Have you prepared a scene-by-scene **script** as a guide for your shooting (page 50)?
- Have you, if necessary, **selected other people** to assist with the production and required related services (page 30)?
- Have you considered the **specifications** necessary for your production (page 53)?

Following careful planning, the preparation of a video recording requires attention to a number of production steps:

1. Recording scenes on videotape
2. Preparing titles, captions and art work
3. Recording titles, captions and art work on videotape
4. Recording narration, music, and sound effects on audiotape
5. Editing and mixing all above elements to a composite master video recording
6. Making duplicate copies from the master recording
7. Developing any correlated printed and visual materials for use with the video recording

The information in Part III, Chapters 10, 11, and 12, on photography, graphic techniques, and sound recording is basic to the successful preparation of a video recording. As necessary, refer to the page references indicated with the following topics. In addition, give attention to the perception and learning theory principles enumerated in Chapter 2, and the research on production factors summarized in Chapter 3.

VIDEO PRODUCTION EQUIPMENT

Three essential components comprise the videotape recording system:

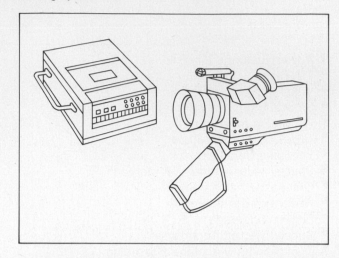

- **Camera**—receives visual images through the lens that are focused on the light-sensitive surface of the image pickup tube (or a microchip), where they are converted into electrical impulses that are sent through a cable to the video recorder.
- **Microphone**—receives sound waves and converts them into electrical impulses that are sent through a cable to the video recorder.
- **Video recorder**—through its recording heads acting on the passing tape, converts electrical impulses into a magnetic force of varying frequency and intensity, resulting in a precise magnetic pattern on the tape surface.

The Camera

Most video cameras in use today transmit images in color. The camera receives visual images through the lens. The images are focused on a light-sensitive surface within the camera. This surface is part of one or three "image-pickup" tubes (or an imaging microchip) which convert the light waves into electrical impulses. These in turn, are sent through a cable to the video recorder.

The camera has a viewfinder through which the camera operator composes and then observes the scene being shot. If this is an electronic viewfinder, on playback of recorded scenes, the action can be observed through this

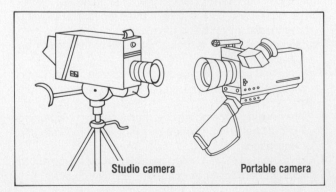

Studio camera   Portable camera

viewfinder. Since the image in the viewfinder commonly is black-and-white, a color monitor attached to the video recorder is necessary for judging color.

In a television studio, large cameras are mounted on pedestals or tripods. But for most shooting, small, compact portable cameras are used. Their relative light weight and flexibility allow recordings to be made at almost any location under normal or low light levels.

A zoom lens is more commonly used with a video camera than is a set of fixed-focal-length lenses. (See the discussion of lenses on page 75.) The zoom allows for flexibility in selecting a view from a range of magnifications without having to move the camera closer to or farther from the subject. The lens should have a zoom ratio of at least 10 to 1, be motorized, and include automatic setting of the f/number with manual override.

Although fixed-focal-length lenses are infrequently used for general shooting, close-up or wide-angle needs can often best be met with them. Especially at these extremes, good quality fixed-focal-length lenses are superior to the great majority of zoom lenses.

Just like film, a video camera reacts to the color temperature of different light sources. (See page 82.) Therefore, it is necessary to adjust for the quality of incoming light since shooting may take place under various light sources or combinations of sunlight, fluorescent lights, and floodlights. This is done by pushing a button or turning a dial to place a filter in the light path behind the lens or to make other camera adjustments. There is an "outdoor/indoor filter" setting. Also, a **white balance** setting must be made which fine tunes the camera for the particular lighting conditions under which the scene will be recorded. This may be done automatically or manually while the camera is aimed at a white card. Each time the camera is turned off or moved to a new location, the white balance should be reset.

The Microphone

The microphone receives sound waves, converting them into electrical impulses that are sent to the video recorder through a cable separate from the one carrying the picture images. With video equipment, sound may be recorded on the videotape at the same time as the picture is being taped. This is known as **synchronized** sound.

Some portable cameras include a built-in microphone and also an external microphone jack for a separate microphone. The built-in microphone is nondirectional and can pick up extraneous sound from outside the picture area. Therefore, it is desirable to use a separate microphone when possible. Refer to the discussion of microphones for audio production on page 144.

Three microphone types are commonly used in video production. A **lavalier,** designed to be hung around a speaker's neck or clipped to clothing, is unidirectional with a cardioid pattern, and picks up little sound other than the speaker's voice. A **fixed** microphone, attached to a stand that is placed on a table or suspended on a boom above the scene, is nondirectional to record remarks from a number of persons grouped around it. A third type is a unidirectional **shotgun** microphone, which is positioned away from the scene and is aimed at the speaker. It is highly directional and its main use is in documentary-type recordings that wish to capture sound in a natural, realistic way.

In addition to these types of microphones, a **wireless** or **radio** microphone may be used when wide or long shots are to be taken or the subject will be moving across a large area and the microphone itself, or its connecting cord, would be visible. This microphone acts as a radio transmitter, broadcasting the audio signal to a receiver placed out of the scene which is attached to an input on the video recorder. A wireless microphone requires careful handling to insure that a proper signal is being consistently recorded.

It is advisable to use an earplug or headphones to monitor sound as it is recorded. Then check the sound through a speaker when the recording is played back.

The Video Recorder

Picture and sound signals, generated at the camera and microphone, are received by the video recorder through its recording heads. The electrical impulses are converted into precise magnetic patterns, representing picture and sound, on the tape surface.

Videotape width is one factor that determines the format of a video recorder. Available widths are 2 inch, 1 inch, $\frac{3}{4}$ inch, $\frac{1}{2}$ inch, and $\frac{1}{4}$ inch or 8mm. Of these, $\frac{3}{4}$ inch and $\frac{1}{2}$ inch are the most widely used sizes for low-budget education and training programs. Commercial broadcasts and high-quality professional recordings require the wider tapes. The $\frac{1}{4}$ inch and 8mm formats can be expected to have extensive uses in the future.

The professional-level 2 inch and 1 inch machines use videotape on **open reels.** In appearance, they resemble a reel-to-reel audio tape recorder. Each of the other sizes is on reels placed in a sealed cassette, thus the designation **videocassette recorder** (abbreviated as **VCR**). There are two $\frac{1}{2}$ inch VCR formats—*Beta* and *VHS*. Older-type $\frac{1}{2}$ inch video recorders are of the open-reel format.

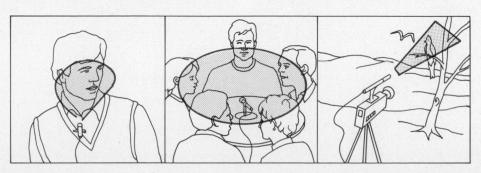

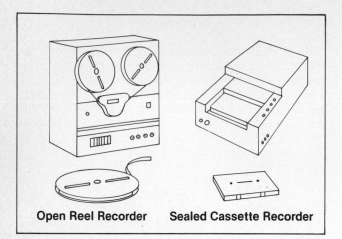

Open Reel Recorder   **Sealed Cassette Recorder**

None of these formats is compatible. A tape can *only* be used with a recorder or playback machine which accepts that particular format tape. Some VCR's include a tape-speed switch so that tapes in the same format may be operated at various speeds. This results in different maximum recording times.

Helical scan or **slant-track** recording is used with both the open-reel and cassette formats. In each one, the tape moves *diagonally* across the video recording head drum. The video picture is recorded as a series of diagonal stripes across the tape, called "slant tracks." In a cassette, one or two **audio tracks** are recorded along the upper edge of the tape after it leaves the drum. Thus, lip synchronous sound, narration, or sound effects can be recorded. Later during playback music, the video playback heads in the drum, acting on the passing tape, convert the magnetic patterns for picture into electrical impulses, sending them through a cable to the television monitor. At the same time, the audio heads convert the magnetic patterns along the upper edge of the tape into electrical signals for sound to the monitor. Another recording head places very precise signals at regular intervals along the lower edge of the tape. This becomes the **control track** to regulate tape speed on playback.

These are the essential controls found on videocassette recorders and players:

- Power ON/OFF switch
- Cassette EJECT switch
- Tape movement controls—PLAY, STOP, PAUSE, REWIND, FAST FORWARD, RECORD

In addition, various models of recorders and players include the following:

- Audio level vu-meters and volume control (automatic and manual) for each track
- Tape counter to help locate a scene on the tape
- Skew or tension control to adjust tightness of tape around the videohead drum to avoid a wavering picture
- Tracking control to adjust video head speed for minor picture distortions when playing back a tape that was recorded on a different make of VCR
- Remote control with a separate hand-operated control unit which allows tape to be played (at various speeds), stopped, rewound, and paused
- Input jacks as connecting points at which camera, microphone, separate audio recorder, or another VCR can be attached
- Output jacks as connecting points at which another VCR, a television receiver or monitor, or an audio speaker can be attached

VCR models in both $\frac{3}{4}$ inch and $\frac{1}{2}$ inch formats, with their accompanying cameras, require a power source that is either 120-volt AC (a wall outlet in the United States), or a battery pack for use at any location. The battery pack permits 30 minutes or more of recording and playback time.

Supplementary Production Equipment

The three items of equipment described above—video camera, microphone, and videocassette recorder—are the

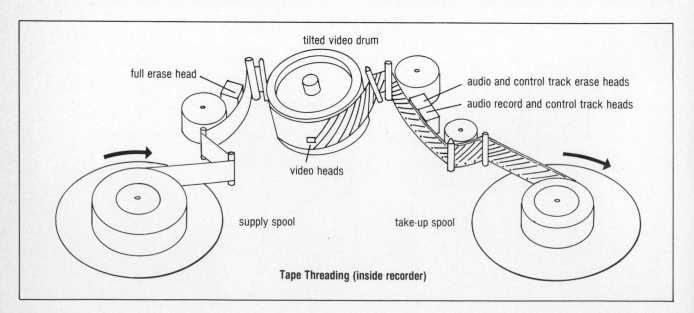

Tape Threading (inside recorder)

basic tools for production. With this equipment you can satisfactorily record scenes. But there are limitations. Scenes must be taped in sequential order. Very likely there will be a discontinuity in the picture—a **glitch**—appearing on the tape between scenes when you stop and then restart the recorder for the next scene. With some cameras, titles can be superimposed and special visual effects may be added at the time of recording. If this cannot be done, titles must be shot as separate scenes or added when the tape is edited (see page 236).

Certain items are of value to supplement the basic production equipment:

- **Video monitor** or **television receiver**: With portable, battery-operated equipment, the black-and-white camera viewfinder serves as the picture monitor when a tape that has just been recorded is replayed. The sound is heard in an earphone. It may be better to use a color video monitor or television receiver for viewing scenes after they are shot. Of the two, a monitor is electronically simpler and offers a superior picture. The video and audio outputs are wired directly into the audio and video circuits of the monitor. A television receiver, which is used in a home, receives television signals through an antenna or cable connection. The electronic signals are converted into video and audio impulses through a process called "demodulation" and then proceed through the circuits of the receiver to become the visible picture with sound.
- **Tripod** or **cart**: For steadiness during use, a camera should be attached to a sturdy support. The video recorder, unless light weight to carry, can be placed on a small cart and rolled from one location to another.
- **Lights:** Floodlights should be available to illuminate indoor scenes. Quartz halogen lamps mounted on heavy-duty stands are preferred. The lights may be used to supplement available light or to provide the main illumination following the key-fill-back-accent lighting pattern. See page 88 for information about lights and lighting procedures.

Maintenance

In order to ensure satisfactory performance from your video equipment, careful handling is necessary.

- Do not aim the camera at the sun or other bright light since this can damage the video tube.
- Turn the camera OFF when it will not be used for a period of time.
- Cap the lens when the camera is not in use.
- Keep the lens clean by dusting with a soft camel's hair brush.

If high-quality tape is used, the videocassette recorder will require little maintenance. Protect it from dirt, dampness, and direct sunlight. Cleaning internal parts and demagnetizing the recording heads should only be done periodically by a qualified technician, or when a malfunction is observed.

SELECTING AND HANDLING VIDEOTAPE

The technical quality of a recording depends to some degree on the characteristics of the videotape that is used. Poor-quality or worn tape may produce a distorted picture. Wrinkled tape can damage recording and playback heads in the recorder. To ensure that tape contributes to a satisfactory recording, keep these facts in mind as tape is chosen and used:

- Most videotape will record both black-and-white and color signals.
- Videotape consists of a magnetic oxide coating on a polyester base. Iron oxide had been the standard material for many years. Now chromium dioxide and other substances, including a high concentration of metal (8mm tape), are used as the coating for high-quality tape. They offer better picture and sound quality and improved image sharpness.
- If small black or white horizontal flashes are visible on the monitor when tape is played back, there has been momentary loss of head-to-tape contact. Such imperfections are called **dropouts** and can be caused by dust or dirt on the tape, flaking of magnetic oxide because of tape wear, a tape manufacturing defect, or dirty or worn heads in the recorder.
- Tape, in its container, stores best when placed in a vertical position.
- Do not expose tape to dust, high humidity, or temperature extremes.
- Store tapes away from TV receivers or other equipment that could generate a magnetic field. This can cause erasure of recorded signals on a videotape.
- Consider the effect of carrying video recordings (also audiotape recordings and undeveloped film) through an X-ray inspection station at an airport. A single passage may not be harmful, but repeated exposure to the X-ray can disturb electronically-created images. Either request hand inspection or place the cassette in a special lead-foil container.

Because of the variety of video recorder formats available, you must carefully choose tape cassettes that will be compatible with the equipment you use. For example, a VHS tape cannot be played on a Beta or 8mm format machine, nor can a $\frac{3}{4}$ inch U-matic tape operate in either a VHS or Beta recorder. When selecting tape, make two important decisions. First, identify the format of the recorder. Second, choose a suitable maximum tape recording time. Playing times may vary from about 30 minutes to 8 hours. Playing time is based on both the amount of tape in the cassette and the speed of tape movement in the recorder. For highest quality picture and sound, record at the fastest speed. Check the instructions with your recorder for the required tape format and acceptable playing time videocassette tapes.

If the program to be recorded will be a long one, have at hand a sufficient number of videocassettes so you can shoot segments of the program on separate cassettes. This procedure will make it quicker to locate scenes when editing.

Review What You Have Learned About Video Equipment and Videotape:

1. Video has replaced film for most media production uses. What *three* major reasons justify this statement?
2. What are the *three* necessary items of equipment for shooting videotape?
3. What *two* settings relative to lighting should be made on a video camera in preparation for using it to record a specific scene?
4. Why is the nondirectional microphone built into the camera not the best one to use?

5. Which videotape sizes are commonly used in education and training programs?
6. Identify these terms:
   a. Videocassette
   b. Helical scan recording
   c. Control track
   d. Glitch
   e. Dropouts

RECORDING THE PICTURE

Planning a video recording is similar to planning other instructional media. But a major difference becomes apparent when you start recording. This difference lies in the word *motion.* Movement is basic to successful video recordings, whether it is in the subject, caused by camera movement when shooting still subjects, or created by editing.

The mere fact that someone or something moves in a scene is not enough to supply this motion. The *ideas* of the recording must move along their planned development. The audience must be given the sense of this kind of motion, even when seeing such stationary things as a mountain or a building on the screen.

Motion can be accomplished by action within a scene, by change of camera angles, by camera movement, by varying the length of scenes, by movement created as you edit and arrange scenes, or by combinations of these elements.

In addition to the concept of motion, you should keep in mind the element of *time,* which is related closely to motion. A photographic print can be studied as long as the viewer wishes to "read" from it the information desired. Slides and filmstrip frames can also be held for any length of time for prolonged examination. But a video scene is displayed only as long as it takes for it to pass the playback head.

Therefore, you must control the period of time the viewer will see a scene. You do this by the length of time the scene is recorded and then projected. You must develop a *feel* for how much screen time is required for the viewer to (1) *recognize* the subject being visualized, and (2) *comprehend* the message or the impression desired.

To shoot and then edit scenes that are of proper length —not too long or too short—requires experience and practice. Some of this can be gained by viewing and studying educational and entertainment films and video recordings. Also watch documentary-type television programs for subject treatment and scene length.

In handling this matter of time, the video medium is unique because it permits you to condense real time by eliminating unnecessary action or extraneous details while the topic being treated remains logical and understandable

to the viewer. A subsequent section on transitions explains how this can be done.

Today most people are visually sophisticated. They have been extensively exposed to the dynamics of television and theatrical films. Their visual orientation has been shaped in great measure by the rapid action of commercials and dramatic shows. Therefore people can accept and even do expect the use of techniques such as showing only essential action with quick scene changes and straight cuts between scenes in place of conventional effects such as dissolves (page 232).

If you want to express your ideas imaginatively with a camera, you are encouraged to use it creatively, but be cautioned that too many unusual effects, like a continually moving camera, out-of-focus shots, double-exposed scenes, and the like, can result in a distracting and boring film. Steady, unobtrusive camera work is often the best policy.

Shots, Scenes, Takes, and Sequences

A video recording is made up of many scenes, shot from different camera angles and put together into sequences to carry the message of the recording.

The terms **shot** and **scene** will be used in this section interchangeably. Each time you start tape moving past the recording head and then stop it, you have recorded a scene or a shot. You can shoot a scene from the script two or more times, calling each one a separate **take** of the particular scene. A **sequence** is a series of related scenes depicting one idea. It corresponds to a paragraph in writing.

When shooting, start the camera just before the action begins (the proper directions are, "Camera start" . . . "Action start") and keep shooting for a few seconds after the action ends unless you are filming to edit in the camera (page 232). In this way you not only make sure of getting the complete action, but have some additional footage needed when editing. With a video camera, the "lead-in" and "lead-out" of a scene should be extra long (at least ten seconds each). This extra time on each side of the main action is mandatory when editing electronically.

Also, because of the "time" factor it may be better to

226

have a scene that is too long than one that is too short and may have to be reshot. Remember that it may take the viewer somewhat longer to identify and comprehend the subject than it does you, because you are very familiar with the subject.

A scene may run from 2 seconds to as long as 30 or more seconds; average scene length is 7 seconds. There is no set rule on scene length; the required action and necessary dialogue or narration must determine effective length. A scene with much detail and activity may require a longer viewing time than a static scene of only general interest. So keep scenes long enough to convey your ideas and present the necessary information, but not so long as to drag and become monotonous.

The manner in which you visually treat the action within scenes, and then relate the scenes, will determine the coherence and effective reality of your video recording.

TYPES OF CAMERA SHOTS

The appearance of a scene can be judged only when viewed through the camera lens. How the camera sees the subject is important, not how the scene appears to the director, to a person in the scene, or to others.

Camera shots are defined according to the way the subject is framed in the viewfinder. This can be done in four ways:

- **Distance**—long, medium, close-up
- **Angle**—high, low
- **Point-of-view**—objective, subjective
- **Camera movement** (or changing lens)—pan, tilt, dolly, zoom

In the section Camera Shots and Picture Composition, in Chapter 10, the first three of these shot groups are explained and illustrated (page 91). Study them carefully. The last category, the moving camera shots, is used when a subject moves and you follow the action. Also, when a subject is too large to be included in a single, set shot, or when you want to visually relate two separate subjects, a moving camera shot is acceptable. These possibilities include:

- **Panning**—a horizontal movement of the camera
- **Tilting**—a vertical movement of the camera
- **Dollying**—a movement of the camera away from or toward the subject (dollying the camera *parallel* to the subject is called *trucking*.)

- **Zooming**—a continuous change in focal length of the camera lens during a scene that simulates the effect of camera movement toward or away from the subject (page 76)

These techniques are generally overused, are often unnecessary, and are frequently poorly done. If a subject in a scene moves, the camera might logically follow the action. This is a good use of camera movement. Or, if a scene is too broad to be caught by the motionless camera, a *pan* (panorama) may show its size and scope. Or you may pan or tilt if it is important to connect two subjects by relating them visually. But do not pan across nonmoving subjects which can be handled satisfactorily by a longer still shot or by two separate scenes.

Closely related to pans and tilts is the use of a zoom lens—*zooming*. Here the same cautions apply about overuse. A series of straight cuts (MS to CU) is often more effective. Save the zoom shot until you feel a real need and one that makes an important contribution to the continuity of your film.

The dolly shot is the most difficult of the moving-camera shots to perform smoothly and effectively. A scene shot as the camera is held in a moving car is one example of a dolly shot. Or the camera, on a tripod, can be attached to some device with wheels—a wagon, an office chair, a grocery cart, or a motorized factory truck—and slowly pushed or pulled in relation to the subject.

When panning, tilting, dollying, or zooming, apply these practices:

- Attach your camera to a tripod, make sure it is level, and adjust the head for smooth movement (a long handle on the head is desirable for good control).
- Always start a moving shot with the camera held still for a few seconds and end the shot in the same way.
- When shooting a moving subject, try to "lead" the subject in the frame slightly (page 95).
- Always rehearse the shot a few times before starting to record.

In summary there is a variety of filming shots at your disposal:

- **Basic shots**—long shot, medium shot, close-up
- **Extremes**—extreme long shot, extreme close-up
- **High-angle** and **low-angle** shots
- **Objective** and **subjective** camera positions
- **Pan, tilt, dolly,** and **zoom** shots

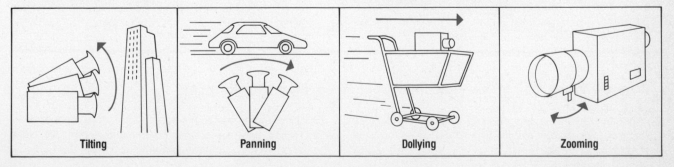

| Tilting | Panning | Dollying | Zooming |

Suggestions as You Plan Your Shots

When there are many scenes in a sequence that include a number of close-ups, *reestablish* the general subject for the viewer with an LS or MS so orientation to the subject is not lost. An establishing shot (LS or MS) is also important when moving to a new activity before close-up detail is shown.

If the viewer has to try and determine where the camera has suddenly shifted or why an unexplained change has occurred in the action, then you have done something wrong. Always plan your scenes to keep the viewer oriented.

For each scene, the camera angle and the selected basic shot (LS, MS, or CU) determine the *viewpoint* and the *area* to be covered. Thus, as you choose camera position and lens you must answer two questions: What is the best viewpoint for effectively showing the action? How much area should be included in the scene?

To help in visualizing the content of key scenes and the placement of the camera in relation to the subject, consider making simple sketches to show the subject position and the area to be covered by the camera. Two examples are shown. These sketches will aid other people (including members of the production crew) to understand what you have in mind. These sketches do not serve the same purposes as the general pictures made from the storyboard

Scene 10 MS Child Reading

Scene 22 CU (High Angle) Inserting Drill

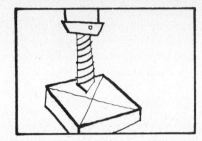

Visualization Sketches in a Shooting Script

during planning (page 48; they apply more specifically to the actual scene being shot.

LS

MS

CU

MS

CU

LS

Review What You Have Learned About Recording the Picture:

1. What are the two concepts that differentiate a video recording or motion picture from most other forms of instructional media?

2. Differentiate the terms *scene, shot, take, sequence.*

3. Relate the use of these techniques—pan, tilt, dolly, zoom—to the following situations:

a. Follow a person climbing stairs
b. Smooth change from an MS to a CU
c. Relate an object on a table to other nearby objects
d. Filming a spread-out farm scene
e. Filming as the camera passes by a number of buildings

4. When preparing to record a scene, how do you best determine whether the shot is set up as an MS or CU?
5. What is a *reestablishing* scene?
6. Label the types of camera shots illustrated by each scene in the sequence that follows.

(a)

(b)

(c)

(d)

PROVIDING CONTINUITY

Selecting the best camera positions, the proper lens, and the correct exposure under the best light conditions may result in good scenes, but these do not guarantee a good video recording. Only when one scene leads logically and easily to the next one do you have the binding ingredient of smooth *continuity*. Continuity is based on the thorough planning and an awareness of a number of factors that must be taken care of during filming.

Matching Action

As you record scenes within the sequence, the subject normally moves. Shoot adjacent and related scenes in such a way that a continuation of the movement is evident from one scene to the next. Such continuation *matches the action* between scenes.

In order to insure a smooth flow from the MS to the CU, match the action at the end of the first scene to the beginning of the second one. Accomplish this matching by having some of the action at the end of the MS repeated at the beginning of the CU.

Then, when editing, from this overlap of action select appropriate frames at which the two scenes should be joined. This transition is most easily accepted by the viewer if the scenes are joined at a point of change—in direction of movement, when something is picked up or shifted, or similar change—rather than in the middle of a smooth movement where the natural flow of action may be disturbed.

In this way "action is matched" and a smooth flow results. As you plan matching shots remember to:

- Change the angle slightly between adjacent shots to avoid a "jump" in screen action.
- Match the tempo of movement from an MS to a CU. Since action in a CU should be somewhat slower than normal or it will appear highly accelerated and disturbing to the viewer, the movement in the previous MS should be slowed down.

MS

CU

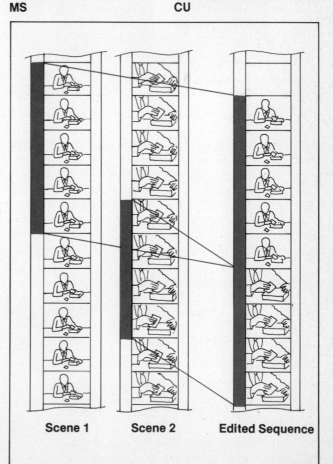
Scene 1   Scene 2   Edited Sequence

● Notice where parts of the subject or objects within the first scene are placed, which hand is used and its position, or how the action moves. Be sure they are the same for the second and any closely following scenes. Be particularly observant if scenes are filmed out of script order.

It is not always essential to match action unless details in adjacent scenes are easily recognizable to the viewer. An LS of general activity (for example, a sports activity scene) does not have to match an MS of a team in action, but when the camera turns from this group to an individual the action should be matched.

For some purposes action can best be matched by using more than one camera, one taking longer shots and the other taking close-ups, the two shooting scenes simultaneously while the action is continuous.

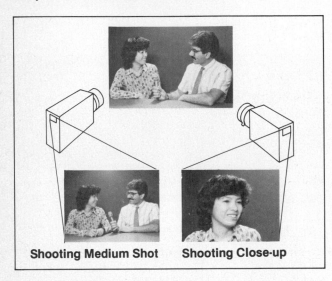

Shooting Medium Shot   **Shooting Close-up**

This **multicamera technique** speeds up recording and ensures accurate matching of action, even though it may require the use of more tape than when shooting with only one camera. Waste can be minimized by starting the second camera on signal just before the first one stops in order to provide the overlap. Multicamera shooting is particularly useful when documenting events like athletic meets, meetings, teleconferences, demonstrations, and news events. Assign camera teams to work together and coordinate their filming to cover related activities at the same time from

positions that will *cut* (edit) smoothly together. Keep a careful record of scenes shot.

Screen Direction

Another kind of matching action is required when a subject moves across the frame. If a person or object moves from left to right, make sure the action is the same in the next scene.

When action leaves the frame on one side, it must enter the next scene from the *opposite* side of the frame for proper continuity.

If directions must change between two scenes, try to show the change in the first scene by having the subject make a turn. If a turn is not possible, include a brief in-between scene of the action coming straight toward the camera. Then a new direction in the following scene seems plausible to the viewer. See the example below.

Be careful when recording parades, races, and similar activities with more than one camera unless someone is assigned to show the action changing direction (going around turns, or moving directly toward or away from the camera). If you do not do this and cameras are located on opposite sides of the activity one will show action from left

to right and another from right to left. The audience will be confused and receive the impression that the moving subject has turned around and is returning to the starting point —and no one will ever get to the finish line!

If screen direction is to be reversed intentionally, include one or more scenes inserted between the two directions that show how the action reversed.

Protection Shots

Careful though you may be, there are times when things go wrong with your planning or filming. Possibly you did not quite match the action between scenes; the left hand instead of the right one was used in the CU, the screen direction between two scenes changes, or you create a *jump* by momentarily stopping the camera during a scene and then starting it again while the action is continuing. Often such situations are not discovered until the editing, and then you may be in trouble. Therefore, by all means protect yourself from possible embarrassment and poor practice.

Protection is afforded by shooting **cut-in** and **cut-away** shots, even though these shots are not indicated in the script. A cut-in is a CU (or ECU) of some part of the scene being shot. It also is called an **insert.** Examples of common cut-ins are faces, hands, feet, parts of objects, and other items known to be in the scene.

A cut-away is opposite to a cut-in in that it is a shot of another subject or separate action taking place at the same time as is the main action. This other subject or action is not in the scene (hence *away*) but is related in some way to the main action. Examples of cut-aways are people or objects that complement the main action; faces of people watching or reacting to the main action are commonly used.

Both cut-ins and cut-aways serve to distract the viewer's attention momentarily and thus permit acceptance of the next scene even though the action may not match the preceding one accurately. By using cut-ins and cut-aways in this way, two scenes can be given continuity even though they would be illogical in direct succession.

Keep these points in mind as you shoot cut-ins and cut-aways:

- Always shoot a number of them, as you may not know until editing whether you will need one or more.
- Make them long enough: a minimum of 10 to 15 seconds, although you may only use 2 or 3 seconds.
- Make them logical, in keeping with the appropriate sequences (watch expressions and backgrounds so they will be consistent with the main action).
- Make them technically consistent with related scenes (lighting, color, exposure).

Transitions

One of the strengths of a video recording is that it allows the use of techniques to take viewers from one place to another or to show action taking place at different times— all this presented in adjacent scenes within a very short period of actual screen time, and accepted quite realistically by the audience. How is this acceptance accomplished? It is achieved through the use of **transitional devices,** which bridge space and time.

The simplest method for achieving smooth pictorial tran-

  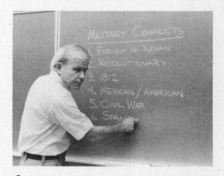

1　　2　　3

Cut-away Shot (reaction shot)

1　　2　　3

Cut-in Shot

231

Transition Scene (title)

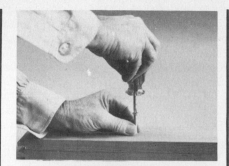

Transition Scene (cut-in)

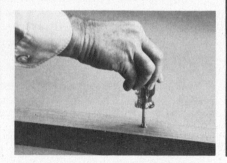

Transition Scene (exit/entrance)

sitions is by use of printed titles or directions placed between scenes (see example above).

Cut-ins and cut-aways can give the impression of time passing and may reduce a long scene to its essentials without the audience realizing that time has been compressed. An example of this may be showing the start of an activity, then a cut-in, then the completion of activity, with the insert scene 2 to 3 seconds in length (see example above).

Exiting from a scene and subsequent entrance at another location creates an acceptable transition in time as well as in space. (Always film the scene for a few seconds after the subject has left the scene to insure an acceptable period of time for the subject to appear in the next scene at a different location.)

A **montage** is a series of short scenes connected by cuts, dissolves, or possibly wipes, used to quickly condense time or distance or to show a series of related activities and places. The audience accepts this rapid series of scenes as if the complete operation or various places had been seen in their entirety.

Visual transitions are common ways of expressing time and space changes. They include fades, dissolves, and wipes.

A **fade-out** and **fade-in** serve to separate major sequences in a recording. They consist of the gradual darkening of a scene to complete blackness (fade-out) followed by the gradual lightening (fade-in) of the first scene of the next sequence. Fades are also used at the beginning of a recording (fade-in) and at the end (fade-out). Use fades sparingly, as too many may produce a disturbing effect which disrupts the flow of the recording.

A **dissolve** commonly indicates lapse in time or a change in location between adjacent scenes. It blends one scene into another by showing the fade-out of the first scene and the *superimposed* fade-in of the next scene. A dissolve is sometimes used to soften the change from one scene to another that would otherwise be abrupt or jarring, due possibly to poor planning or incorrect selection of camera angles in adjacent scenes.

A **wipe** is a visual effect in which a new scene seems to push the previous scene off the screen. The wiping motion may be vertical, horizontal, or angular. This effect is infrequently used but may be important for situations that are closely related, like the start of an experiment followed immediately by the result. A group of scenes connected by wipes may give the impression of looking across a row or a series of objects.

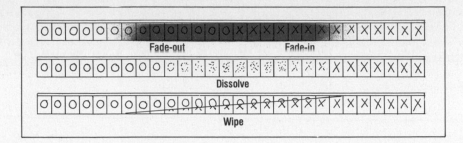

Fade-out   Fade-in

Dissolve

Wipe

When producing a video recording, a **special effects generator** or a **switcher** (which includes a special effects generator) would be used to add visual effects to a tape.

SPECIAL VIDEO PRODUCTION METHODS

There are situations in which you must shoot a program *without* previously preparing a script, or you need to shoot scenes in specific script order *without* any editing. In such instances special methods may be used.

Documentary Recording

As explained on page 56, there may be times when a topic is to be recorded without first writing a detailed script. Your understanding and application of all the aforementioned production techniques relating to camera shots and continuity are essential in documentary recording.

To end up with scenes that can be edited together and result in a coherent product, you must keep in mind what you have shot as well as try to anticipate what may be next. Because you cannot anticipate all the action, you may miss something of importance, or what you expect may not happen. Therefore, always shoot a number of cut-aways and cut-ins to be used as protection shots for insertion between scenes that may not go together naturally.

The editing stage for a documentary recording becomes especially important. Depending on the footage you have, editing may be a straightforward task or require careful decisions that can lead to alternative treatment or interpretation of the subject. See page 58 for suggestions on how to proceed with the editing of a documentary recording.

Editing-in-the-Camera Technique

You may want to prepare your recording without having to remove or rearrange any of the scenes shot. When you shoot one scene after another in sequence, from the first title through the END title, the recording is edited in the camera. As opposed to the documentary approach, this technique requires careful planning and following the script scene by scene. Obviously, when scene locations require shifts from one place to another and back again, shooting in sequence can be very costly in time.

A beginner in video production may believe that, like in motion picture filming, one scene can be recorded, the video camera and videotape stopped, the next scene set up and shot, and so forth through the scenes of the script. The expected result is an edited-in-the-camera tape immediately ready for showing to an audience. When this procedure is used in video, the result may be unacceptable as explained on page 225. Although adjacent scenes may have proper visual relationship and be of proper length, the electronic signal becomes discontinuous, and each scene change is evidenced by a rolling of the picture and other disturbance (called a glitch).

To overcome a glitch between scenes, place the control in PAUSE position instead of in STOP. The continued contact of the recording head with the tape, especially on newer model recorders, will hold the picture and eliminate the image breakup. But the PAUSE control should not be used for more than one minute. It causes wear on the tape and on the video heads; also, it uses up battery time.

In using the editing-in-the-camera technique, each of the following is important:

- You are certain of the sequence of scenes.
- You can control all action to be recorded and know how long each scene will be. Rehearsals are important.
- Titles and illustrations are prepared and ready for recording. You must have the titles set up so you can turn the camera to them for recording in proper order and then go back to the next action scene.
- Light balance and camera movement will be correct for each scene.

Review What You Have Learned About Providing Continuity and Special Video Production Methods:

1. What is "matching action"?
2. Why should a constant screen direction be maintained in a sequence?
3. How can you provide for acceptable direction change?
4. What protection shot might you use in recording a sequence showing a woman sewing a lengthy seam?
5. What transition device would you use in each of these situations:

a. A board being painted and then used after it has dried
b. A man leaving home and arriving at his office
c. A scoreboard showing the score after the first inning and then at the end of the game

6. What purposes are served by visual effects?
7. What are the *three* common visual effects?

8. To what matters should you give particular attention when preparing a documentary recording?
9. Why can you *not* successfully shoot a scene, stop the recorder, then shoot the next scene, and so on? What should you do to effectively shoot all scenes in sequence?

OTHER PRODUCTION FACTORS

The following are important matters for preparing successful video recordings. Give attention to each one at the appropriate time during production.

Video Reference Marks

In order to set a standard for the color quality of the video image record at least 30 seconds of **color bars** on the beginning of the first videotape for a production. This may be done with a setting on the camera. All scenes then appear in color relationship to these color bars. During editing and subsequent use of the videotape, the color bars serve as reference for adjusting players and television receivers for optimum color reproduction as created by the program director or camera operator.

If color bars cannot be created in your camera, shoot a color chart, a commercial television station's color-bar-test pattern from the receiver, or a well-lighted scene that contains a range of colors. Be certain the color temperature and white balance in the scene are all properly adjusted.

In addition to the color bars, an audio tone should be recorded on the tape, also for reference purposes. This can be done with a simple tone generator (usually 400 cycles) connected to the microphone input on the video recorder. It is recorded along with the color bars. The tone permits the proper setting of the audio levels during playback.

Composition

Although composition is a matter of personal aesthetics, there are accepted practices to follow. Note the general suggestions for good composition on page 94.

The proportions of the video frame should always be kept in mind when selecting subjects and composing scenes. Since the frame has the proportions of 3 to 4 (page 107), preference should be given to subjects with horizontal rather than vertical configurations.

Special care must be taken in composing scenes involving motion. Always lead a moving subject with more area in the direction of motion, and carefully rehearse anticipated movement so that proper framing or camera motion can be planned.

When a person handles an object or performs an operation, the camera should be situated for the best view, although it may appear unnatural from the subject's viewpoint. Exaggeration and simplification are acceptable in informational recordings, or in recordings displaying a skill in order to convey a correct understanding or impression.

Lighting Scenes

The light levels necessary for color video cameras, as well as for color film, have been reduced to the point that much recording can take place under available room light levels. When necessary, a general boost in lighting can be obtained by aiming one or more floodlights at a light-colored ceiling.

If an indoor location must be lighted, follow the information on using key, fill, and back lights (page 88). A close-up scene containing one or more faces requires particular care when being lighted. Also, to enhance a sense of depth for the subject, carefully control shadow areas (page 89).

Titles and Illustrations

After the live action has been recorded, and before editing starts, all titles, captions, art work, slides, and other items to be added to the videotape should be prepared. Preparing them at this time helps to ensure that they will be properly placed and effectively worded.

Similar techniques described for preparing titles and illustrations for slides on page 199 can be used now. Review these procedures:

- Word each title so it is brief but communicative
- Keep materials simple, bold, and free from unnecessary details
- Use suitable, clear-cut lettering
- Limit outlines and other lists to four or five lines, each containing no more than three to five words
- Adhere to recognized legibility standards for ease of reading—minimum letter size one twenty-fifth the height of the frame area for video
- Prepare the lettering and art work (page 106), keeping in mind the correct 3 to 4 proportions of the video image area, using simple design features (page 108), and selecting appropriate backgrounds (page 119)
- Then attach each prepared title and illustration to a wall or easel and photograph it with your camera on a tripod (page 96), or use a copy or titling stand
- Record each title or other printed information for at least as long as it takes you to read the material completely twice

For video, carefully consider the placement of content within the format size. Provide for loss of one-sixth the marginal area on each side of the screen. Thus, keep important parts of a visual within the middle two-thirds portion of the screen. This restriction is necessary because of variations in the adjustments of television receivers. For

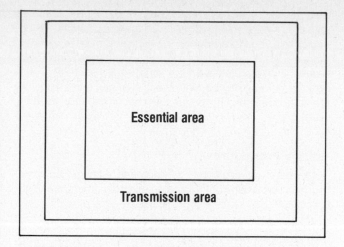

Essential area

Transmission area

planning titles and art work, an outline of the open area of the video frame is printed inside the back cover of this book.

Aside from preparing and shooting titles as described above, a **character generator** can be used. Titles, captions, labels, arrows, and other symbols can be created on a keyboard and transferred, in color, to videotape or added to already-prepared scenes during editing.

Scheduling and Record Keeping

Follow the suggestions on page 55 for preparing a shooting schedule, making arrangements for recording, checking facilities, and keeping a record (a log sheet) of all scenes as shot. To facilitate editing, identify each scene and make notes about the suitability of each "take." When viewing the original footage, the scenes and takes can be identified and selected for final editing.

Be mindful of the need for obtaining permission from those appearing in scenes or from owners of copyrighted materials used. For a sample release form, see page 57.

RECORDING THE SOUND

As part of a video recording, sound can take various forms:

- Synchronous voice or voices of persons appearing in a scene being recorded on the videotape at the same time as the picture is recorded
- Sounds integral to the subject (for example, animal, street, or machinery sounds) recorded on videotape at the same time the picture was recorded
- Sounds integral to or typical of the subject but recorded

on audiotape separate from the picture (may also be obtained from a *sound effects* record)
- Narration that may be recorded by one or more off-camera voices either at the same time the picture is recorded, or preferably on audiotape and added to the picture after the latter has been edited
- Music as an introduction, as background, or between segments of the program, recorded on audiotape and then transferred to the videotape after editing

Any of these sound sources, alone or in combination, may be essential to your video recording. The realism and immediacy so important to the effect of a video program can be attributable to the way the sounds are handled, as well as to the pictures. The basic information in Chapter 12 about recording and mixing sounds applies when creating and handling sound for video. Multitrack audiotape recorders and a mixer are necessary for preparing off-camera recordings before they are transferred to the videotape and combined properly with the pictures.

Here are suggestions when recording sound for a video recording:

- Preferably use a microphone separate from the one built into the video camera. Place it close to the subject for best quality—not over 3 feet away from a person speaking.
- Consider using a shotgun or wireless microphone (page 223) if it is difficult to keep the regular microphone, or its cord, out of the scene.
- If more than one microphone will be used in a scene, attach each one to an input on the audio mixer (page 147), which in turn is connected to the line input on the VCR. Adjust volume levels before recording starts.
- Start recording at least 10 seconds before the sound and action are to start and continue recording for at least 10 seconds after the sound and action are ended. The blank lead-in and lead-out will be necessary during editing.

When complex actions or detailed how-to-do-it sequences are to be recorded and explained, it may be desirable to record the narration or accompanying explanations as scripted on audiotape *before* each scene or sequence is shot. Then play back the sound as each scene is filmed so actions can be performed at appropriate times and at suitable length corresponding to the required narration. This can ensure sufficient action to cover necessary narration and reduce work during editing.

Review What You Have Learned About Other Production Factors and Recording Sound:

1. For what reason are "color bars" put on a videotape at the beginning of a recording period?
2. When preparing titles for a video recording, what *five* procedures should receive attention?
3. What are *four* kinds of sound that may be used on a video recording?
4. Why is it necessary to have blank recorded tape before and after the actual recorded signal?

EDITING VIDEOTAPE

The selection of visuals and sound elements for the script, the choice of camera positions when shooting, the recording of sound, and finally the editing procedure for pictures and sounds make up the main creative aspects of a video production. If your program has been shot in sequence, using the "editing-in-the-camera" technique (page 233), then there is little need for editing. Since this is infrequently the case, you must be prepared to do some picture and sound editing.

Editing is the process of selecting, arranging, and shortening scenes, as well as inserting pictures and sound so that the final result satisfies the objectives you originally established. In order to accomplish this, a person must have a feeling for how to manage the "cinematic" medium. A creative talent is required as well as patience and intuition so that suitable decisions will be made as editing proceeds.

With a staged studio production, editing takes place as the director selects shots and controls their sequential placement on the videotape by means of a switcher. The resulting recording, at the completion of shooting, is the final **master tape** with all scenes, titles, and special effects in correct position.

As opposed to multicamera studio productions, videotape recordings made with a single camera at various locations, or with two cameras shooting independently, are a standard practice. The subject can be anything from a simple how-to-do-it technique at a machine to a complex documentary or drama requiring sophisticated shooting at many locations. Editing the variety of scenes, through electronically copying the desired scenes onto another tape, to form the final master, then follows. This procedure has little effect on footage of the original tape. It is not cut or spliced as is motion picture film.

Preparation for Editing

Editing consists of two tasks. First, decisions concerning content, pace or flow, and continuity must be made. This requires: ordering scenes; selecting takes; determining scene length or the usable portion of a scene; mixing narration, lip synchronization, and other sound elements; and reviewing the need for and placement of titles, captions, and special inserts.

The second task is technical, essentially a copying function. Each scene from the original tape, along with sound elements, is accurately transferred, in proper order, to the final master tape.

Start by viewing the tape or the set of tapes that were shot in order to familiarize yourself with the material. Review the script that was used to guide the shooting, and the log sheets of shots and takes that were prepared during shooting. Note changes that should be made in the script (unscripted scenes that were shot, reordering of scenes, scenes that were combined or eliminated). If you documented an activity without having a script, now is the time to prepare a listing or plan from the footage you view.

Mark on the script the cassette number and the approximate footage location of each scene and which take to use.

Also check where narration may replace synchronous sound, or where sound effects might be added. If the editing equipment you have permits the inclusion of titles, special effects (fades, dissolves, wipes, superimpositions, split frames) indicate places on the script where such titles and effects are to be placed. With these changes and additions, the previous shooting script now becomes your **editing** script.

Editing Procedures

There are various levels of videotape editing. Each one requires certain equipment and an exacting procedure. Reference will be made to three levels of editing.

Pause-edit method

This is the simplest method of editing. It is similar to editing audiotape as you copy from one recorder to another. It employs two VCRs—one to play back the original tape and the other to record scenes into a final program. The method is similar to the editing-in-the-camera technique in which the PAUSE button is used while shooting a sequence of scenes (page 233). It is a simple procedure of stringing scenes together. This is called **assemble editing.**

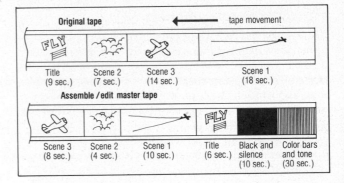

Here is the procedure:

1. Set up a VCR *player* (with monitor) and a VCR *recorder* (with monitor). Attach video and audio *outputs* on the player to video and audio *inputs* on the recorder.

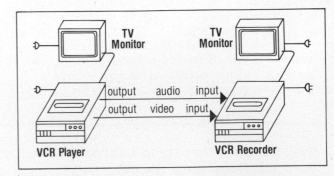

2. Preset the counters on both units to zero (although the counters are not accurate, they will closely indicate location of specific scenes).
3. Put a fresh videocassette, prerecorded with 30 seconds of color bars, in the recorder VCR.

4. Review the original tape on the player VCR, selecting scenes needed and note counter numbers for start and end points of each scene.

5. Dub first scene to recorder VCR. Exactly at the end of the scene put the recorder VCR into PAUSE. Do *not* STOP the recorder.

6. Find the start of the next scene on the player VCR.

7. Rewind the original tape for ten seconds of playing time ahead of the scene's starting point.

8. Start the playback VCR. At the end of the ten seconds (or when the image at the beginning of the scene appears on the monitor) release the PAUSE button on the recorder VCR. Return to PAUSE at the end of the scene.

9. Locate the next scene to be copied and repeat the process.

The pause-edit method is limited as to accuracy in assembling scenes at proper beginning and ending points. If an error is made while transferring, it may be necessary to go back and repeat one or more scenes. But if the equipment is operated carefully, there should be little, if any, breakup of the image between shots. The PAUSE control will stay engaged for about 5 minutes, although after about 1 minute the tape can be damaged. This would be due to the continual rotation of the recording head drum, rubbing against the tape.

The availability of more sophisticated equipment would allow more accuracy and speed in editing, and more flexibility for inserting additional audio and visual elements.

Manual backspace method

This is another assemble-editing procedure as scenes are put together in script order. Two VCRs are also required. One is a regular **playback** machine and the other a special **edit/record** VCR. With the edit/record VCR you can control precisely (without a glitch) where scene 1 ends and scene 2 starts. The procedure requires: (1) that tapes on *both* machines be backed up; (2) that both tapes be played forward; and (3) that, when the edit point is reached, the EDIT button on the edit/record VCR be depressed. This starts the transfer of scene 2 from the playback machine to follow the end of scene 1 on the edit/record VCR.

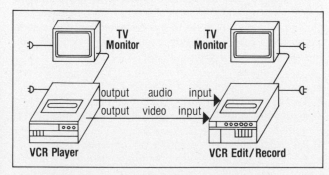

The process takes place in this way:

1. Before starting to edit, attach the video camera to the edit/record VCR and put about 30 seconds of color bars and an audio tone on the front of a fresh videotape. See page 234 for an explanation and the importance of this procedure.

2. Set up a VCR *player* (with monitor) and the *edit/record* VCR (with monitor). Connect video and audio outputs from the player to video and audio inputs on the edit/record VCR.

3. Transfer scene 1 from the player to the fresh tape on the edit/record VCR, following the color bars. End the transfer somewhat after scene 1 ends.

4. Locate the point at which you want scene 1 to end.

5. Find the start of scene 2 on the playback VCR (called the "edit-in" point). Stop the tape at the precise edit-in point.

6. Rewind both tapes the *same* amount (at least 10 seconds, using a stop watch). This will allow both machines to reach speed and stabilize, when playing and recording, before the edit is made.

7. Start both machines simultaneously. Watch the edit/record VCR monitor. At the precise edit-in point, press the EDIT button. The machine automatically changes from PLAY to RECORD mode, recording scene 2 from the playback VCR.

8. Let the tape run past the point at which you plan to make the next edit. Then stop both VCRs.

9. Start with step 4 for making the next edit, and so on.

In this manual backspace method, accuracy depends first on properly cueing the two tapes when backspacing, and then on the reaction time of the person doing the editing to push the EDIT button when performing the edit. If the timing for either action is off, the result will be an improper matching of action from one scene to the next.

Automatic backspace method

This method adds an **editor controller** between the player VCR and the edit/record VCR. The editor controller is a programing device that accepts editing location instructions for selected scenes (edit-in and edit-out points as numbers on the counter and placed in the memory of the controller). The editor controller then automatically has the two VCRs backspace, stop, play, and record to ensure frame-accurate edits.

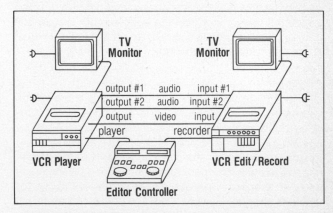

Although available models of editor controllers may have somewhat different operating procedures and control button labels, the procedure for **assemble-mode** editing is:

1. From the player VCR connect audio and video outputs to audio and video inputs on the edit/record VCR. Connect each VCR to a monitor. Connect from the remote

237

control on the player VCR to the player VCR plug on the controller, and from the remote control on the edit/record VCR to the recorder VCR plug on the controller.

2. Use a fresh videocassette with color bars and audio tone already in place as on the master tape in the edit/record VCR.

3. Using the *player* controls on the editor controller, locate the start of scene 1 on the player VCR. Press the PAUSE/STILL button on the controller to freeze the picture on the first frame of the scene. On the numerical keyboard of the controller, enter the counter number for this frame.

4. Using the *edit/record* controls on the editor controller, locate the point on the master tape at which scene 1 should be entered. Press the PAUSE/STILL button on the controller to freeze the picture at this point. Enter the counter number into the controller's memory.

5. If you wish, preview the edit on the edit/record VCR monitor to be certain it appears at the proper spot. If it is wrong, reprogram the edit points.

6. Press the EDIT/RECORD button on the controller to effect the edit. The controller will automatically back-space both tapes on both VCRs the *same* distance, start forward, and at the proper point start the transfer of scene 1 to the master tape. Stop the two machines after the scene ends.

7. Prepare to transfer scene 2 by starting with step 3 above and repeating the procedure.

During this editing procedure, the controller senses the position and movement of each tape according to pulses that were recorded on the **control track** along the lower edge of the videotape. By counting these pulses, which are like sprocket holes on motion-picture film, the controller can back up a tape for a certain distance, such as 300 pulses for 10 seconds. Then the controller starts both tapes playing forward at the same time. The controller will initiate the record function on the edit/record VCR when the last of the 300 pulses is reached. The transfer will continue either for the number of pulses (frames) that were programed into the controller's memory or until the edit/recorder is manually stopped. Such an edit is accurate to within 1 to 3 frames of where you want it to take place.

Insert edit

In addition to assembling scenes in order, there will be times when you want to insert a brief close-up or a cut-away (page 231) to replace part of a scene already on the master tape, or to add narration or other sound in place of synchronous sound. In such instances, you perform an **insert edit** onto the already-recorded master tape. You may transfer a complete scene or either new picture or new sound, *leaving* the other component unchanged. The process requires that the new material be fitted into a certain space on the master tape. In this procedure it is necessary to locate *both edit-in* and *edit-out* points on the master tape. Recall that in assemble-mode editing *only* the edit-in point was necessary. Now you want to be certain not to erase any of the master tape beyond that portion in which the new material will be placed. Once you have located the editing points on both

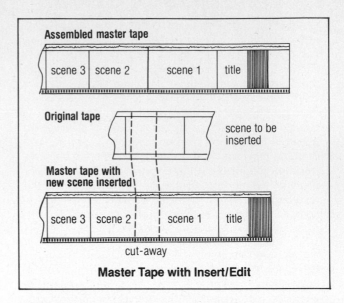

Master Tape with Insert/Edit

the player tape and the edit/record VCR tape, follow the above procedure. Use the appropriate control (RECORD PICTURE or RECORD AUDIO) to transfer the new segment as replacement of what is now on the master tape.

In order to better identify each video frame and for even more accurate editing, a **time code** system may be used. An eight-digit code number (hour, minute, second, and frame number) is recorded on an audio track of the videotape which identifies each frame of the tape. This is done either when the original tape is being shot or is added to the tape prior to editing. A **time-code generator** is used for this purpose. During editing, each VCR relays the code to the editor controller. You select frame numbers for editing and program the controller with them. As described above, the controller backs up both tapes and then carries through the editing operation accurately.

Editing for Special Effects

If you wish to superimpose a title or a symbol over a scene, or create a visual effect like a dissolve or a wipe from one scene to the next, it is necessary to have the two scenes playing and being transferred to the master tape at the same time. This adds to the complexity of the editing process, but can be very effective. To record such effects, you will need this equipment:

- Two player VCRs with monitors
- Two time-base correctors to synchronize scenes and adjust the quality of images when mixing scenes together
- An audio mixer with amplifier and speaker
- A switcher to control which scene or scenes are played
- A special effects generator as part of the switcher to create visual effects
- An edit/record VCR with monitor
- An editor controller capable of handling all equipment.

The two original videotapes containing the scenes to be combined are designated as **roll A** and **roll B.** As before,

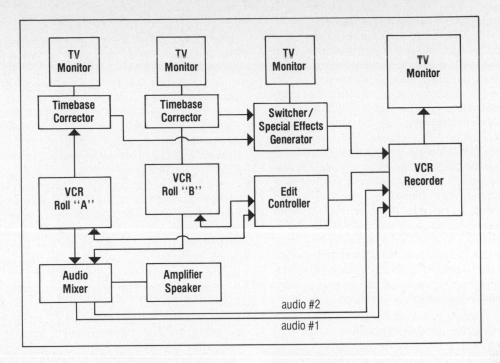

the player VCRs are connected to the edit/record VCR. Any special equipment is in circuit with the editor controller. For example, the **switcher** functions like a "vision" mixer to permit video signals from either one or both A and B tapes to be transferred to the recorder VCR. The **special effects generator** creates the fade-in and fade-out of scenes, the wipe of a scene across the screen, and other effects. The **time-base correctors** stabilize and adjust each scene on the two tapes to achieve consistent, acceptable picture quality. A **character generator,** as another piece of equipment, can be added to the circuit for creating titles and simple graphics which are transferred as video signals to the master tape.

The procedure for A and B roll editing is similar to the programed method using the editor controller described previously. The two player VCRs and the edit/record VCR must all be programed to move together according to time codes. Each edit should be rehearsed and accepted before it is actually made.

In the above editing methods, only generalized proce-

dures have been presented. Check the operational manual with your equipment for specific details and precautions needed when making assemble, insert, and special-effects edits.

Suggestions When Editing

- Avoid interruptions; try to complete a whole sequence during one work period.
- Keep your audience in mind, anticipating its lack of familiarity with the subject and the need for continuity of action, which you might overlook as a consequence of your own knowledge of the subject.
- Be alert to all motion media techniques—LS–MS–CU relationships, establishing scenes, the use of cut-ins and cut-aways, screen direction, matching action, and transitions.
- Use music and sound effects purposefully to create moods and to add a sense of realism to your production.

Now Review What You Have Learned About Editing Videotape:

1. What is the purpose for editing videotape?
2. When scenes are put together in script order, this is known as _____ editing.
3. What equipment is required for each of the *three* editing methods discussed in this section?
   a. pause-edit
   b. manual backspace
   c. automatic backspace
4. Which one of the above editing methods will result in the most accurate editing points?
5. What is meant by *insert* editing? When would the proce-

dure be used?
6. What is the importance of each of the following?
   a. control track
   b. edit-in point
   c. A and B rolls
   d. time code
7. What purpose is served by each of the following items of equipment?
   a. time-base corrector
   b. character generator
   c. special effects generator

SPECIAL FILMING TECHNIQUES

At the beginning of the chapter it was indicated that video recording is the primary method for treating subjects when motion and related purposes require attention. But, although special video equipment may be used, there are certain special techniques (such as time-lapse and animation) that, at present, can more suitably be carried out with a motion-picture camera.

The descriptions that follow are limited to the procedures required for accomplishing specific filming tasks. Fundamentals of motion-picture photography, including selecting film, using a motion-picture camera, editing and splicing film, and combining sound with the picture have been treated adequately in former editions of this book and in other references. Refer to them for assistance.

If the resulting film footage is to become part of a video recording, it can be converted to videotape by a video studio or film laboratory and then inserted into the master tape. Narration also can be added to the videotape.

Compressing and Expanding Time

Because a motion-picture camera can be set to operate at various speeds, it is possible to slow down or speed up normal action. This permits certain subjects to be viewed and studied that otherwise might not be possible.

To record action that normally takes place *too rapidly* for ease of study, use a **slow-motion** technique. Set the camera speed to *exceed* the projection rate—a motion-picture speed of 32, 48, 64 frames per second (fps) instead of 18 or 24 fps. When the film is projected at normal speed, a slowdown effect results because the action on the screen takes a longer period of time than the action took before the camera. Saying this another way, the projected image moves more slowly than the original action. This allows for detailed analysis of movements which in real time would be too rapid for the eye to follow.

Remember to increase the illumination on the subject and compensate for change in exposure on a manual-setting camera when using faster speeds. This becomes necessary because there is a reduction in the amount of light reaching the film, caused by shorter exposure time per frame. Refer to your camera instruction booklet or to a motion-picture handbook for guidance.

To record action that normally takes place *too slowly* for ease of study, use a **rapid-motion** technique. By setting camera speeds *slower* than the projection rate and then projecting the film at normal speed, action is visually compressed or speeded up. Recall how rapidly the people and cars move in "old-time" movies that were filmed at 8 or 16 fps and are now projected at 24 fps.

The slowest setting on most cameras is 8 fps, so you do not have a wide choice of slower than normal filming speeds for increasing action. But the 8 fps rate may be suitable for speeding up some types of action you wish to study or use for a special effect.

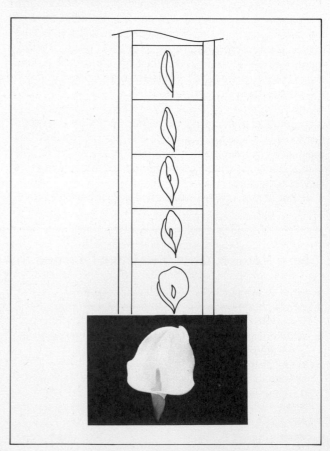

A variation of the rapid-motion technique for accelerating action is **time-lapse** photography. When individual frames are exposed one at a time, this procedure is called **single-framing.** Your camera may be equipped with a special control that permits single framing. Some video equipment includes this feature. In this way you can record subjects having a very slow movement, like the opening of a flower bud, the formation and movement of clouds, and color changes or growths in laboratory experiments. Individual frames are exposed at a predetermined rate, for instance 1 per second or 1 per minute or 1 per hour. Find out how long the action to be filmed normally takes. Then decide how much film time you want to use and determine how often a picture should be taken. Here is a problem:

Purpose: To film a color reaction change in a test tube
Normal time for reaction to take place: 8 hours (480 minutes)
Film time to be used: 10 seconds at 24 fps (240 frames)

$$\frac{480 \text{ minutes}}{240 \text{ frames}} = 2 \text{ minutes per frame}$$

This means that for a period of 8 hours a frame should be exposed every 2 minutes. But first run about 2 seconds of screen time to familiarize the viewer with the subject before the time-lapse action starts.

Result: Eight-hour change is shown in 10 seconds.

Motor-driven commercial equipment is available for such automatically timed single framing. See the references on page 245 for further details and techniques on time-lapse photography.

Filmography

A technique that employs single-framing and relates closely to the next topic, animation, is the filming of still pictures directly onto motion-picture film. The result is called a **filmograph.** Magazine pictures, historical photographs, drawings, and any other flat, still pictures can be placed under the camera on a copystand and the necessary number of frames exposed according to the narration or effect to be created. As previously mentioned, this technique can be used to transfer slide/tape programs to videotape or film.

In filmography, motion can be simulated by moving or zooming the camera into a picture to film a close-up of part of the subject, or by moving out from a detail to a broader view. The picture can also be moved across the lens to create the effect of panning or tilting. In either situation, the camera position, lens focal-length setting, or picture shift should only be changed slightly as each frame is exposed. When this sequence of separate movements is projected continuously, the illusion of motion—inward, backward, or across—results.

One tendency in making a film this way is to shoot the still pictures for too short a length of time—sometimes 2 to 3 frames per scene. This may be too brief a time for the viewer to grasp the meaning of the picture. You may have to experiment with filming various numbers of frames to realize both the understanding and effect you want to create.

Animation

The single-framing procedure can be extended to filming things that normally cannot move to make them appear as if they are moving on the screen. This is **animation.** It resembles time-lapse in that single-framing is used to film sequences and the method to determine movement per frame is similar. But here the sequence consists of successive drawings, slight movement of three-dimensional figures and objects, or progressively developed parts of a diagram. Animation is a fascinating area of motion picture production, but one that takes patience and attention to numerous details.

Animation requires careful planning, particularly storyboarding (page 48). A storyboard helps you establish the movements and time required for each action. Because of the repetition of details (movement, expose, movement, expose, . . .), develop a checklist of steps to follow—action, frames to expose, completion, etc. Keep a careful record (log sheet, page 56) as you shoot animation. Any mistake made when filming may require that you start again from the beginning, so plan and prepare completely and carefully.

Some of the common animation techniques, with examples, are:

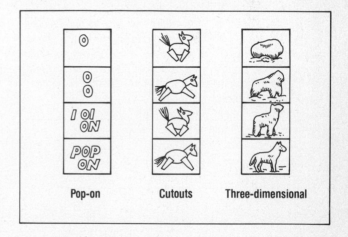

Pop-on   Cutouts   Three-dimensional

- **Pop-on**—Set the first letter of a title in place, shoot three or four frames, add the next letter, shoot frames, and so on as each letter is placed in proper position; draw a line, a segment at a time, shoot three or four frames, add the next portion, and so on; prepare a series of drawings on a process, each of which adds a small detail to the previous one, then film each drawing sequentially, on a few frames.

- **Cutouts**—Figures and objects cut from colored paper or cardboard can be set under the camera, exposed, moved slightly, exposed, and so on; other figures can be constructed with hinged parts so that walking, throwing, and other humanlike motions can be animated.

- **Three-dimensional**—All kinds of objects, including figures and shapes constructed from modeling clay, can be placed on a table and moved slightly between exposures, or altered in shape; or use people to simulate motion by single-framing such things as skating on grass, sliding down a hill, and so on.

241

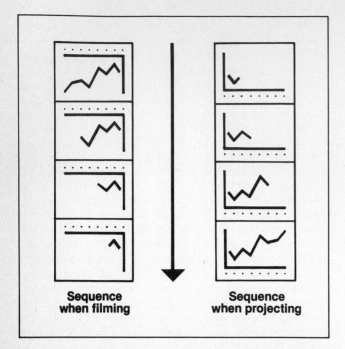

**Sequence
when filming**   **Sequence
when projecting**

- **Wipeoff,** or **scratchoff**—prepare a diagram or drawing on clear acetate with water-based inks or felt pens. Small segments of the diagram are progressively removed as a few frames are exposed. When the film is turned end-for-end and projected, the diagram appears to develop and grow. Use 16mm double-perforated film for sprocket holes to be on the proper side for projection when film is turned.

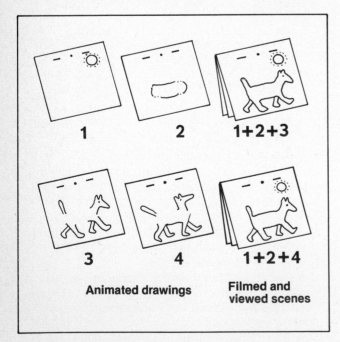

1   **2**   **1+2+3**

3   **4**   **1+2+4**

Animated drawings   **Filmed and
viewed scenes**

- **Cel**—A series of drawings made on clear acetate sheets (called "cels") of figures or objects that can "move" by altering their positions slightly from one drawing to the next; these are placed in careful *registration* under the camera and photographed with appropriate cel changes between frames.

- **Computer animation**—It is now possible to create animation using computer graphics systems similar to those described in Chapter 11 (page 115). Instead of creating a single still image, you design a series of images which vary in such a way that they will give the illusion of motion when projected in rapid sequence. These images are then recorded on motion-picture film in a film recorder or on videotape in a specially designed videotape recorder. Some systems are so sophisticated that if you create the first and last image in an animated sequence, the computer will fill in all of the images which should appear in between. However, this procedure requires an expensive system that is available only in a well-equipped computer animation studio.

As you prepare to try one or more animation techniques, keep these suggestions in mind:

- Necessary camera features for animation work include: through-the-lens viewing and focusing, close-up attachment if lens does not focus close enough, single-frame control, and preferable override on automatic-exposure setting (use a light meter).
- Use a sturdy copystand or other frame to hold the camera immovable in a fixed position. Any slight camera movement will cause subjects to suddenly jump and usually will require that you start filming the whole sequence again.
- Prepare art work and select objects large enough so you can handle them easily and have a large enough field for shooting (minimum of 6×9 inch frame area); see page 106.
- Expose just a few frames per movement for smooth motion. One-frame exposure with very small increments of motion is preferred, but the time and care required become very demanding. Increments greater than four frames can result in erratic and sudden movements.
- Linear movements per exposure should be very slight ($\frac{1}{4}$ inch or so for flat copy work). As with the number of frames to expose, if a segment of movement is too great, the subject will seem to move in a sudden and jerky rather than smooth and continuous manner.
- Determine length of time the animation sequence will be projected and then calculate the number of movements required. For example, if a twelve-letter title is to unscramble and move into position in 20 seconds, then: $20/12 = 1.7$ seconds of movement per letter at 24 fps—40 frames required for each letter; if 2 frames per movement—20 movements required per letter.
- Some actions normally accelerate and decelerate, like a moving ball. More rapid motion can be shown by having distance increments close together at the beginning and gradually increase their size as the subject speeds up.
- Select a time for filming animation when you will not be disturbed or distracted. Keep careful records to check-off each movement, exposure, camera setting, and so on.
- Be sure the art work or subject is placed right side up when you stand in a normal position for filming and sighting through the camera viewfinder.

There are many variations to the process of film animation and other problems you will need to solve if you become a serious film animator. For further discussion of animation, see the references on page 245.

Review What You Have Learned About Special Filming Techniques:

1. Why use a motion-picture camera for filming animation?
2. You wish to show the form of a diver by slowing down action on the diving board. What camera settings would you change from the normal and what are possible new settings?
3. What attachment on a motion-picture camera is essential for *time-lapse?*
4. How often would a frame be exposed if you wished to show the changes in clouds across the sky from 10 A.M. to 4 P.M. and covering 20 seconds of film shot at sound speed (24 fps)?
5. What is a *filmograph?*
6. What type of animation is represented by each of the following?

a. Make a graph using colored tapes for the column bars; with a blade, remove a tiny section of a column bar and shoot a few frames; remove another section and shoot frames; and so on.
b. Place part of a device in position and shoot a few frames; add a new part of the device, shoot frames; and so on.
c. Make sections of a diagram on separate sheets of clear acetate; start with the first section in place and shoot a few frames; add another section to the first one and shoot frames; and so on.
d. Place small models in position; shoot a few frames; move the models to a slightly new position, shoot frames; and so on.

DUPLICATING A VIDEOTAPE RECORDING

Once the master recording is completed, copies may be needed for distribution purposes or because the playback machine to be used requires a tape format different from that of the master ($\frac{3}{4}$ inch to $\frac{1}{2}$ inch VHS). In either case, duplicate copies must be prepared.

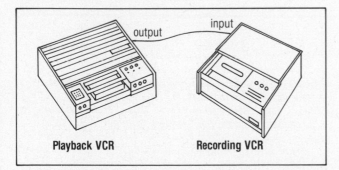

Playback VCR   **Recording VCR**

Duplicating, like editing, requires two VCRs—one to play the master tape and the other to record the duplicate tape. If a format change is required the recording unit should match the appropriate format. The units are connected (from the *output* of the unit with the master tape, set in the playback mode, to an *input* on the duplicate-tape unit, set in the recording mode). If proper care is given to audio and video level settings and machine alignment (tracking control), a good-quality recording can result.

The original camera recording is termed the **first generation** tape. The edited master recording, made from the original camera footage, becomes the **second generation** copy. From the master, one or more **third generation** copies may be made. These latter ones usually will be the tapes to be used with audiences. If the master is to be protected and not played, and if a large number of final copies must

be made, then a third generation copy will serve as the tape from which distribution copies (fourth generation) would be prepared.

With each successive generation, there will be a slight reduction in image quality. This is particularly true when $\frac{1}{2}$ inch format copies are used to prepare further copies.

PREPARING A VIDEODISC

Instead of duplicating videotape for use at various locations, multiple copies of the master tape can be prepared commercially as **videodiscs.** This is financially feasible if over 100 copies will be required and if the content will not need revision for a long time. Any changes will require the preparation of new discs. In appearance, a videodisc is like a long-play ($33\frac{1}{3}$ rpm) audio record. Up to 30 minutes of moving pictures with sound, or 54,000 individual frames, or combinations of the two, can be placed on each side of the disc.

The image and sound are read by a laser stylus in a videodisc player. Picture and sound quality will be outstanding if the video master is one- or two-inch tape. Videodisc "mastering" and disc reproduction can only be done at a few videodisc service centers. There are detailed specifications and requirements for preparing the premaster videotape that is to be used for videodisc duplication. See references for videodisc service centers on page 245.

There is a growing use of both videotape and videodisc in conjunction with microcomputers. This is called **interactive video** instruction. See the discussion in Chapter 20 (page 255) for information on this technology.

PREPARING TO USE YOUR VIDEO RECORDING

If the videotape is to be shown to a group, either video monitors or a video projector can be used. The placement and adjustment of the equipment are essential. Give attention to these matters:

● Provide a sufficient number of monitors so no member of the audience will be seated beyond the anticipated

legibility limit of words and numbers on the screen—generally 8 feet distance for a 21 inch picture tube or 30 feet for a 4 foot screen.

● Make sure each person in the audience has a clear, undistorted view of the monitor or screen and can hear the sound. This may require that monitors be placed on tall stands and a video projector with its screen on a stage.

● Control stray light that may fall on the monitor or screen by closing window drapes. If necessary, turn the monitor slightly so that any stray light that cannot be controlled will be reflected away from viewers.

● Maintain a low light level in the room during projection.

● If someone other than you will operate the video player, meet and instruct the person as to your plan for use.

● Adjust all sets or the projector controls to obtain a good-quality picture (steadiness, color, contrast) and an adequate sound level. Cue the tape to its beginning. Then turn off the player (leave the monitors on) until ready for use.

Review What You Have Learned About Duplicating and Using Videotape:

1. For what reasons should a copy of a videotape be used from a generation that is closest to the master recording?
2. If it were necessary to prepare 200 copies of a videotape for distribution and use within an organization, in what form might it be best to make the duplicates? What is

one requirement for the videotape from which the copies would be made?
3. On what principle do videodiscs operate?
4. What is one major advantage for using a video projector over a video monitor when showing a videotape to a large group?

REFERENCES

Video Equipment and Maintenance

Kerr, Robert J. "Selecting a Video Recorder. Part 1." *Audio Visual Directions* 4 (March/April 1982):30–35.
———— "Selecting a Video Recorder. Part 2." *Audio Visual Directions* 4 (May 1982):52–56.
Noronha, Shonan. "Non-Technical Maintenance Procedures for Video Professionals." *Photomethods* 25 (August 1982):15–16, 57.
Straub, Ken. "The Focus is on Camera: How To Select One for Video." *Audio Visual Directions* 3 (August/September 1981): 29–32.
Sullivan, Sam L., and Long, Hugh. *Video Fundamentals: An Operational and Minor Maintenance Manual for Non-Technical Personnel.* Huntsville, TX: KBS, Inc., P.O. 2812 SHSU.

Video Recording

Basic Fundamentals of Video. Series of Eight Instructional Videotapes on Production and Other Aspects of Video. Compton, CA: Sony Video Communications, 1979.
Bluck, John. "Production Tips for One-Camera Shoots." *Educational and Industrial Television* 13 (December 1981):60–65.
Fuller, Barry, et al. *Single-Camera Video Production Handbook.* Englewood Cliffs, NJ: Prentice-Hall, 1982.

Gayeski, Diane. *Corporate and Instructional Video Design and Production.* Englewood Cliffs, NJ: Prentice-Hall, 1983.
LeBaron, John. *Making Television: A Video Production Guide for Teachers.* Totowa, NJ: Teachers College Press, 1981.
LeBaron, John, and Miller, Philip. *Portable Video: A Production Guide for Young People.* Englewood Cliffs, NJ: Prentice-Hall, 1982.
Utz, Peter. *Video User's Handbook.* 2nd ed. Englewood Cliffs, NJ: Prentice-Hall, 1982.

Video Editing

Browne, Steven E. *The Videotape Post-Production Primer.* Burbank, CA: Wilton Place Communications, 1982.
Shetter, Michael. "Primer on Computer-Assisted Videotape Editing. Part 1—The Architecture of the Editing System." *Audio Visual Directions* 4 (January 1982):14–16.
———— "Part 2—Time Code." *Audio Visual Directions* 4 (February 1982):35–37.
———— "Part 3—Basic Editing." *Audio Visual Directions* 4 (March/April 1982):21–27.
———— "Part 4—Making the First Cut." *Audio Visual Directions* 4 (May 1982):28–32.
———— "Part 5—Getting Fancy." *Audio Visual Directions* 4 (June 1982):11–16.

———— "Part 6—Sound, the Forgotten Power." *Audio Visual Directions* 4 (July/August 1982):9–11.

———— "Part 7—More on Sound." *Audio Visual Directions* 4 (September 1982):39–41.

———— "Part 8—List Management." *Audio Visual Directions* 4 (October 1982):29–33.

———— *Videotape Editing: A Guide to Communicating with Pictures and Sound.* Elk Grove Village, IL: Swiderski Electronics, Inc, 1200 Greenleaf Avenue, 1982.

Utz, Peter, and Sauder, Joseph. "Not All Edits are Created Equal." *Audio Visual Directions* 5 (May 1983):31–35.

Videodiscs (and service)

Crowell, Peter. "The Disc Program Development Team." *Educational and Industrial Television* 14 (June 1982):45–49.

News, Videodisc Design/Production Group, KUON-TV/University of Nebraska—Lincoln, P.O. Box 83111, Lincoln, NE 68501.

Optical Recording Project/3M, 223-5S 3M Center, St. Paul, MN 55144.

Motion Picture Techniques

Cheshire, David. *The Book of Movie Photography.* New York: Knopf, 1979.

Kemp, Jerrold E. *Planning and Producing Audio Visual Materials.* 4th ed., Chapter 23. New York: Harper & Row, 1980.

Lefkowitz, Lester. "Time-Lapse Photography." *Industrial Photography* 31 (May 1982):29–33.

"Time-Lapse Photography: Putting the Squeeze on Reality." *Audiovisual Notes from Kodak.* Publication T-91-8-1. Rochester, NY: Eastman Kodak Co., 1978.

Whitaker, Harold, and Halas, John. *Timing for Animation.* New York: Focal, 1981.

The World of Animation. Publication S-35L. Rochester, NY: Eastman Kodak Co., 1983.

Video Journals

Audio Visual Directions, Montage Publishing, 5137 Overland Ave., Culver City, CA 90230.

Educational and Industrial Television, C.S. Tepfer Publishing Co., 51 Sugar Hollow Rd., Danbury, CT 06810.

Industrial Cine/Video, United Business Publications, Inc. 475 Park Avenue South, New York, NY 10016.

Videodisc News, Videodisc Services, Inc., P.O. Box 6302, Arlington, VA 22206.

Videography, United Business Publications Inc. (See above.)

Video News, Phillips Publishing, 7315 Wisconsin Ave., Bethesda, MD 20814.

Video User, Knowledge Industry Publications, Inc. 701 Westchester Ave., White Plains, NY 10604.

Chapter 20

COMPUTER-BASED INSTRUCTION

- Uses for CBI
- Planning for CBI
- CBI Equipment
- Designing CBI
- Writing the Program
- Graphics for CBI
- Completing the Process
- Computer-Controlled Interactive Video
- Preparing to Use CBI

The development of computer technology has led to the emergence of computer-based instruction (CBI). This method of teaching applies many of the principles from learning theory described in Chapter 2 (page 14) to ensure that satisfactory learning will take place. As students study, each one "interacts" with the material being presented on the computer screen, often by responding to questions and making choices in order to proceed.

USES FOR CBI

The ways in which the computer can be utilized in the instructional process can be categorized as follows: (See illustrations on facing page.)

Drill and Practice

The easiest and most common CBI task is to provide practice for reinforcement of a concept or skill. The computer is programed to provide the learner with a series of questions or exercises typical of those found in a workbook. The practice exercises might include working simple math problems, estimating the size of an angle in degrees, or identifying geometric shapes. An example of the latter type is shown below and represents the simplest form of drill and practice. In this sequence a problem is posed, a response is solicited, the response is judged, and feedback is given before the next problem is posed. More elaborate programs will begin with questions or a pretest to assess the entry level of the learner and then use that information to provide practice at the most appropriate level of complexity.

Some programs maintain a record of student responses which are reported to the student or instructor at the end of

the exercise. The record of performance serves as a basis for prescribing additional instruction.

Tutorials

Tutorials attempt to emulate a human tutor. Instruction is provided via text or graphics on the screen. At appropriate points a question or problem is posed. If the student's response is correct, the computer moves on to the next block of instruction. If the response is incorrect the computer may recycle to the previous instruction or move to one of several sets of remedial instruction, depending upon the nature of the error.

Simulations

Programs which attempt to emulate dynamic processes are called **simulations.** In some of the common examples, students use microcomputers to simulate flying an airplane, running a small business, or manipulating the controls of a nuclear power generator to prevent a meltdown. Such programs offer the opportunity to experience "real world" problems without the associated risks such as a fatal crash, a bankruptcy, or a nuclear disaster. In addition, a well designed simulation can radically reduce the costs of instruction, while expanding or compressing the timeline to one more appropriate for learning.

Simulations are one of the most powerful instructional applications of computers, but they are also the most difficult to achieve. To be effective teaching tools, they must accurately reflect the processes they model.

Games

If properly designed, computer games can use the learn-

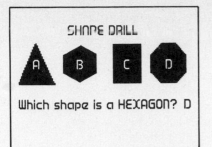

SHAPE DRILL

Which shape is a HEXAGON? D

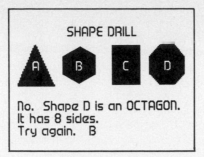

SHAPE DRILL

No. Shape D is an OCTAGON.
It has 8 sides.
Try again.   B

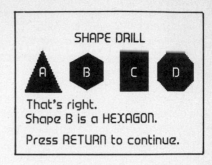

SHAPE DRILL

That's right.
Shape B is a HEXAGON.

Press RETURN to continue.

Drill and Practice

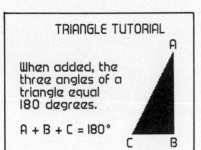

TRIANGLE TUTORIAL

When added, the
three angles of a
triangle equal
180 degrees.

$A + B + C = 180°$

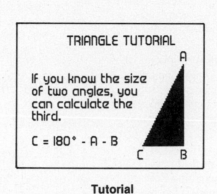

TRIANGLE TUTORIAL

If you know the size
of two angles, you
can calculate the
third.

$C = 180° - A - B$

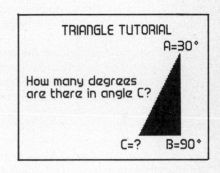

TRIANGLE TUTORIAL

$A=30°$

How many degrees
are there in angle C?

$C=?$ $B=90°$

Tutorial

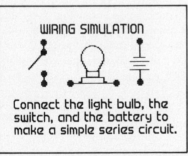

WIRING SIMULATION

Connect the light bulb, the
switch, and the battery to
make a simple series circuit.

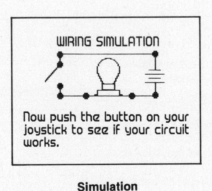

WIRING SIMULATION

Now push the button on your
joystick to see if your circuit
works.

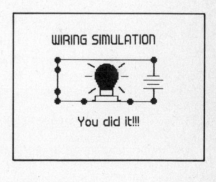

WIRING SIMULATION

You did it!!!

Simulation

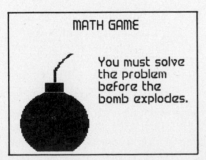

MATH GAME

You must solve
the problem
before the
bomb explodes.

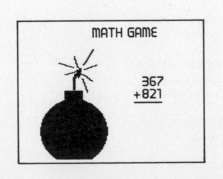

MATH GAME

367
+821
——

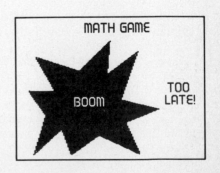

MATH GAME

BOOM

TOO
LATE!

Game

247

ers' competitive nature to motivate and to increase learning. One of the more successful educational games combines video arcade action with typing exercises to improve typing skills. The user must rapidly and correctly type the words that appear in the corners of the screen in order to keep from being zapped by the ever-menacing alien ships.

As with simulations, good instructional games are difficult to design and designers must ensure that the integrity of the learning objectives is not lost in the attempt to provide a gaming atmosphere.

Controlling Other Media

Computers are excellent devices for controlling the display of other media. Microcomputers are used extensively to operate banks of slide projectors and related equipment for multi-image presentations (see page 218). With a computer program 35mm slides or video images can be selected and randomly accessed.

Interactive video combines the realism and impact of video images with the interactive nature of the computer. For example, an interactive video tutorial which describes how to modify a piece of machinery could begin by using a CBI tutorial to provide background information and assess the previous knowledge of the learner. At the appropriate times, still pictures could be displayed on a video screen and overlaid with computer-generated labels to identify critical components. A video clip would demonstrate how the learner should handle the equipment and make final adjustments. The computer could then present questions to assess the learning which has occurred and review the concepts which the learner has not mastered.

PLANNING FOR CBI

The programing for CBI, like other media, requires systematic planning and preparation. Before developing a CBI package, consider this planning checklist:

- Have you clearly expressed **your idea** and limited your topic (page 28)?
- Will your program be for **motivational, informational,** or **instructional** purposes (page 28)?
- Have you stated the **objectives** that should be served by your program (page 29)?
- Have you considered the characteristics of the **audience** which will use the materials (page 30)?
- Have you prepared a **content outline** (page 32)?
- Have you considered whether **CBI is the most appropriate medium** for accomplishing the objective and handling the content (page 42)?
- Have you developed a **treatment** and a **storyboard** to help organize and visualize the content (page 48)?
- Have you developed a **script** which indicates how the materials will be presented (page 50)?
- Have you, if necessary, **selected other people** to assist with the preparation of the materials (page 30)?

In addition, the development of computer-based instruction requires the following steps:

1. Assembling the equipment
2. Designing and flow charting the sequence of instruction
3. Developing the questions to be used for review and branching
4. Planning the images on the screen
5. Writing the computer program
6. Developing the graphic images and sound effects
7. Preparing accompanying printed materials
8. Testing and revising the program

If you plan to use the computer to control other media such as a video player or slide projector you will also need to include two additional steps:

9. Procuring the appropriate interconnection device
10. Writing the computer instructions which control the additional equipment

CBI EQUIPMENT

In order to develop computer-based instruction you will need the following basic equipment:

- A **microcomputer** or a **terminal** which is connected to a mainframe computer.
- A **cathode ray tube** (CRT) which displays the image.
- A **data storage system** such as a disc drive.

A description of these and other computer equipment can be found in the section on computer hardware in Chapter 11 (page 115).

Most CBI systems are microcomputer based. Typically a single learner, or small group of learners, interacts with a single computer. The entire capability of the computer is focused on the teaching task, with only one instructional dialog occuring at a time.

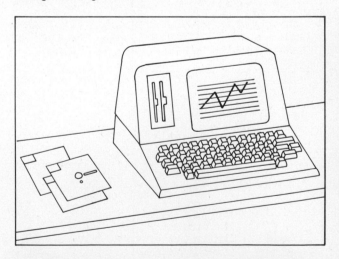

On the other hand, large mainframe computers can support several simultaneous interactions. Many stations (terminals) can be connected to a large computer which stores a large number of CBI programs. Several users, each working at a terminal, can simultaneously interact with those programs and be involved in very different instructional activities.

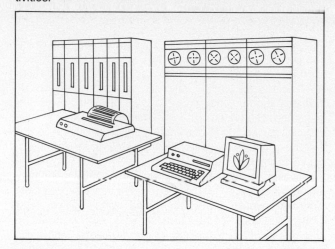

DESIGNING CBI

Most instructional media contain information or instruction, presented in a fixed sequence, which will be the same for all learners. This is called a **linear program** and does not allow for individual differences among learners.

A computer-based instructional program may also be linear as it advances according to a predetermined sequence. But more often, the program is controlled by the learner as a choice is made or an option is selected in response to a question. This leads to an appropriate following sequence of instruction. A number of options may be available at any given time. This is called a **branching program** and can result in much more individualized learning.

This branching capability offers a new degree of complexity and your planning should clearly reflect how you

intend to use this characteristic of CBI. The development of a flow chart can assist in this process.

Flow Charting

The purpose of a flow chart is to illustrate the various sequences of steps that a learner might follow in a CBI lesson. Fairly complex flow charts can be constructed with a few symbols. Arrows indicate the direction of movement through the flow chart. The arrows should point down or to the right, but may point in other directions when required branching is back to an earlier part of the program.

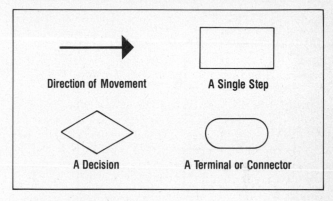

The rectangle represents a single step in the process, such as the presentation of a single frame of information or the manipulation of data. A rectangle usually has one line flowing into it (input) and one line flowing out (output).

The diamond indicates a decision point at which a branch will occur. For example, the computer may ask learners if they wish to continue with the lesson, then either proceed to the next block of instruction or stop, depending upon the response. Or the computer may judge the accuracy of a learner's response to a question and then offer new instruc-

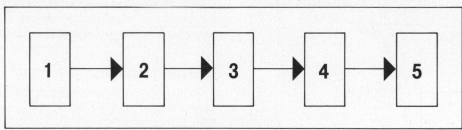

Linear Program

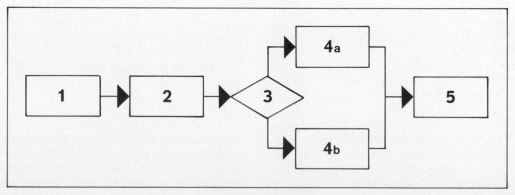

Branching Program

tion or recycle through previous instruction. A diamond usually has only one input and two or three outputs.

The oval is used to indicate a terminal or connecting point. It marks the beginning and end of a flow chart and indicates where it connects with other flow charts or reconnects with itself at another point. Terminals usually have only one input or one output.

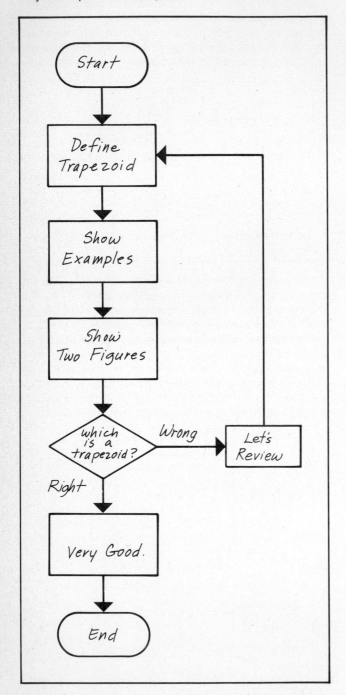

Develop a flow chart for CBI by first listing all of the steps in the instructional process. Then identify those steps which are decision points and will result in branches. Draw the steps (boxes) and the decision points (diamonds) in sequence on a sheet of paper and use lines with arrows to show how the program flows. The process will help you get

a much clearer understanding of how the program should be constructed.

Questions

The branching capability of CBI is a teaching tool which can permit learners to tailor the instruction to their own specific needs. A well-designed computer-based instructional activity will provide options which let the instructor or individual learner determine the level of difficulty and the sequence of instruction. Furthermore, the student's responses can be used to continually shape the instruction so that the unmastered content is stressed, rather than treating all content as equal.

However, these capabilities are only available when they are designed into a CBI program. As the designer, you must determine where branches should occur. Since most branches result from questions, those questions should be carefully selected to ensure that they clearly reflect the learning objectives.

Questions can be stated in a variety of forms. However, some forms of questions are much easier for a computer to score. Multiple choice questions are the easiest to judge since the learner must select one of a few predetermined answers.

On the other hand, fill-in-the-blank questions are much more difficult to score since the correct answer can often be stated in a variety of equally acceptable ways. The developer who designs CBI questions of this type must specify *all* acceptable answers in advance. This is complicated by the fact that a student might know the right answer, but be incapable of spelling it correctly. Only the more sophisticated computer-based instructional systems are capable of correctly judging misspelled responses to open-ended questions.

Whatever form is used, your questions should be clearly stated and simple in structure so they will be understood by the learner. In addition they should indicate the type of response required (true-false, short-answer, or whatever).

Branching Patterns

Computer-based instruction can be designed to branch in any number of ways. However, five types of branching patterns are common.

Linear format with repetition

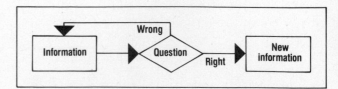

This is very similar to standard linear programing, except that questions are inserted as steps in the program between segments of information. If a learner answers a question correctly, he or she proceeds to the next segment. If not, the computer has the learner review the previous segment and asks the same question again.

Pretest and skip format

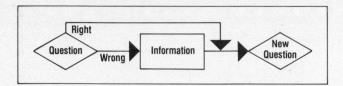

In this case the learner is given a pretest prior to any instruction. The program then skips over any instructional segments which the learner has previously mastered. As a result the instruction becomes much more efficient for the individual.

Single remedial branch

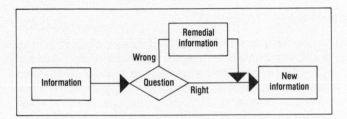

In this case the learner is provided with some information and then asked a question. If the answer is correct, positive feedback is received and the learner advances to the next segment. All incorrect answers receive the same remedial instruction.

Multiple remedial branches

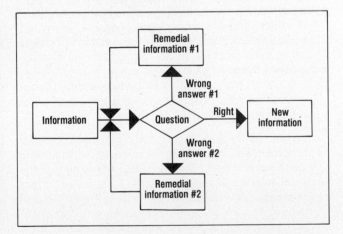

This is similar to the previous type except that each incorrect response results in a different type of remedial instruction. In this situation the learner is often allowed to continue selecting solutions until the correct answer is obtained. In some programs a predetermined number of incorrect responses, with remedial feedback, is allowed before the correct answer is revealed by the computer, and the instruction moves to new information.

Compound branches

Patterns of this type are commonly used in diagnostic or problem-solving situations where the questions can be answered with a simple yes or no. Each branch leads to a new question until a final solution is reached. Such programs can have very different results depending upon the answer to a question at a branching point. (See example on the following page.)

Each of these five patterns might be used separately and repeated with progressive content sequences, examples, questions, and answers to form a complete instructional unit. Or a variety of patterns can be combined into a single unit as required to meet the specific instructional objectives of that unit.

Screen Image

The design of the screen image which appears on the CRT can affect the learners' response to CBI and should be well planned. The screen should *not* be thought of as a printed page to be filled with sentences. Information can be presented in any arrangement on the screen and it can be a dynamic display. This means that diagrams, words, phrases, or sentences can appear instantaneously, slowly, or in rapid succession, and placed anywhere on the screen. The element of time can be used to provide pauses for emphasis or to give suitable time for a person to read the message and then react.

A good approach is to develop a sketch for each screen display in much the same way you would design a storyboard for a slide presentation (page 48). You may find it useful to use a coding form such as that shown below. Such a form lets you specify the exact location of each character on the screen and provides you with additional information to simplify the programing process.

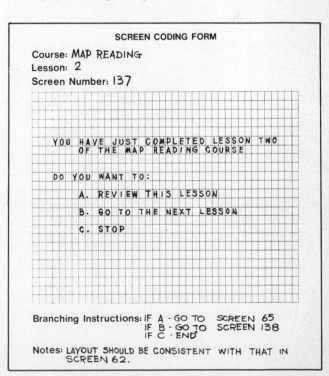

SCREEN CODING FORM

Course: MAP READING
Lesson: 2
Screen Number: 137

YOU HAVE JUST COMPLETED LESSON TWO OF THE MAP READING COURSE

DO YOU WANT TO:

A. REVIEW THIS LESSON

B. GO TO THE NEXT LESSON

C. STOP

Branching Instructions: IF A - GO TO SCREEN 65
IF B - GO TO SCREEN 138
IF C - END

Notes: LAYOUT SHOULD BE CONSISTENT WITH THAT IN SCREEN 62.

251

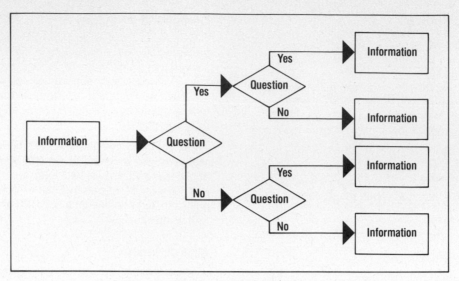

Compound Branches

Ideally, each frame (screen image) consists of a single concept or idea and the learner should control the rate at which each one is presented. This is usually done by pressing the RETURN key or space bar when ready for the next frame. Some computers utilize a **touch-screen** technique in which the user touches a finger or a **light pen** to a place or an item on the screen. This action directs the computer to respond accordingly. The sensitive screen consists of a special membrane or a touch panel, one type of which emits narrow beams of infrared light in a crisscross pattern in front of the screen. These invisible beams form a grid that acts like a computer keyboard when touched. In touching the screen, the user actually breaks the beams at X, Y coordinates, thus directing the computer to access information.

When writing the text for a screen, use short sentences and include only that information which is required to meet the learning objectives. Be certain to maintain the flow from one screen image to the next, and consider the continuity for all possible branches of the program.

The same graphic principles of design, legibility, and so forth, as discussed in Chapter 11, and for printed media in Chapter 13, apply to computer images in CBI. Due to the low resolution of computer systems, you must pay particular attention to legibility and should design illustrations which will appear simple to the user, thus to better ensure they are clearly understood.

Special attention should be given to the following (see reference by Heines):

- **Readability**—Use san serif lettering of medium weight and suitable upper and lower case size; limit line length; justifying lines (right margin alignment) not necessary; break text at natural phrasing points.
- **Space**—Blank space helps readability; group items carefully and for aesthetic purposes.
- **Color**—Use to highlight and give emphasis; avoid "hot" colors that pulsate on screen (pink, magenta); use colors in combination that complement and not clash (yellow and blue, for example, rather than green and blue).
- **Special effects**—Use sparingly and purposefully for attention and emphasis; consider reverse lettering (black on white or color box), underlining or boxing-in words and phrases; and flashing (on–off display).

While you will want to maintain a consistent style for the screen, you should plan for some variety in style to maintain user interest. Variety can be added through the use of graphics and sound effects. A computer permits the manipulation of graphic symbols in a way that can increase interest and improve learning simultaneously. Such programing should be relevant and meaningful in support of the learning objectives.

Review What You Have Learned About Using Equipment, and Designing Programs for Computer-Based Instruction:

1. To which instructional use for computers is each situation related?

a. Guiding a student through a basic electronics program with readings, answering questions, giving feedback on answers, reviewing as necessary, and allowing the student to progress successfully.

b. Learning chemical formulas for pharmaceutical products by having to identify many examples that are shown repeatedly.

c. Presenting a realistic financial planning problem in which numerous variables are introduced and have to be considered in reaching a solution.

d. Engaging in competition as a history lesson is studied with points being awarded as the student proceeds through the program.

e. Programing six projectors for a 3-screen slide presentation.

2. What are the three necessary pieces of equipment for a microcomputer-based CBI system?

3. Differentiate between a *linear* and a *branching* CBI program.

4. What are the *four* symbols used to construct a flow chart and what does each represent?

5. What form can questions take in a CBI program to ensure standardization of replies?

6. What type of branching pattern would you select for each situation?

a. When a student answers a question incorrectly further information is provided, then the question is repeated.

b. Based on the results of a test taken before studying the program, the student is directed to those parts of the program with which she is not already knowledgeable.

c. Questions are interposed periodically as the student goes through the program, but they do not affect the sequence of instruction.

d. When the student answers a question incorrectly, as based on the answer selected, appropriate review is given and the question is repeated.

e. Each student is led through the program along an individual path based on answers given at decision points.

7. What are seven matters that should receive attention when designing the image that will appear on a computer screen?

WRITING THE PROGRAM

A computer program (called **software**) is a series of instructions which tell the computer exactly what is to occur at all times. The components of a program, as they appear on the CRT, will include many of the following:

1. Orientation information (contents of the program)
2. Directions and procedures for carrying out the learning activities
3. Textual information and graphic displays
4. Questions, problems, and other ways of testing understanding and making applications of the knowledge learned
5. Feedback on learner response (most often as evaluation of answers)
6. Further explanation of content related to questions and answers
7. Choice of concept or path for continuing program

Computer Language

A computer language is the complete set of **commands** which can be used to write a program, as well as the rules (syntax) which tell you how the commands must be used. There is a variety of computer languages which can be used to write programs. BASIC, Pascal, and FORTRAN are common, general-purpose languages.

When writing a program, you must recognize that a computer requires very specific instructions. These instructions are written as **statements** in the program. Usually each line in the program is a separate statement. A statement often includes the following elements:

- **Line Numbers**—(Required for BASIC programs, but not required for most other languages.) Inform the computer

of *the order* in which instructions are to be carried out. The numbers are often multiples of 10 so as to leave space for adding more lines at later times.

10

20

30

40

- **Operators**—(Also called **commands** or **key words**.) Tell the computer *what to do.* The short words or phrases immediately after the line number prepare the computer to do something very specific with the information that follows in the rest of the statement, or to the program as a whole. These are the *action verbs* in the syntax of a computer language.

PRINT

GOTO

RUN

LIST

IF..THEN

FOR . . . NEXT

END

- **Operands**—Tell the operator in the statement *what to act upon.* Depending upon the operator used, an operand may be required, optional, or not allowed. Operands may be:

253

Literals—Strings of characters or numbers which are translated literally by the computer. These are often enclosed in quotation marks.

"YOUR NAME"

"WELCOME TO THE PROGRAM"

"THE AREA IS 3 FEET BY 3 FEET"

Variables—Memory locations in the computer which have specific names, but whose value can be changed. This is roughly equivalent to variables in algebra.

Variable name	Possible variable values
NAME	JIM, JOHN, JERRY, GEORGE
APPLES	1, 6, 25, 108

Expressions—Mathematical equations which the computer can calculate as it runs the program. The calculations are performed under rules which are very similar, but not identical to those of algebra.

$2 + 2$

$27 + (3*A)$

$A + B/C$

The following is an example of a short CBI program written in BASIC:

```
10 PRINT "THIS IS A SHORT MATH PROGRAM"
20 PRINT "HOW MUCH IS 8+2+4-3?"
30 INPUT A
40 IF A=11 THEN PRINT "THAT IS CORRECT"
50 IF A < > 11 THEN PRINT "THAT IS NOT"
   CORRECT. THE RIGHT ANSWER IS 11.
```

Authoring Languages

To increase the efficiency of the programing process, a number of specialized **authoring languages** have been developed which are designed specifically for the production of CBI. PILOT is an example of an authoring language which is available in versions for a number of computer systems. The following is the above sample math program written in PILOT:

```
T:This is a short math program
T:How much is 8+4+2-3?
A:
M:11
TY: That is correct.
TN: That is not correct. The right answer is 11.
```

Due to the specific and repetitive nature of CBI, a simple authoring language command may replace several commands in a general purpose language such as BASIC. Thus, an authoring language can reduce the time required to develop CBI.

While authoring languages such as PILOT are easier to learn than most general purpose languages such as BASIC, they still require considerable knowledge on the part of the programer. In response to this situation, a number of authoring systems have been developed which assume little or no prior programing experience. When turned on, these systems often display lists of options, called **menus,** from which the user makes a selection. If the user decides to present a question, the system asks if the question is to be multiple choice, true/false, or short answer. Once this is determined, the system asks the user to specify the question, the correct answer, possible incorrect answers, and what should occur when each response is given. Upon receiving this information, the system generates the segment of the computer program required to ask the question.

The following sequence illustrates how an authoring system might work:

SAMPLE AUTHORING SYSTEM PROGRAMING SEQUENCE	
PROMPT BY COMPUTER	RESPONSE BY USER
Type your text.	This is a short math program.
Type your question.	How much is 8+2+4−3?
Is the question A. Multiple choice B. True/false C. Short answer	C
Enter the correct answer.	11
Enter the response to a correct answer.	That is correct.
Enter the response to an incorrect answer.	That is incorrect. The correct answer is 11.

Such an authoring system is relatively easy to use. However, it can limit the flexibility of CBI design since it will only let you design materials which fit a narrow range of predetermined formats.

While some computer languages may function on a number of different microcomputers, any given version of a specific language is likely to function only on a single type of microcomputer. For that reason, you should design CBI using the type of computer on which the software will be used and should use only a language which is designed to operate with that specific system. Several references and sources are listed at the end of the chapter to assist you in locating additional information on authoring languages.

254

GRAPHICS FOR CBI

As in other media, graphics can increase the appeal and effectiveness of computer-based instruction. Graphic images in CBI allow you to increase interest by adding visual variety to the presentation of information on the CRT and break up the monotony caused by using text only. In addition, they allow you to illustrate a structure such as a human organ with a drawing, or use animation to demonstrate a motion such as the flow of electrons through a wire.

Computers vary considerably in their graphics capabilities. Some systems are designed to offer text only. Some provide graphics, but only in monochrome, with a very limited selection of colors, and with restrictions as to how they can be used. Most offer a maximum resolution that is considerably coarser than that found in a video image. Some systems which have considerable potential for graphics are difficult to program in order to take advantage of that potential. Therefore, it is important to carefully consider graphics capabilities when selecting a CBI system.

Just as authoring languages can increase the efficiency of developing CBI programs, sophisticated languages or programs called graphics editors can radically increase the speed by which graphic images are constructed. They include additional programing which lets you draw common shapes with much fewer instructions.

Some graphics editors permit the use of a joystick, light pen, or graphics tablet to identify positions on the screen and eliminate the need to type in X and Y coordinates. With such a system, you type a single command or key stroke to indicate to the computer that a shape such as a box is desired. You then point to any two diagonal corners and the box is drawn.

For further details about computer graphics systems, review the information on page 115 in the Graphics chapter.

Computer Animation

The graphics capabilities of most common CBI systems permit simple animation which can be used to illustrate a motion, provide interest, or draw attention to some portion of the screen.

As with traditional animation, a series of individual images must be constructed which create the impression of motion. For example, the images below, when alternately displayed and moved across the screen, will create a little person pushing a hand truck.

To create the impression of motion, the computer is instructed to draw the first image at the starting location. That image is held on the screen momentarily and then erased before the second image is drawn. The process is repeated until the animation sequence is completed.

COMPLETING THE PROCESS

Once the program has been written and programed, two activities remain to be done before it is ready for use.

Documentation

No CBI program is complete without **documentation.** Documentation is the descriptive material which accompanies a program and tells the user how the program is to be used. At a minimum the documentation should include (1) a description of the computer system(s) upon which the software will run, (2) a list of the objectives, and (3) instructions as to how the program is started.

In addition, clear and consistent directions should be provided for all parts of the program. These may be in the form of printed materials, or may be incorporated into the software itself. Self-documenting materials often use lists of options (menus) which allow the user to select what comes next at critical points in the program. Some materials have a help command which will call a summary of instructions to the screen whenever they are needed.

Testing

Unlike human beings, computers are very particular about the accuracy of the instructions they receive. A single misplaced letter or symbol can render a computer program useless as an instructional tool. Therefore it is very important that CBI materials be thoroughly tested before they are released for broad use.

Errors such as these are called "bugs" and the process by which they are located and removed is called **debugging.** This is best accomplished by letting a variety of people try the materials to see what types of problems might occur. Ideally these people should be representative of the learners for whom the materials are designed and should go through the materials under the anticipated circumstances for their use. They should be asked to work through the program several times, trying all of the options, so that each branch can be tested. In addition, they should judge the effectiveness of the instructions and the clarity of the documentation.

COMPUTER-CONTROLLED INTERACTIVE VIDEO

Various types of media equipment can be connected (interfaced) with a computer. Most common are videocassette players and videodisc players, although as previously indicated, slide projectors, audiotape players, and other

media also can be controlled by a computer. The computer provides directions, textual information, and manages testing of student learning. The media equipment presents audiovisual material in short segments as directed by the computer. The two units of equipment function in coordination, rather than as separate entities.

There are a number of options for controlling the operation of the video player:

- **Level 1**—A remote control pad, containing a series of function keys, is attached directly to the video unit and operated by the user. A specific frame, anywhere in the program, can be located and images viewed individually as well as in slow motion or regular speed, with or without sound.
- **Level 2**—A microprocessor in the video player accepts a computer program previously placed on the videotape or videodisc. The presentation is controlled by this program and also by input from the remote control pad.
- **Level 3**—The video player is connected to an external microcomputer. The input from the user is entered through the computer's keyboard or other input device. In addition to controlling the presentation, text, graphics, questions, and answers will appear on the computer screen.

The Interface and Its Functions

The link between the computer and the image source is through an **interface,** an electronic circuit which translates computer commands into signals that direct the video player to locate segments of the videotape or videodisc. These segments may be still pictures, moving picture segments (with or without sound), or sound without any picture.

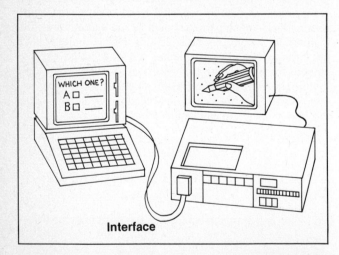

Interface

To achieve maximum interactive capabilities the equipment should provide electronic remote control access to all functions including forward, reverse, fast forward, and rewind. In addition, with video equipment it is desirable to have still frame capability and multiple audio channels. On a videocassette system, an audio channel (designated the "control track") is used to record a series of signals by which the computer determines the location of images on the tape. After indexing the control track frame numbers,

they are placed into computer memory. Then a search can be made by shuttling the tape fast forward or in reverse to locate and then show a desired frame or a motion sequence. This movement may be by choice of the learner or automatically directed by the computer in response to a decision the learner has made.

Interfaces require software in order to function properly. Ideally, the software should permit the user to program events in a number of general purpose and authoring languages. The video control commands should be fairly simple and easily embedded in the CBI program. The following is an example of a BASIC language command sequence which tells a videodisc player to display frames 5286 to 6492 of a videodisc:

1000 PRINT V$"PLAY(5686,6492);VIDEO"

The program which contains the commands used to control a specific instructional sequence is usually stored on a floppy disk which is inserted in the computer's disk drive. On self-contained videodisc units, this information can be stored on the videodisc itself and loaded directly into the microprocessor of the videodisc unit when it first starts to play.

Some computer systems have interfaces built in when they are manufactured. Others permit the installation of one at a later date. This may be as simple as inserting a circuit board in an empty slot in the computer. Some vendors offer units which are external to the computer and communicate with it through a standard computer port. As indicated previously, certain video equipment (primarily videodisc units) contain their own microprocessors which let them be programed directly and used in an interactive manner without the assistance of an external computer.

The capabilities of interfaces vary considerably. Although some universal interfaces are available, most are designed specifically for a given computer and a limited number of video display units. Most systems offer the ability to switch between a video and computer display on the same CRT and some allow you to superimpose the two images.

It may be desirable for some applications, or because of study space limitations, to use a single viewing screen that displays all images.

Tape Versus Disc

While both videocassette and videodisc units can be used effectively for interactive video applications, both have their unique advantages (see page 243).

Programs in videocassette format can be produced relatively inexpensively, especially if only a few copies of the final program will be required (less than 100). In addition, they can be rapidly revised as programing problems are discovered or when instruction is modified. However, the linear nature of videocassette operation is not as adaptable to using still-picture frames and can cause delays of up to a minute or more as the video player fast forwards or rewinds to locate the next segment. Such delays can destroy the pacing of an instructional sequence and detract from learning. Careful sequencing of visual segments can minimize this problem, but efficient random access capabilities are limited.

Most videodisc images, on the other hand, can be accessed in 2 to 5 seconds, thereby reducing or eliminating this problem. However, most commonly available videodisc systems require that discs be produced to precise standards, and once manufactured, they cannot be modified. Developments are underway to make available new-type videodiscs that permit greatly simplified local production.

An additional advantage of the videodisc is that the second audio track, when coupled with computer-controlled audio switches, can be used to provide alternative soundtracks for the same visual sequence.

If efficiency and economy of production are your primary concerns you should use a videocassette system. Careful sequencing of visual segments can minimize, but not eliminate, the access delay problems. If, on the other hand, these conditions are important—the program will require a large number of copies, each unit will have a high frequency of use (100 or more times), rapid access time is important, playback equipment is available to the user, and you can afford the disc manufacturing process—then consider a videodisc system. However, all video sequences and programs should be thoroughly tested in videocassette format before manufacturing the disc.

Planning

The same preliminary elements of planning and considering the variety of branching patterns as previously described for designing CBI (page 250) can be applied in computer-video interactive programing. In addition, because of the variety of media alternatives that can be used —textual, visual (photographic or graphic still picture, or motion at regular speed, slow motion, or accelerated speed), audio, or combinations of media—planning is much more complex than for either a CBI program or a conventional media program. Therefore, once the objectives and subject content have been selected, these planning matters should receive attention:

- Treatment of a concept can be more direct, thus dispensing with the normal pattern of introduction, content presentation, and summary.
- Develop a flow chart which specifies how each element

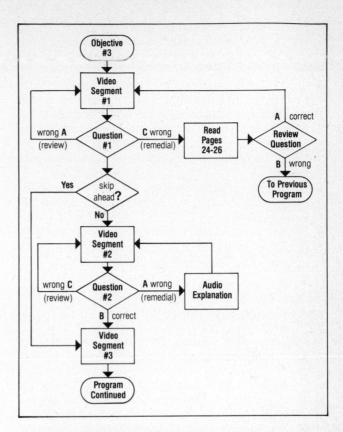

of the program should be treated—information, simulation, response choices, and so forth. This is a multidimensional form of a treatment (page 48). Options and alternative paths should be identified within the flow chart.

- Decide on media form for each program component (CRT display, audio, still, motion).
- Locate existing media material that may be considered for use in portions of the program. Give attention to obtaining permission for use from the copyright holder (page 57).
- Develop a detailed storyboard and script from the flow chart to guide production and editing. Because of the nonlinear sequencing and branching possibilities, scriptwriting becomes very complicated when placing all pieces in proper order and relationship to insure continuity.

Production

The complexity of interactive video planning carries over into the production phase. Both computer programing skills and a variety of media production skills are required. The original audio and visual materials may be prepared on either 16mm film or on one inch videotape. Use standard production techniques described in Chapter 19. In carrying out the production work, attention should be given to these matters:

- If the student will be allowed to restudy materials, the pace of information being presented can be more rapid than usual.

- Since the tape will consist of many short segments that can be used by students in various arrangements, normal transitions (dissolves, fades) for continuity purposes would not be necessary.
- Because of the various possible arrangements of video sequencing, care should be given to maintaining consistent overall production style and image quality.
- Allow 4 to 6 frames of still picture content on tape (rather than using a single frame) in order to overcome any limited tolerance in the equipment for locating a specific frame.
- When editing, place items on the tape in anticipated order of the intended sequence for use, thus reducing tape shuttle time.
- Special requirements for preparing videotapes to be converted to videodisc should be obtained from the company that will provide the service.

PREPARING TO USE CBI

As with any other medium, computer-based instruction must be used properly to be effective. Because of its interactive nature, most CBI occurs on an individualized basis, or with small groups of two or three. Of course, large group presentation of CBI is also possible, but usually less effective.

When designing an individualized CBI station, be certain to provide an environment which is comfortable and conducive to learning. Make certain that:

- The keyboard and monitor are placed in positions which are comfortable for the students.
- Lighting is at or near normal levels, but that no distracting glares or reflections strike the screen.
- Space is provided for writing or working with additional materials.
- Instructions and documentation are readily available.
- Earphones are provided if audio materials are used and the space must be shared with others.

Care should be taken to see that the equipment is maintained on a regular basis. Although the current generation of computer equipment is much hardier than previous generations, avoid extremes in temperature or humidity. Since dirt and moisture can rapidly destroy equipment and software, keep the area clean and discourage drinking, eating, or smoking while at the workstation.

Most computer-based instruction which is delivered by a CRT is designed for viewing at a very close distance. As a result, the text is frequently much smaller than that used in video materials. When using CBI for large-group instruction, you must allow for this difference, and be certain that all viewers are much closer to the screen. This means using additional monitors or a large-screen projection system. In addition you should consider the other environmental factors affecting any form of large-group instruction.

As with other forms of media, CBI does not occur in a vacuum. Whether the instruction occurs individually or in a large group, the success of your CBI will depend not only on its content and the quality of production, but also on the manner in which it is introduced and reinforced.

Review What You Have Learned About Writing the Program, Completing the Process, and Controlling Video With a Computer:

1. Refer to the sample BASIC computer program on page 254). Identify each element in the *second* line and give its name as used in programing.
2. How does an *authoring* language differ from a *general purpose* language?
3. Describe how you might use animation with a computer to change a *square* into a *circle*.
4. As a minimum, what *three* items should have attention in "documentation" of software?
5. What is meant by "self-documentation"?
6. To what phase of instructional media planning does the "debugging" step in CBI development relate? (*Hint:* See Chapter 7.)
7. What is an *interface?*
8. What five functions should an interface control?
9. Under what circumstances is it wiser to use *videocassettes* over *videodiscs* in an interactive video system?
10. When preparing to use CBI with individual students, what *four* matters should receive attention?
11. What are 3 planning matters and 3 production techniques that are unique to computer-video programing?

REFERENCES

Computer-Based Instruction

Burke, Robert L. *CAI Sourcebook.* Englewood Cliffs, NJ: Prentice-Hall, 1982.

Coburn, Peter. *Practical Guide to Computers in Education.* Reading, MA: Addison-Wesley, 1982.

Heines, Jesse M. *Screen Design Strategies for Computer-Aided Instruction.* Billeria, MA: Digital Press, 1983.

Jones, Aubrey B. Jr., *I Speak BASIC to My Apple.* (Atari, IBM, PC, and other products). Rochelle Park, NJ: Hayden, 1982.

Kearsley, Greg. *Computer-Based Training: A Guide to Selection and Implementation.* Reading, MA: Addison-Wesley, 1983.

Interactive Video

Dargan, Thomas R. "Five Basic Patterns to Use in Interactive Flow Charts." *Educational and Industrial Television* 11 (June 1982):31–34.

———— "Flow-charting—The Key to Interactive Video." *Educational and Industrial Television* 13 (November 1981):60–64.

DeBloois, Michael L., ed. *Videodisc/Microcomputer Courseware Design.* Englewood Cliffs, NJ: Educational Technology Publications, 1982.

Floyd, Steve, and Floyd, Beth, eds. *Handbook of Interactive Video.* White Plains, NY: Knowledge Industry, 1982.

Ics Journal. Mountain View, CA: Interactive Communications Society, P.O. Box 4520.

Interactive Video. Sony Video Communications, 9 West 51st St., New York, NY 10019.

"Interactive Video." *Performance and Instruction Journal* 22 (November 1983): total issue.

McEntee, Patrick, and Rosenfeld, Mark. "Interactive Production: Tape vs Disc. What's Best to Use in a Nonlinear Production?" *Video Systems* 8 (October 1982):22–28.

Sippl, Charles, J., and Dahl, Fred. *Video/Computers.* Englewood Cliffs, NJ: Prentice-Hall, 1981.

CBI Journals

Apple Education News, Apple Computer, Inc., 10260 Bandley Drive, Cupertino, CA 95014.

Atari Special Additions, Atart, Inc., Home Computer Division, 1196 Borregas Ave., Sunnyvale, CA 94086.

Compute! Magazine, Compute!, Box 5406, Greensboro, NC, 27403.

The Computing Teacher, International Council for Computers in Education, University of Oregon, Department of Computer and Information Science, Eugene, OR 97403.

Creative Computing, P.O. Box 789-M, Morristown, NJ 07960.

Computers and Electronics, P.O. Box 13877, Philadelphia, PA 19101.

InfoWorld—The Newsweekly for Microcomputer Users, Circulation Department, 375 Cochituate Road, Framingham, MA 01701.

PC, P.O. Box 2443, Boulder, CO 80321.

Personal Computing, Hayden Publishing Co., Four Disk Drive, Riverton, NJ 08077.

TRS-80 Microcomputer News, Radio Shack, Tandy Corporation, 1600 One Tandy Center, Fort Worth, TX 76102.

SOURCES FOR CBI HARDWARE AND SUPPORTING SOFTWARE

Allen Communications, 140 Lakeside Plaza II, 5225 Wiley Post Way, Salt Lake City, UT 84116 (interactive video interfaces).

Apple Computer, Inc., 10280 Bandley Drive, Cupertino, CA 95014 (microcomputers and software).

Atari, Inc., 1265 Borregas, Sunnyvale, California 94086 (microcomputers and software).

Bell & Howell Co., 7100 McCormick Road, Chicago, IL 60645 (CBI authoring systems for the Apple microcomputers).

CAVRI Systems, Inc., 26 Trumbull Street, New Haven, CT 06511 (interactive video interfaces).

Commodore Business Machines, 901 California Avenue, Palo Atlo, CA 93404 (microcomputers and software).

Control Data Corporation, P.O. Box O, Minneapolis, MN, 55440 (PLATO system and related software).

Digital Equipment Corporation, 146 Main Street, Maynard, Massachussetts 01754 (microcomputers, mainframe CBI systems, interactive video systems, authoring systems, supporting software).

Hazeltine Corporation, 7680 Old Springhouse Road, McLean, VA 22102 (mainframe CBI authoring systems).

IBM Corporation, 1133 Westchester Avenue, White Plains, NY 10604 (microcomputers, mainframe systems, authoring systems, and software).

Island Graphics, Box V, Bethel Island, CA 94511 (graphics software for microcomputers).

Penquin Software, 830 4th Avenue, Geneva, IL 60134 (graphics software for the Apple microcomputers).

Radio Shack, a Division of Tandy Corporation, Fort Worth, TX 76102 (microcomputers and software).

Sony Information Center, CN-02450, Department K, Trenton, NJ 08659 (microcomputers and interactive video systems).

Texas Instruments, Inc., 8600 Commerce Park Drive, Houston, TX 77036 (microcomputers and software).

Part Five

MANAGEMENT

Chapter 21

MANAGING MEDIA PRODUCTION SERVICES

- Extent of Services
- Personnel
- Financial Support and Budgeting
- Facilities
- Equipment, Supplies, and Services
- Working with Clients
- Production Controls and Procedures
- Evaluating Production Services

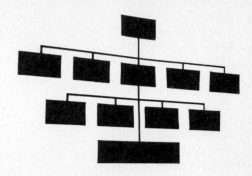

Media production services, whether a part-time assignment for one person in an organization, or a wide range of production activity handled by a large staff, should be operated on a businesslike basis. Because production services represent a sizable investment by the organization, there must be a return in the form of specific benefits to the organization.

The return is indicated by the accomplishment of purposes established by the organization. These purposes may be **motivational, informational,** or **instructional.** This means that media are designed for use in teaching and training, for public relations or other organizational communication needs, or to encourage some form of action by an audience.

Once purposes are identified, then the level on which the services will be provided needs attention. There are three levels as explained on page 8—**mechanical preparation, creative production,** and **conceptual design.** These levels are interrelated to the extent that the simplest techniques, practiced on the mechanical level, will be utilized on the higher levels. This hierarchy is useful when considering the types of activities in which a media production center is likely to engage.

A service that functions only at the mechanical level requires a different staff and facility than does one operating at the creative or conceptual design levels. The emphasis in mechanical preparation is solely on the equipment and techniques required to produce a specific product. This work can be handled by technicians, trained students, or possibly by the clients themselves. At the creative production level, attention shifts to planning and producing a relatively sophisticated media product and using effective communication techniques. A more talented staff is required. At the conceptual design level, professionals, who are aware of the instructional development process (page 4) in addi-

tion to the mechanical and creative aspects of production, should have the ability to convert instructional plans into meaningful visual and verbal materials.

This rationale, relating to purposes and levels of media production services, provides a framework within which management decisions can be made concerning the many aspects of a production operation.

EXTENT OF SERVICES

Media production services may include the following categories and the specific activities within each one:

- **Production planning**
  Preliminary planning (objectives, content, media selection)
  Media planning (treatment, storyboard, script)
- **Graphic arts**
  Design and layout for graphic materials
  Illustrations and art work, including charts, graphs, and titles prepared by hand and with computer technology
  Lettering
  Mounting
  Laminating
  Pasting up original copy for reproduction
  Duplicating single copies of materials on paper with or without size change (includes diffusion-transfer and high-contrast film processes)
  Preparing overhead transparencies
- **Still photography**
  Picture taking in black-and-white and color, on location and in studio
  Lighting
  Copywork

263

Slide duplication
Processing black-and-white and color film
Printing black-and-white and color negatives
Mounting slides
Making filmstrips
Duplicating filmstrips
Planning and producing slide/tape programs
Planning and producing multi-image presentations
Preparing overhead transparencies

- **Video recording and motion picture photography**
  Recording picture and sound on location and in studio
  Lighting sets
  Creating special effects including titles, animation, time lapse, and slow motion
  Editing pictures and sound
  Preparing final master on tape or film
  Making duplicate copies
- **CBI production**
  Flow charting
  Writing programs
  Creating graphics
  Designing screen images
  Developing computer/video interactive programs
- **Audio recording**
  Recording narration
  Recording music and sound effects
  Editing sounds
  Mixing sounds to master recording
  Adding synchronizing signals to slide programs
  Duplicating tapes
- **Duplication and reproduction**
  Preparing headings and captions
  Making litho negatives and printing plates
  Reproducing copies on printing press
  Duplicating originals or masters on copy machine
  Collating and binding
- **Models and displays**
  Planning models, exhibits, and displays
  Making models in cardboard, plastic, wood, and metal
  Designing displays and exhibits
  Constructing two-dimensional displays and exhibits
  Constructing three-dimensional displays and exhibits

A person directing a media production service will select activities from the above categories to provide the necessary services within an organization. These are chosen in terms of the established purposes and the production levels previously indicated. Also, consideration must be given to services that might be available from other departments or offices within the organization (reproduction or print shop, model-making shop, and so forth) so as not to duplicate services.

PERSONNEL

Each category of service requires one or more staff members having capabilities relating to the activities enumerated above. A single individual may fulfill many of these roles. The major responsibilities are handled by the following personnel positions:

- Instructional designer (to cooperate with subject matter experts for planning media)
- Writer (for script preparation and developing printed media)
- Graphic artist
- Still photographer
- Media production specialist (for producing slide/tape and multi-image programs)
- Audio recording technician
- Video and motion picture producer/director
- Video engineer
- Lighting technician
- Video and film editor
- CBI developer/programer
- Printer
- Model maker

A staff consisting of capable, trained technicians and professionals will provide the best service. Below this level would be students or volunteers who can be taught basic skills as described in this book. Also, some instructors and other clients would be enthusiastic about, or receptive to, preparing their own materials. While they may have the necessary skills, the need for staff supervision can be time consuming and the resulting quality of the media product may not justify the efforts.

FINANCIAL SUPPORT AND BUDGETING

The cost for media production services may be handled in various ways. The service may be allocated a certain sum of money for a given period (generally for a fiscal year) as budgeted by the parent organization. Under this plan, the costs for all services and projects are absorbed through these budgeted funds. These would include personnel salaries, supplies, services, and often new equipment purchases.

At the other extreme, some media production services are entirely self-supporting by charging clients or departmental budgets for work performed. This is known as a **chargeback** system. Standard, widely used services like making slides on a copystand would have a per unit pricing schedule. This is based on a calculation of time and materials required with a percentage amount added for overhead and equipment depreciation. In other situations, estimates for a firm bid price may be made for designing and producing special materials (video recordings, slide/tape or multi-image presentations) which are more complex, requiring extensive amounts of staff time. Within this category of recovering costs would also be income generated by selling services and products beyond the organization.

Between these two limits of *complete funding* and *fully reimbursement* would be charges for a portion of the services. This is often for materials used to complete a project or job. The production department's own budget absorbs the personnel, overhead, and depreciation charges. Thus, there is a sharing of expenses.

When there is a complete or partial form of chargeback, there can be a tendency to accept, with little question, any production work requested. Income must be generated to keep staff busy. Few questions about objectives or quality of original materials might be asked of the client. Frequently, quick and simple techniques, that may not be appropriate in terms of need or quality, are all that is requested. Since the client or sponsoring department is paying the bill, and often wants to keep costs to a minimum, it may be difficult to insist on the acceptance of adequate consultation.

On the other hand, when there is no chargeback, the production department can set standards and would be more likely to produce high-quality materials. However, without a charge, a client may request more service than is required for a job, resulting in wasteful effort and the use of excess materials. Examine these alternatives when considering how to establish the financing for a production program.

FACILITIES

While some related services can share the same working areas, most categories of services require special facilities for their activities. The space required to house production functions may be as follows:

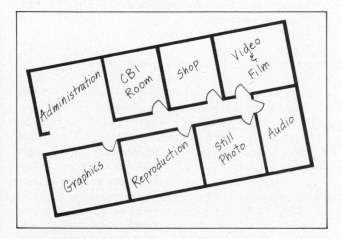

- **Graphic arts**
  Area with drafting tables as work stations, supported by space for required equipment and storage of supplies; requires pure water, extensive electrical power, and controlled ventilation
  Light-controlled area for process camera use (diffusion-transfer process)
  Space for computer graphics system (may require special electrical and environmental needs)
- **Still photography**
  Studio for picture taking (high ceiling and sufficient electrical outlets and circuits)
  Darkroom with suitable ventilation and temperature-controlled pure water for film processing and printing
  Workroom for copywork, slide duplication, slide editing, and slide mounting
- **Video recording and motion picture photography**
  Sound-controlled studio for video recording and filming (of sufficient size for possible multi-camera use), re-

quires high ceiling and sufficient electrical circuits (note: although this facility is included, the requirements for a full-scale video studio are limited since most production work is by single-camera, on location)
Control room area for directing studio programs
Quiet, separate area for videotape and film editing
Workroom for equipment used to create special effects, make masters, and duplicate tapes
- **CBI production**
  Clean, carpeted workroom
  Wall for developing and displaying flow charts
  Space for microcomputer, printer, and possible terminal to mainframe computer
- **Audio recording**
  Soundproof room for recording narration
  Workroom with suitable space for audiotape editing and sound mixing
  Projection area for synchronizing and previewing programs
- **Duplication and reproduction**
  Graphic-type workroom for preparing copy
  Darkroom for processing film
  Camera room for using process camera
  Area for plate-making
  Press room for operating printing press
  Finish area for completing job (collating, binding)
  Storage area for paper supplies
- **Models and displays**
  Workroom with shop equipment requiring sufficient electrical power and adequate ventilation
  Large floor area for laying out and assembling displays and exhibits
  Storage area for supplies
- **Administrative area**
  Office space for management, instructional design, and secretarial functions
  Conference room for planning meetings with clients and staff
  Space for computer functions, unless available in separate production areas

EQUIPMENT, SUPPLIES, AND SERVICES

Each aspect of media production requires certain equipment and supplies. These are obtained through separate budget categories. Equipment is termed a **capital expenditure** because it is considered to be a tangible asset, having use for a period of years. Some major pieces of equipment may be leased or rented for short-term use. Once purchased, items of equipment are depreciated (the value is proportionately reduced) over their lifespan period. Depending on the item, this may be for 5, 8, or even 10 years. At the end of this time, with available budget, replacement equipment can be purchased.

Supplies, as consumable items, are obtained through **operating budget** categories. They are bought from vendors on open accounts, through contracts, or as the result of purchase orders being issued after bids for lists of supplies have been submitted and then evaluated for price and item quality.

Outside services frequently are necessary to supplement those services being normally offered by the production department. Outside services may be used for processing film, for preparing materials beyond the capability of a production operation (like making a master filmstrip from slides, converting 16mm film to videotape, or obtaining computer-generated transparencies), or for employing special personnel talents (a writer, narrator, editor, or program evaluator). An organization may also find it necessary to contract with another production group for preparing materials or producing complete programs beyond the abilities of the facility.

The decision of whether or not to use outside services is made by considering these factors:

- Need is beyond the competencies and equipment capabilities available within the organization
- Less costly to fulfill the need outside than to provide it as an in-house service
- Time constraints or the size of the job make it difficult or impossible to complete the work as requested through normal procedures
- Need is occasional and the cost to maintain fulltime capability is not justifiable

The technologies that have caused changes in media production are evident by the new equipment and procedures that increasingly become available. It seems obvious that such changes will continue to occur. Therefore, keeping alert to further developments to evaluate their benefits and, when appropriate, to integrate them into a production program is a necessity.

WORKING WITH CLIENTS

In the expression *media production services,* the emphasis should be on the word **services.** Individuals, teams, or groups in an organization or from the community at large may come to a production department or company voluntarily, requesting a service. In other instances, the use of media may be mandated for a communication or instructional need ("Overhead projectors will be used when teaching in this room!").

The individuals or groups for whom service is performed are the instructors and other staff members in an organization. We frequently call them **clients.** The satisfactory working relationship between production staff members and clients is very important for ensuring successful service and also for projecting a positive image of production services to others.

Here are some behavioral guidelines that staff members should employ in order to communicate properly and to work successfully with clients:

- **Listen carefully**
  Be alert to important things said (purposes for materials, time requirements, funds available)
  Give full attention to client
  Avoid distractions (telephone calls, interruptions by other persons)

- **Ask questions**
  In what form should materials be made?
  With what group will they be used?
  Where will they be used?
  When are they needed?
  How will costs be handled?
- **Be sensitive to client's manner so as to successfully interact**
  Formal or informal
  Talkative or reserved
  Serious or humorous
  Nonverbal clues (facial expression, mannerisms, body language)
- **Include client when decisions are to be made**
  Approvals concerning design and color selection
  Time frame for preparation
- **Admit you don't always know or understand**
  Media choice may be unclear
  Requirements may not be clear or specific
  Details are vague or left out
  Deadline is not clearly stated
- **Ask if you should contact someone else**
  For authorization
  For further information
- **When necessary, inform the client that production cannot take place as requested**
  Point out workload
  Refer to higher level administrator
  Suggest alternatives
- **Recognize an impossible situation**
  What is requested may not be a media form that can be conveniently prepared
  Client is unreasonable in expectations
  Time frame is too short
  Techniques required are not available
  Work will be too expensive to complete

If difficulties arise when working with a client, request help from the production department manager.

PRODUCTION CONTROLS AND PROCEDURES

At the beginning of this chapter it was noted that a media production service should operate on a businesslike basis. This has been evident when considering financial support, budgetary matters, and decisions relative to the purchase of equipment, supplies, and outside services. Also a number of other procedures can make the operation more businesslike and efficient. They include:

- Monitoring the quantity and quality of work produced
- Making the best use of individual staff members' abilities and their working time
- Meeting agreed time schedules for work completion
- Avoiding waste in use of supplies
- Keeping expenses within budgetary limits
- Abiding by existing copyright regulations for materials to be duplicated and for those produced

Computer Functions

Ono way to attend to many of the above procedures is to make use of a computer. For certain management functions, prepackaged programs are available and can be adapted to local needs, or special programs can be developed for a microcomputer. With such equipment and software, you can accomplish the following:

- Prepare budgets, monitor expense accounts, and make future estimates
- Maintain inventories of equipment, accessories, and supply items
- Print out periodic reports (weekly, monthly, annually), listing media prepared, time required, costs for materials and services
- Develop a data base of suppliers and outside service agencies, including items and price information, services provided, address and characteristics of individuals to serve as actors, narrators, and to fulfill other essential functions
- Bill clients for services performed while also recording the information for reporting purposes
- Organize negative and slide files under various "descriptor" headings for quick retrieval (see discussion on page 206)
- File information on special or infrequently used graphic and photographic procedures
- Visualize the change in production services over periods of time expressed in chart form for such numerical items as products prepared, personnel time, budgets provided, expenses, and clients served
- Prepare estimates for proposed projects as based on data from previous projects

From this list, it is evident that the computer can significantly simplify and extend aspects of a production manager's job in many ways. The speed and versatility with which information, in a variety of formats, can be assessed are other advantages for having the assistance of a computer prior to making management decisions.

Human Functions

As specified above, a computer can be programed to collect and display data that will inform a manager, as well as provide assistance in evaluating aspects of a production program. There are other functions that require human attention beyond the competencies of a computer. With reference to the six management procedures listed at the beginning of this section, the following are suggested:

- Assign job or project numbers to all work as received. Set up a visible control board so that progress on a job can be seen, especially when coordination among areas of production will be required. A modification of the planning–storyboard procedure, with cards, can be used (see page 32).
- Develop production service forms to be used for gathering data which will be entered into computer memory.
- As a manager, spend time observing and visiting with

staff members while at work. Look over the materials developed to assess their quality. See use of materials in classrooms and in other situations to judge their suitability.
- Establish policy and procedures for handling copyrighted materials. Emphasize importance to staff.
- Set a plan for copyrighting major productions as completed (see page 67).
- Establish procedures for informing persons in the organization about availability of prepared materials that can be used for purposes other than those for which they were initially requested.
- Hold periodic staff meetings to discuss budgets and operational matters, to provide information on new equipment and procedures, and to invite suggestions from staff members.

EVALUATING PRODUCTION SERVICES

The value of any service is judged on how well it serves the needs established by the organization of which it is a part. Here are three questions that can help you (and others in the organization) to evaluate the value and success of production services:

- **How productive has the service been?**
  Number of items prepared and programs produced per year, possibly in relation to the size of the production staff
  Number of clients or separate departments served
  Growth in these two measurements over a period of years
- **What has been the contribution of the materials and media programs to the organization?**
  Learning time reduced and/or amount of learning increased
  More students or trainees handled
  Personnel working time reduced or transferred to other activities
  Higher productivity, increased sales, better services provided, improved safety record
  Improvement in attitudes of employees toward job and organization
  Image of organization enhanced
- **What is the opinion of instructors, clients, and other key individuals within the organization about the services provided by the production department?**
  Opinions expressed by individuals on rating forms
  Informal comments received
  Other indications of change in level of support

Data for answering the first question would be obtained from computer records as explained in the previous section. The second question requires careful followup after materials or programs are completed and are being used. Training department records, interviews with training staff, public information staff, and others who have used production services, should be able to provide some evidence for answering this question. It is not easy to directly equate the use of media with measurable results. (Recall the values for using media as stated on page 3 in Chapter 1.)

The third question requires that a special effort be made to obtain feedback about the production services. Requesting completion of questionnaires with rating scales, as well as being alert to reactions by persons outside the production area will reveal feelings and provide subjective evaluative data. How the production services are "treated" at budget time, when additional personnel are requested, or when the purchase of special equipment is approved, are further indications of the value and importance of the service to the organization.

Finally, do not ignore the importance of telling your own story and keeping the organization informed about your services, special projects that have been completed, and other benefits offered to the organization. Promote your services through such means as these:

- Distribute a brochure or circular describing the services to instructors and potential clients.

- Invite individuals and groups to tour the production facility.
- Offer to demonstrate to groups within the organization the extent and variety of services and products offered.
- Set up exhibits in prominent places illustrating the extent of services offered.
- Distribute a periodic newsletter reporting on projects, recognizing instructors or clients who have made extensive use of services, describing new techniques, and providing other newsworthy information.
- Prepare an annual report tabulating details about the services offered and uses made of them during the year.

Handle public relations activities, like the above, carefully. If too much effort seems to be spent on promotional efforts, then time is being taken from regular production work. Soundly planned efforts such as these can pay off in generating instructor and client interest as well as in contributing to a positive image of the media production services.

Review What You Have Learned About Managing Media Production Services:

1. What are the eight categories into which media production activities can be grouped?
2. To which production personnel role is each of the following related (an item may apply to more than one category)?
   a. duplicating slides
   b. editing pictures
   c. using a copy machine
   d. preparing titles
   e. script preparation
   f. mounting pictures
   g. assembling a bulletin board
   h. lighting a location
   i. mixing music and narration

3. What are the *two* usual ways of financing a production service and an advantage or limitation of each one?
4. What are the *three* budget categories and cost items in each one?
5. What are *five* classes of behavior that a production staff member should practice when working with a client?
6. In what *five* ways can a production service be operated in a businesslike way?
7. What *six* production-management functions can be supported by using a computer?
8. For what *three* questions should answers be obtained in evaluating a media production service?
9. What are *four* ways for obtaining recognition and promoting a media production service?

REFERENCES

Burlingame, Dwight F., et al. *The College Learning Resource Center.* Littleton, CO: Libraries Unlimited, 1978.

Dennison, Linda. "The Chargeback System: Useful Tool or Useful Hindrance." *AV Video* 6 (March 1984):37–40.

Larish, John J. "Computerizing Your Photo Unit." *Technical Photography* 15 (September 1983):34–38.

Merrill, Irving R., and Drob, Harold A. *Criteria for Planning the College and University Learning Resources Center.* Washington, DC: Association for Educational Communications and Technology, 1977.

Schmid, William T. *Media Center Management.* New York: Hastings House, 1980.

Wadsworth, Raymond H. *Basics of Audio and Visual Systems Design.* Indianapolis: Howard W. Sams, 1983.

APPENDIXES

A. Answers to Review Questions

B. Glossary

Appendix A

ANSWERS TO REVIEW QUESTIONS

Chapter 1 (page 9)

1. (a) 7, (b) 2, (c) 4, (d) 3, (e) 1, (f) 6, (g) 8
2. Nature of students
   Objectives to be accomplished
   Teaching/learning methods and activities
   Evaluation procedures
3. Selection of media and its development is influenced by such instructional planning elements as: nature of student group, objectives to be served, content to be treated, and methods of instruction
4. Presentation to groups
   Individualized or self-paced learning
   Instructor–student interaction
5. a. Individualized learning
   b. Instructor presentation
   c. Student interaction
   d. Instructor presentation
   e. Individualized (self-paced) learning
6. Mechanical level—simple preparation
   Creative level—production of a complete instructional media
   Design level—selection and production of media within an instructional design framework
7. a. Mechanical
   b. Design
   c. Creative
   d. Mechanical
   e. Creative
   f. Design
8. Developing skills in interpreting, judging, responding to, and using visual representations of reality
9. Take responsibility, share within a group, become perceptive and analytical of visual world, become fluent in expressing ideas verbally and visually

Chapter 2 (page 16)

1. Perception—the internal awareness a person develops for recognizing an event or object in the environment. Gaining a person's attention, holding his or her interest, and making sure the correct message is received are important considerations in designing audiovisual materials relating to perception.
2. Fleming and Levie, *Instructional Message Design.*
3. a. Perceptual element
   b. Basic principle
   c. Perception and cognition
   d. Attention

   e. Perceiving picture and words
   f. Perceptual capacity
   g. Basic principle
   h. Perceptual distinguishing
4. The transmission-channel step
5. Behaviorism
6. a. Instructional resources
   b. Learning objectives
   c. Feedback
   d. Participation
   e. Individual differences
   f. Prelearning
   g. Practice
   h. Application
   i. Reinforcement
   j. Organization of content
   k. Motivation
   l. Learning sequence
   m. Motivation
   n. Organization of content
7. Cognitive, affective, psychomotor; affective domain most difficult to serve
8. a. Psychomotor
   b. Cognitive
   c. Psychomotor
   d. Affective
   e. Psychomotor
   f. Cognitive
   g. Affective
9. a. Higher
   b. Low
   c. Low
   d. Higher
   e. Higher
10. a. Concept
    b. Principle
    c. Fact
    d. Application
    e. Concept

Chapter 3 (page 21)

1. Agree
2. Disagree (depending on value you attribute to finding 26)
3. Agree
4. Agree
5. Agree

6. Agree
7. Disagree
8. Disagree
9. Disagree
10. Agree
11. Agree
12. Agree
13. Agree
14. Agree
15. Disagree
16. Disagree
17. Disagree
18. Agree
19. Agree
20. Disagree
21. Agree
22. Agree

Chapter 4 (page 34)

1. Purpose or idea, objectives, audience, content outline
2. Subconscious thoughts
3. a. objective
   b. content outline
   c. purpose
   d. objective
   e. audience
4. Motivation, information, instruction
5. a. information
   b. instruction
   c. motivation
   d. information
   e. instruction
6. Subject specialist, communications specialist, technical staff
7. Action verb, content reference, performance standard
8. b, c, d, g, h, i
9. a. (1) cognitive
      (2) cognitive
      (3) cognitive/affective
      (4) psychomotor
      (5) affective
      (6) affective
   b. (3) To *understand* their role . . .
         (reword) To *explain* to other employees their role . . .
      (6) B. To *know* how youth activities help . . .
         (reword) To *cite* four instances of how youth activities help . . .
10. Age, education, maturity level; skills in reading, writing, math computations; attitude toward topic; present knowledge or skill relative to topic
11. May offer useful ideas
    Materials already produced might serve objectives and be suitable for use
12. Write information on cards and display cards on a planning board
13. Reader's activity

Chapter 5 (page 45)

1. a. Overhead transparencies, multi-image presentations
   b. Display media
   c. Printed media, audiotape recordings, computer-based instruction
   d. Slide series, filmstrip, video, film
2. a, b, d, f, g
3. (Note: alternative answers for following are possible)
   a. Media required?—yes
      Visual form only?—yes
      Graphic only?—yes
      Overhead transparencies
   b. Media required?—yes
      Audiovisual technique?—yes
      Still only?—yes
      filmstrip/tape (preferable to slides as costs lower for multiple copies required to serve 2,000 students)
   c. Media required?—yes
      Audiovisual technique?—yes
      Motion necessary?—yes
      video recording
   d. Media required?—yes
      Multi-image technique?—yes
      multi-image—slides and motion picture footage
   e. Media required?—yes
      Audiovisual technique?—yes
      Still only?—yes
      paper/tape (audio-notebook)
4. a, b, c, e, f
5. Reader's activity

Chapter 6 (page 53)

1. **Treatment:** a verbal description of how the content will be sequenced. The treatment starts you toward visualization of the content. Thus, you are moving from verbal to visual, a normal thought-process sequence.
   Storyboard: a series of pictures with narration notes which illustrate how each sequence will be treated. The storyboard shows visual situations which prepare you mentally for the specifics you will need when making actual pictures.
   Script: the plan that describes each scene to be photographed or drawn along with necessary sound. This is the map or blueprint which indicates the details of the instructional media to be produced.
2. Move from things known to the unknown or new
   Start with an introduction, then develop the details, finally summarize or review
3. The student or trainee has to either mentally or actively respond to a part of the content or to a situation presented. This may require answering questions, other written or verbal activity, or some type of physical performance.
4. Allows for more effective learning of content
5. Expository, personal involvement, dramatic
6. Simple sketch, detailed sketch, 2×2 color slide or photographic contact print, instant photograph
7. Frame or sequence number, visual, narration information, special notes

8. (a) False, (b) True, (c) True, (d) False, (e) False, (f) True, (g) False, (h) True
9. Allow sufficient time to develop the subject and for the audience to comprehend it, but not so long that interest will be lost
10. They allow you to see what will have to be done in order to produce the materials, the assistance needed, the probable costs, the time required, and so forth
11. Type of media, materials required, sound features, length, facilities and equipment, special techniques, special assistance, completion date, budget
12–16. Reader's activity

Chapter 7 (page 64)

1. (a) 7, (b) 4, (c) 10, (d) 1, (e) 3, (f) 12, (g) 11, (h) 6, (i) 9, (j) 2, (k) 13, (l) 5, (m) 8
   The sequence for production may vary as some of the minor elements are handled at different times.
2. (a) True, (b) True, (c) False, (d) True, (e) True, (f) True, (g) False, (h) True, (i) False, (j) False, (k) False, (l) True, (m) False, (n) False, (o) True, (p) True, (q) True, (r) True
3. Positively: a, b, d, g
4. Does the material serve the objectives?
   Does it exhibit a smooth continuity?
   Does the narration support the visuals?
   Is the material of suitable length?
   Has anything important been left out or unnecessary material been included?
   Are all pictures satisfactory for communicating the content?
   Is the material technically acceptable?
5. a. To obtain reactions, suggestions, and help that might improve your instructional media.
   b. Checkpoints: Content outline, storyboard review, script, formative evaluation, or try-out time
6. Reader's activity

Chapter 8 (page 67)

1. Check the room; arrange for equipment; provide for physical comfort of audience; provide handouts; instruct person(s) who will assist; rehearse use of materials; prepare group for viewing
2. Reader's activity
3. Put copyright notice on material—© Your Name, Year (for sound recording)—℗ Date, Your Name
   Obtain application form for legal copyright protection

Chapter 10

Camera Types (page 75)

1. Reader's activity
2. Reader's activity
3. a. Self-processing film camera
   b. Twin-lens reflex camera
   c. 35mm camera
   d. Sheet-film camera

Camera Lenses (page 77)

1. Reader's activity, page 76
2. Millimeters
3. Wide angle, normal, telephoto
4. (a) Normal, (b) wide angle, (c) telephoto, (d) telephoto
5. Reader's activity

Camera Settings (page 80)

1. Twice as much
2. (a) $\frac{1}{500}$ second, (b) $\frac{1}{250}$ second, (c) $\frac{1}{125}$ second
3. $f/22$ and $\frac{1}{30}$ second
4. $3\frac{1}{2}$–5 feet
5. Less

Film (page 82)

1. Type of light source (daylight or tungsten light)
2. Kodacolor negative or color reversal film
3. Indicates the relative light sensitivity of the film; higher number for greater sensitivity (faster film)
4. For use under lower light levels or to film faster action
5. A filter passes light of its own color, while blocking out light of the complementary color.

Exposure (page 86)

1. Film speed, $f/$stop, shutter speed; $f/$stop and shutter speed; film speed
2. $f/32$ at $\frac{1}{15}$ second
3. Film speed and light level indication
4. $f/$stop and shutter speed
5. When light is coming toward the camera, not over your shoulder; when photographing an unusually bright or dark area that is not important to the subject in the picture
6. $f/32$ and $\frac{1}{8}$ second
7. Incident light meter

Lighting (page 90)

1. Small areas
2. $f/16$
3. Flat shadowless lighting on the subject with heavy background shadows
4. Incident meter reading of key light, measured at the scene, is four times that of the fill light. No, the ratio is too high.
5. Describe use of either a reflected or incident light meter.
6. a. Key—main light on subject; 45° to side of camera and placed higher than camera.
   b. Fill—lighten shadows created by key light; at camera position on other side from the key.
   c. Background—lighten background, reduce shadows on background and give some depth to scene; aimed at background from side.
   d. Accent—highlight or outline the subject and separate it from background; above the subject, aimed down.
7. To avoid reflections.

Picture Composition (page 95)

1. Generally (c) or (e) would be preferable because of the careful framing.

2. (a) Low angle, (b) high angle, (c) subjective, (d) objective, (e) either high or low angle
3. (a) MS, (b) XCU, (c) LS, (d) LS, (e) CU high angle, (f) CU, (g) MCU low angle, (h) MS low angle
   Sequence: c-a-g-b-h-e-f-d (your sequence may vary and still be acceptable to an audience)

Close-up and Copy Work (page 99)

1. All cameras that have separate window-type viewers
2. No. Calculation is necessary if a bellows is used.
3. Macro lens
4. Larger number (that is, smaller opening); slower; tripod essential
5. Divide 40 by 5. Set meter at film speed of 8.

Processing Film and Making Prints (page 104)

1. Developer—causes chemical change to form black silver image on film.
   Rinse—stops action of developer.
   Fixer—sets the image on the film.
   Wash—removes all chemicals.
2. See list on page 100.
3. Paper developer instead of film developer in first step, others the same.
4. You can examine photographs to select best negatives for enlargements.
5. Faster processing and requires less print handling and darkroom space.
6. A single developer step is used in color negative processing, while with color reversal processing, after the first developer, a color developer is used which brings out the positive image. Eventually the negative image is bleached out.
7. Either shoot color negative film or color slide film of the operational steps. Either the negatives or positive slides can be printed on paper as color photographs.

Chapter 11

Planning Art Work (page 110)

1. 6×9 inches or 9×12 inches
2. To allow sufficient "bleed" space when filming because a camera may record a larger total area than is seen through viewfinder
3. 8×8 inches—126 slides; 8×10 inches—photographs and overhead transparencies; 12×18 inches—standard 35mm slide; 9×12 inches—filmstrip, motion picture, and video
4. When a number of layers comprising a visual must be accurately aligned.
5. (a) simplicity and line; (b) balance and space

Graphic Tools and Materials (page 112)

1. 14-ply cardboard
2. Nonreproducible blue pencil
3. T-square
4. Technical drawing pen
5. Compass
6. X-acto knife
7. Triangle

8. 4H–9H pencil
9. Tracing paper

Illustrating Techniques (page 114)

1. Opaque projector
2. Ready-made pictures (tearsheets)
3. Photo modifier
4. Clipbook pictures

Statistical Data (page 115)

1. Bar or column
2. Circle
3. Line
4. Bar or column
5. Surface

Using Computer Graphics Systems (page 118)

1. Art on paper, 35mm slides, overhead transparencies, video
2. Increased speed of production, ease of making revisions, increased emphasis on design
3. (a) storage, (b) input, (c) output, (d) output, (e) input, (f) output, (g) output
4. a. Modem: device which permits transmission of output from computer via telephone to another location.
   b. Software: instructions for a computer as a program.
   c. Menu: list of options in a program as they appear on the video display terminal.
   d. Pixels: rows of dots that make up the computer-generated image.
5. Resolution (sharpness of image)
6. Partial system—design materials on your own computer, transmit to a production center, receive completed visuals.
   Production services—visual totally designed and produced by a commercial company.

Coloring, Shading, and Background for Titles (page 120)

1. Spray paints are quicker to use and provide a more even coloring, but time is required for masking and protecting areas not to be sprayed with a color.
2. Richness of tones and type of paper surface to be reproduced reliably on film.
3. It is too hard to separate the color sheet from the backing sheet unless a "lip" remains for insertion of a blade. If a piece is cut to exact size it is very difficult to line it up exactly over area to be covered. Cutting it larger allows some margin when placement is made.
4. Airbrushing
5. b and c
6. (a) black overlay, (b) white overlay

Legibility Standards (page 124)

1. So the lettering can be read easily by audience at an anticipated maximum distance.
2. Legibility decreases as quantity of information increases.
3. Use capital letters for short titles, lowercase letters for six words or more.

Space letters optically (as they look proper to you). Allow $1\frac{1}{2}$ letter widths between words and twice as much between sentences.
Separate adjacent lines about $1\frac{1}{2}$ times a lowercase letter.
Contrast letter color with background color.
Light letters against dark background have greater visibility.
Dark letters on light background should have wide strokes.
Avoid script letters.
Moderate bold line is preferred to thin or thick line width.
4. 1 inch
5. About $\frac{1}{5}$ inch high
6. Larger
7. Larger
8. 6 feet

Lettering for Titles (page 127)

1. Hold pen firmly in hand, make only arm movements.
2. Against a T-square or a lightly ruled line.
3. Protects back of letter sheet from dirt and inadvertent pressure that may transfer letters.
4. Preferably a special burnishing tool, but any blunt object like the rounded edge of a pen can be used.
5. Wricoprint, Koh-i-noor Rapidoguide, Leroy, Pressure Machine, and Phototypesetting.
6. Template lettering
7. Pressure machine
8. Reader's activity. Then see Table 11-2 on page 128.

Mounting and Protecting Surfaces (page 135)

1. Coat only one surface and put two surfaces together for temporary mounting while cement is still wet.
Coat both surfaces and allow them to dry for permanent mounting.
2. You can see guide marks through it; it does not adhere to rubber cement.
3. A heat-sensitive adhesive is coated on both sides of tissue paper.
4. So both picture and tissue can be accurately trimmed together
5. Puncture bubble with a pin, reapply heat and pressure.
6. Reader's activity
7. To reduce the temperature so as not to damage heat-sensitive material.
8. The edges of the picture to be joined are trimmed *before* the dry-mount tissue is tacked to the back; tacking the second piece of the picture is done with the picture face up.
9. For cardboard or cloth, size is the deciding factor. Choose rolling or folding on the basis of convenience of use.
10. So there is no exposed adhesive that might stick to the working surface or to the dry-mount press.
11. Extra pressure is necessary, so a sheet of cardboard or masonite is used in the press under the rubber pad.
12. Yes, with a cold mounting-adhesive material.

Making Paste-ups (page 138)

1. Paste-up: attaching each black-line element of a page in proper position.
2. Nonreproducible blue: color of pencil with which to make marks on paste-up sheet; will not show when page is duplicated or photographed with high-contrast film.
3. Continuous-tone photograph: black-and-white with shades of gray.
4. Halftone picture: black-and-white image, consisting of uniformly spaced black dots of different sizes that blend together to convey shades of gray when printed.
5. Halftone negative: result of photographing a continuous-tone photograph through a halftone negative contact screen.
6. Clear window: clear area on high contrast negative where halftone negative will be attached; results from black paper or red adhesive area on paste-up.
7. Registered overlay paste-up: separate paste-up, usually on clear acetate or translucent paper, for a color, aligned accurately over the base paste-up sheet.

Duplicating Line Copy (page 139)

1. Diffusion transfer (PMT) and high-contrast film
2. Use a good quality office copy machine.
3. Diffusion transfer: expose negative paper in camera to paste-up, develop through processor in contact with positive receiver, image transfers from negative to positive paper.
High-contrast film: expose high-contrast film in camera to paste-up; process film through developer, stop, fixer, and wash; dry negatives; remove unwanted marks on negative with opaque solution; enlarge image on negative in enlarger onto photo paper, process paper, and dry.

Reproducing Printed Matter (page 142)

1. (a) Electrostatic, (b) offset, (c) spirit
2. (a) Spirit, (b) electrostatic, (c) offset, (d) electrostatic

Chapter 12

Sound Recording Equipment, Tape, and Facilities (page 148)

1. a. Unidirectional
   b. Nondirectional.
   c. Bidirectional and dynamic
2. Remove oxide from heads, capstan, and rollers; demagnetize the heads.
3. a. Open reel two-track stereo
   b. Open-reel multitrack
   c. Open reel or cassette two-track monophonic
   d. Four-track stereo cassette
4. Volume level, tone, fading sound, blending sound
5. a. 1 mil, 7-inch reel ($7\frac{1}{2}$ ips)
   b. 1.5 mil, 7-inch reel ($7\frac{1}{2}$ ips)
6. C-60 cassette (full recording on one side); or C-30 cassette (14.30 minutes side 1, 10:30 minutes side 2)
7. Soundproofed room, eliminate ambient noise, avoid equipment that may generate 60 cycle electrical interference

275

Sound Recording Procedures (page 154)

1. Uses conversational tone; gives emphasis to certain words; changes rate of speaking; delivery fits mood and intent of words; provides suitable pauses and pacing.
2. Proper microphone placement; select recording tape speed; avoid using AGC; make volume level check on voice with moderately high setting; provide comfortable position for narrator; run test on equipment; have a glass of water at hand.
3. Use a mixer into which both the microphone and record player input in order to combine the sounds before going into the recorder, or record the narration on tape and then mix the recorded narration and the record as they go onto the tape.
4. Return to the last correct sentence and rerecord from then on; or, continue recording by repeating the material correctly, preceded by a verbal indication, such as "take 2."
5. Advantages: Allows setting volume level at start of presentation and as background creates a mood while providing continuity for presentation.
   Drawback: Musical background, if too loud or recognizable, can interfere with narration.
6. Physical editing is cutting and splicing a copy of the original tape, thus eliminating and reordering sections as necessary.
   Electronic editing is omitting and reordering parts of the original recording as a copy is being made by dubbing to another tape.
7. Review page 151.
8. It shortens listening time, reduces boredom, keeps listener alert, may increase learning.
9. Either create an actual sound—musical note, chime, buzzer, or bell, or use a tone generator and select a tone between 100 and 440 cycles per second.
10. Use only one side of the cassette for narration with audible signal and repeat the same recording, with signal, on other side; have narration on one side and inaudible signal on second track.
11. Reader's activity, see page 153.
12. Reader's activity.

Chapter 13 (page 168)

1. Learning aids: guide sheets, job aids, picture series
   Training materials: handouts, study guides, instructor's manuals
   Information materials: brochures, newsletters, annual reports
2. Syllables per 100 words; sentences per 100 words
3. 12th grade level (4 sentences, approximately 164 syllables)
4. Underline, capitalize, boldface, italics, box-in
5. Cropping: marking a photograph to indicate area to be printed.
   Copyfitting: adapting text to fit available space.
   Dummy: final layout of pages that will be used to guide paste-up.
   Duplication: making a few copies.

Galley proof: text set in type to be checked against original manuscript.
Justify: spacing words so that every line is the same length.
Repro copy: final type of text, with all corrections, to be printed.
Reproduction: making many copies.
6. a. Guide sheet
   b. Study guide
   c. Brochure
   d. Handout
   e. Newsletter
   f. Picture series
   g. Instructor's manual

Chapter 14 (page 172)

1. a. Flip chart
   b. Magnetic chalkboard
   c. Chalkboard
   d. Exhibit
   e. Hook-and-loop board
   f. Chalkboard
   g. Bulletin board
2. a, c, d

Chapter 15

Planning Transparencies (page 178)

1. Reader's activity, starting on page 173.
2. Horizontal, because in rooms with low ceilings the lower portions of a vertical transparency cannot easily be seen by all members of an audience.
3. $7\frac{1}{2} \times 9\frac{1}{2}$ inches.
4. Check answers against suggestions on page 174.
5. Vertical format rather than horizontal, quantity of information included, size of lettering, need for copyright clearance.
6. Eliminate unnecessary details; modify visual elements; replace lettering with larger size type; divide a complex diagram into sections; consider using masking and overlay techniques.
7. Clarify elements, give emphasis, increase interest.
8. $\frac{1}{5} - \frac{1}{4}$ inch (page 124)
9. Most suitable would be dry transfer, Wricoprint, Leroy, Koh-i-noor, and pressure-machine lettering (Kroy).
10. Register each overlay drawings with the master drawing by use of guide marks in two corners.

Direct Preparation on Acetate (page 179)

1. No. Marks made by some do not adhere to acetate.
2. Use a solvent, such as lighter fluid.
3. No.
4. Cover with a sheet of acetate.

Reproductions of Printed Illustrations (page 184)

1. Electrostatic process.
2. The writing material used to make the image on the paper must have a carbon base.

3. No.
4. Underexposed. Slow down the machine.
5. With felt pen or colored adhesive.
6. Reader's activity, (page 181)
7. Translucent tracing paper or transparent acetate.
8. Cover the area on the tracing paper with opaque paper.
9. Light source—master drawing—diazo film
10. Ultraviolet exposure
11. Overexposure. A shorter time
12. Full-color electrostatic process
13. Check the picture for clay coating
14. First with a roller; second with dry-mount press
15. Any remaining clay will appear as dark areas on the final transparency as the clay is opaque to light
16. Laminating a picture

Photographic and Computer Methods (page 188)

1. A camera
2. Diffusion-transfer (PMT) and high-contrast photography
3. Kodalith Ortho Type 3 from Eastman Kodak
4. High-contrast subject: line drawings consisting of black ink on white paper.
   Halftone subject: Shades of gray from black to white comprise the illustration.
5. The enlarger is used instead of the camera. A larger sheet of film is used.
6. The developer is a diluted paper developer, and development time may vary.
7. Shoot a number of negatives of the subject according to the number of overlays needed. Opaque out all but the necessary area for the overlay on each negative. Print each negative.
8. All answers can be used.
9. Draw it on acetate with a plotter, as directed by the computer; or use a film recorder to prepare the transparency on a sheet of color film.

Completing and Filing (page 190)

1. On the underside
2. Base on underside; overlays on separate sides, top of mount
3. Sliding mask and hinged section masks
4. Reader's activity, page 189

Chapter 16 (page 195)

1. With an informative recording the listener only receives information. With an instructional recording the listener also should have opportunities to interact with the material presented.
2. See the list on page 192.
3. See the list on page 191.
4. Audio-notebook or audio-tutorial system.
   The audio-notebook generally is limited to correlated printed material, whereas the AT system includes a variety of other activities and resources for students in addition to pencil and paper work.
5. Refer to the list on page 194.
6. Answering questions, solving problems, completing readings, applying the information, making observations, receiving answers to the questions, and so on
7. Informal, conversational, one-to-one
8. Not over 10 minutes
9. Reader's activity
10. See the discussion on page 194.

Chapter 17

Preparing Slide Series (page 208)

1. See explanations on page 197.
2. See Table 17–1, page 197. Many other 35mm reversal color films are available; Justify your answer.
3. To block out unwanted areas; change proportion of a slide; create an unusual shape
4. a. See 5 and 6, page 199.
   b. See 4, page 199.
   c. See procedure, page 199.
5. A film recorder containing 35mm color film.
6. Script is prepared *after* the picture taking of the event is concluded.
7. E-6
8. How well the slide treats the subject and technical quality (exposure, focus, composition)
9. See discussion on page 201.
10. Reader's activity. See methods starting on page 203.
11. Increase in contrast and shift in colors. Use a film that will control contrast (film designed for duplication); balance colors with color correction filters.
12. Reader's activity. See page 205.
13. An image is always on the screen, even when one is being removed (fading out) and the next one starting to appear (fading in).
14. Reader's activity, page 206.

Producing Filmstrips (page 211)

1. Filmstrip—3 to 4; slide—2 to 3
2. No. A 35mm camera with a half-frame aperture is required
3. Copy slides with a half-frame camera or send slides to a film laboratory which will copy them with a half-frame camera (often a 35mm motion-picture camera).
4. Orientation of subject should be in a horizontal format.
5. Have a film laboratory prepare an internegative from the original slides and from it positive filmstrip prints.
6. Audible signal: Requires no special equipment; uses just one track; synchronization lost when cue missed; may be difficult to find suitable device for generating the signal
   Inaudible signal: Special signal generator necessary 50-Hz signal recorded with narration on same track; can put same or separate message on each track; 1000-Hz signal and narration using both audio tracks; synchronization lost if cue missed electronically
7. Reader's activity. See page 210.

Chapter 18 (page 220)

1. Provide a motivational experience for the viewer; process large amounts of information effectively in a short time.

2. See list, starting on page 213.
3. (a) False, (b) True, (c) False, (d) True, (e) True, (f) False, (g) True
4. Put narration on one track and slide-change cues by voice on second track.
5. Reader's activity, page 217.
6. Dissolver
7. Real time: programing in actual clock time required to present the program.
   Leisure time: taking as much time as you wish while programing each image in turn.
8. (a) Flash, (b) freeze, (c) fade-out, (d) wipe
9. See list, page 219.

Chapter 19

Video Production Equipment and Videotape (page 226)

1. Scenes recorded can be seen immediately.
   Videotape is lower in cost than is film and can be reused.
   Synchronized pictures and sound more easily recorded on tape.
   When editing is completed, videotape can be used immediately.
   Duplicate copies are lower in cost on videotape.
2. Camera, microphone, video recorder.
3. Outdoor/indoor filter adjustment and white balance
4. Picks up extraneous sounds not wanted and it would be better to locate microphone closer to subject for many scenes.
5. $\frac{1}{2}$ inch and $\frac{3}{4}$ inch
6. a. Videocassette—tape reels inside a sealed cassette.
   b. Helical scan recording—tape moves diagonally across the video recording drum head and video picture is recorded as slanting tracks across the tape.
   c. Control track—signals recorded along lower edge of tape to regulate tape speed.
   d. Glitch—discontinuity in recorded picture between scenes due to stopping and restarting the recorder.
   e. Dropouts—momentary sound loss on tape or horizontal flashes on videoscreen.

Recording the Picture (page 228)

1. Motion; time
2. Terms *scene* and *shot* have same meaning, namely, the action represented by the exposure of a length of film; *take* refers to a number assigned each time the same scene is filmed; *sequence* is a series of scenes all related to the same idea or concept.
3. (a) Tilt, (b) zoom, (c) pan, (d) pan, (e) dolly.
4. Look through the camera viewfinder.
5. A scene that orients the viewer to the subject after a number of close-up scenes.
6. (a) MS, (b) CU, (c) LS, (d) MS low angle

Providing Continuity and Special Video Production Methods (page 233)

1. Carefully relating the action at the end of one scene with that at the start of the next scene

2. To keep the audience properly oriented
3. Show direction of movement changing or a head-on shot directly toward the camera; cut-away to related action or an observer.
4. Close-up of her face; spool with thread coming off; or other scene you may think of
5. (a) Hour hand of clock moving or title "Allow paint to dry"; (b) few seconds of closed house door, traffic on street; (c) fade-out/fade-in
6. As bridges for time and space
7. Fade-out/fade-in; dissolve; wipe
8. Keeping in mind what has already been shot and anticipating what might be coming next
9. Because a glitch or discontinuity of image appears on the tape each time the recorder is stopped and then started again; Avoid glitches by putting the recorder *briefly* in the PAUSE position between shooting scenes.

Other Production Factors and Recording Sound (page 235)

1. Used as a reference when adjusting video players and receivers in order to show the best color images
2. Word titles briefly; use simple, bold lettering; adhere to legibility standards; proportion the lettering in 3 to 4 ratio; record title for a period of time that allows title to be read twice; provide for one-sixth marginal loss on video screen.
3. Synchronous voices or sounds, narration, sound effects, music
4. Important for use during editing so that a recorder or player can be backspaced and stabilized before the transfer of a scene takes place.

Editing Videotape (page 239)

1. To select, rearrange, and shorten scenes, as well as to insert scenes or sound in place of a portion of a scene.
2. Assemble editing
3. a. Pause edit: one player VCR, one recorder VCR.
   b. Manual backspace: one player VCR, one edit/record VCR.
   c. Automatic backspace: one player VCR, one edit/-record VCR, an editor controller.
4. Automatic backspace method
5. Putting a brief picture or sound, or a complete scene on tape to replace part of a scene that has already been transferred to the master tape.
   Most commonly, insert editing is used to add a cut-in or cut-away shot, also to replace synchronous sound with narration or a sound effect.
6. a. Control track—pulse along lower edge of tape which determines tape playback speed and is used to locate specific places on the tape.
   b. Edit-in point—initial frame of a scene at which transfer from the player tape starts to the master tape being recorder.
   c. A and B rolls—original videotapes that are to be combined or used in special effects (dissolve or wipe) when transfer is made to the master tape.

d. Time code—a series of eight-digit code numbers recorded to indicate each frame on the videotape and used in editing.
7. a. Adjusts picture quality of scenes on A and B rolls
   b. Creates titles and graphics
   c. Creates fades, wipes, and other effects of scenes on the master videotape

Special Filming Techniques (page 243)

1. Ease of single-framing images.
2. Change fps to 32 or even 48; compensate for less light reaching the film by using a larger lens aperture (smaller f/number).
3. Single frame control.
4. One frame every 45 seconds for the 6 hours.
5. A film consisting of copy work from still pictures shot by using the single-frame technique
6. (a) Scratchoff; (b) pop-on; (c) cel; (d) three-dimensional.

Duplicating and Using Videotape (page 244)

1. Image quality, including sharpness and color fidelity would be best.
2. As videodiscs (if equipment available for use). The content is firm and will not be changed for a long period of time. If videodisc not available, use videocassette.
3. Image and sound read by a lazer stylus
4. The image is larger (and more realistic in comparison with motion pictures), thus more people can be accommodated unless more monitors are set up through the viewing room—often a difficult bothersome job.

Chapter 20

Using, Equipment, and Designing Programs for CBI (page 252)

1. (a) Tutorial; (b) drill; (c) simulation; (d) game; (e) controlling other media.
2. Microcomputer, CRT, disc drive, or other storage device.
3. Linear—instruction and responses are in a fixed sequence.
   Branching—based on either student choice or the answer to a question, different sequences may be taken through the program.
4. direction of movement   ⟶

   single step in process   ▭

   decision point   ◇

   terminal or connecting point   ⬭
5. Multiple choice
6. (a) Single remedial branch; (b) pretest and skip; (c) linear; (d) multiple remedial branch; (e) compound branches.
7. Present a single concept or idea on a screen display; recognize the rate at which words and diagrams may appear on CRT; consider arrangement of items on screen; plan for some variety of images; provide for continuity of sequential images; consider legibility standards for words; insure readability of words; use blank space, color, and special effects carefully.

Writing the Program, Completing the Process, and Controlling Video with a Computer (page 258)

1. 20—line number
   PRINT—operator
   "HOW MUCH IS 8+2+4−3?"—operand (literal)
2. General purpose languages are designed for many different functions. They tend to be flexible, but require considerable knowledge, time, and effort to develop an effective CBI lesson. Authoring languages are designed specifically for writing CBI. Authoring languages are usually easier to learn and more efficient to use when developing CBI.
3. Construct a series of graphic images which represent the desired motion. Plot the first image, erase it, and plot the second image in a slightly different location. Continue in this manner until the action is completed.
4. A description of the computer system(s) on which the software will run; a list of objectives; instructions on how to use the program.
5. The documentation is incorporated into the program itself so that the need for written documentation is minimized.
6. Formative evaluation
7. An electronic circuit which controls a peripheral device such as a videodisc player.
8. Forward, reverse, fast forward, rewind, single-frame advance.
9. When economy and efficiency of production are important; when revisions in content or treatment are likely; for testing purposes prior to the development of a videodisc.
10. Comfort for the student when using the equipment; lighting level; working space; availability of instructions and documentation; availability of earphones for audio material.
11. See lists on pages 257 and 258.

Chapter 21 (page 268)

1. Production planning; graphic arts; still photography; video recording and motion pictures; CBI production; audio recording; duplication and reproduction; models and displays
2. a. Still photography
   b. Still photography, video and motion pictures
   c. Duplication and reproduction
   d. Graphic arts
   e. Production planning
   f. Graphic arts
   g. Models and displays
   h. Still photography, video and motion pictures
   i. Audio recording

3. Complete funding—can set standards and produce quality materials; client may request more service than necessary.

   Full reimbursement—must generate service for income; usually leads to acceptance of whatever work is requested.

4. Capital expenditures—equipment purchase and construction work.

   Operating budget—materials and supplies.

   Outside services—supplementary services.

5. Listen carefully; ask questions; be sensitive to client's manner; admit you don't always know or understand; find out if someone else should be contacted; inform client of situation in which production cannot take place as requested; recognize an impossible situation.

6. Monitoring quantity and quality of work completed; using staff members' skills and time properly; meeting time schedules for work completion; avoiding waste in using supplies; keeping expenses within budgetary limits; abiding by copyright regulations

7. See list on page 267.

8. How productive has the service been?

   What has been the contribution of the materials and media programs to the organization?

   What are the opinions of the instructors or clients about the services provided?

9. Distribute a brochure describing the services.

   Conduct tours of the production facility.

   Offer demonstrations about services.

   Set up exhibits.

   Distribute a newsletter.

   Prepare and distribute an annual report.

Accent light   a light that accentuates and highlights an object in a scene

acetate (clear)   a plastic sheet that permits a high degree of light transmission, resulting in a transparent appearance

affective domain   category of instructional objectives relating to attitudes, values, and appreciations within human behavior

animation   a filming technique that brings inanimate objects or drawings to apparent life and movement by single-frame exposure

aspect ratio   proportion or format of an instructional media, such as 2 to 3 for slides, or 3 to 4 for video

assemble editing   putting scenes of a video program in 1–2–3 order according to the script

attributes of media   capabilities of an audiovisual medium to exhibit such characteristics as motion, color, sound, or simultaneous picture and sound

audible picture advance signal   a signal on tape that is heard (bell, buzzer), thus informing the user to change the slide or filmstrip to the next picture

audiocassette recorder   a recorder that uses magnetic tape on spools that are enclosed in a plastic case

authoring language   a computer language which is designed specifically for CBI

authoring system   computer software which takes you step-by-step through the process of developing CBI

background light   the illumination thrown on the background to lighten it, giving the scene depth and separating the subject from the background

BASIC   a general purpose computer language

Beta   a $\frac{1}{2}$ inch video format manufactured by Sony (differs from VHS format)

bidirectional microphone   one that picks up sounds only on two opposite sides

branching program   a CBI program containing options, either selected by the student or to which the student is directed, thus consisting of sequences that may be different for each student

camera-ready   paste-up copy of pages that are ready for reproduction with a process camera

captions   the printed explanations to accompany visuals

cardioid microphone   one that picks up sounds in a heart-shaped pattern, namely, with major sensitivity from one side, some sensitivity on adjacent two sides, and almost none from back side.

cassette   (see audio*cassette recorder* or video*cassette recorder*)

CBI   computer-based instruction

character generator   electronic device used to create titles and simple graphics which can be translated into video signals on videotape

chargeback system   billing a department for services provided

clearance form   (see *release form*)

clipbooks   printed booklets containing a variety of commercially prepared black-and-white line drawings on various subjects

close-up   a concentration of the camera on the subject, or on a part of it, excluding everything else from view

cognitive domain   category of instructional objectives relating to knowledge, information, and other intellectual skills within human behavior

color bars   color pattern put on the beginning of a videotape recording that serves as reference for adjusting players and receivers for maximum color quality when a videotape is played

colored adhesive   a translucent or transparent color printed on a thin acetate sheet having adhesive backing for adherence to cardboard, paper, acetate, or film

command   (see *operator*)

communications specialist   a person having broad knowledge of audiovisual media and capable of organizing the content of instructional media to be produced so that the stated purposes will best be served

compressed speech   (see *variable speed recording*)

condenser microphone   one that transmits sounds when a plate that receives direct current is caused to vibrate with respect to an adjacent fixed plate

contact print   a photographic print the same size as the negative, prepared by exposing to light the negative and paper, placed together

continuity   the logical relationship of one scene leading to the next one and the smooth flow of action and narration within the total instructional media

continuous-tone subjects   illustrations consisting of shades of gray, varying from black to white

control track   continuous series of signals placed along lower edge of videotape when recording to regulate tape speed during playback and also used during editing for locating scenes

copy-fitting   adapting text to fit available space on a page

copy stand   a vertical or horizontal stand for accurately positioning a camera when photographing flat subjects very close to the lens

credit title   a listing of those who participated in or cooperated with the media project

cropping   marking the edges of a photograph to indicate visual area to be printed

cut   instantaneous change from one image to another within a sequence of projected visuals

cut-away shot scene of a subject or action taking place at the same time as the main action, but separate from it, and placed between two related scenes which have a discontinuity of action

cut-in shot a close-up feature of a subject being recorded on videotape and usually placed between two scenes that have discontinuity of action

debugging testing a computer program and correcting any deficiencies

depth-of-field the distance within a scene from the point closest to the camera in focus to the farthest in focus

developer a solution in which the chemicals set the photographic image by acting on silver salts in exposed film that have been affected by light during picture taking

diazo process a method for preparing overhead transparencies requiring film containing one of a possible number of dye colors, which is exposed, in contact with an opaque original prepared on translucent or transparent paper or film, to ultraviolet light, and then developed in ammonia vapor

diffusion-transfer process two-step duplication when a negative sheet is exposed in a process camera to original art or printed matter, then processed in contact with a positive sheet of paper or film with the image developing on the negative and transferring to the positive paper or film

disc drive a device that will transfer program stored on a disc into the computer's memory

dissolve effect involving two superimposed scenes in which the second one gradually appears as the first one gradually disappears

dissolve unit a device that activates one or more projectors to fade out an image on the screen while an image projected by another projector begins to fade in as a superimposed image over the first one

documentary approach a method of taking pictures without detailed script preparation

documentation descriptive material accompanying a computer program and tells the user how to use it

dolly shot a video scene recorded as the camera is moved toward or away from the subject

drill and practice a computer program consisting of questions and exercises for the user to complete

dropouts momentary loss of sound on an audiotape or horizontal flashes on a video screen due to dirt on the tape or loss of head-to-tape contact

dry mounting sealing a picture to a cardboard or cloth backing with a heat-sensitive adhesive material

dry-transfer letters transferring symbols from a sheet to the working surface by rubbing with a blunt tool

dubbing the transfer of a recording from one unit to another; commonly record-to-tape, audiotape to audiotape, or videotape to videotape

dummy final layout of all elements for a page that serves as guide for paste-up

duplication making a few copies (less than 10) of original pages

dynamic microphone one that employs a moving coil in a magnetic field to generate an electrical signal from sound waves

edit controller a machine that permits the precise location of a videotaped scene's beginning and end points for cueing player and recording VCRs

editing the selection and organization of visuals after filming and the refinement of narration or captions; or, the procedure for rearranging elements of an audio or videotape recording, removing bad-sound takes, and lengthening pauses

editing VTR or VCR a recorder that can record video and audio tracks separately or simultaneously, and will smoothly and accurately record scenes without distortion on a master videotape or videocassette

editing-in-the-camera technique recording scenes in sequence according to their listing on the script

electrostatic duplication a process that uses an electric charge and powdered toner to create copies on paper or acetate

equalizer a device for boosting or subduing certain audio frequencies to improve the tonal quality of a recording

establishing shot a medium or long shot that establishes the whereabouts of a scene and serves as orientation

exposure index a number assigned to a film by the manufacturer which indicates the relative emulsion speed of the film for determining camera settings (f/number and shutter speed) according to American Standards Association (ASA) and International Standards Organization (ISO) in terms of required light conditions

exposure meter (see *photographic light meter*)

expressions mathematical equations which a computer can calculate as it runs a program

f/number (f/stop) the lens setting selected from a series of numbers consisting partially of . . . 2, 2.8, 4, 5.6, 8, 11, 16, 32, . . .

fade-in a visual effect in video recording in which a scene gradually appears out of blackness

fade-out a visual effect in video recording in which a scene gradually disappears into blackness

feedback informing a student of success or progress in learning

fill light the secondary light source illuminating a scene, which brightens dark shadow portions created by the key light

film recorder a special camera to reproduce colored images on photographic film as output of a computer

filmograph a sequence of still pictures on a videotape or motion-picture film

fixer a solution in which the chemicals desensitize the developed film image to light and change all undeveloped silver salts so they can be removed by washing

flannel (felt) board a presentation board consisting of a flannel or felt surface to which objects backed with flannel, felt, or sandpaper will adhere

flash an effect used in multi-image presentations consisting of ON–OFF changes of a lamp while a slide remains in projection position

flow chart a diagram that illustrates the sequence(s) of steps that a student might follow in a computer-based instructional program

focal length a classification of lenses, being the distance from the center of a lens to the film plane within the camera when the lens is focused at infinity

frame an individual picture in a filmstrip or video recording

freeze holding a single image on the screen during a multi-image presentation, video or film sequence, or projecting two images together on the same screen

galley proof first copy of the press run of typeset material to be proofread

game a computer program built on competition to motivate the user to learn

glitch interference in the video image that appears on the screen when, during recording, the recorder is stopped and then restarted for the next scene

guide number or exposure guide number a number assigned to a film for the purpose of calculating exposure when electronic flash units are used; it is based upon the film speed, the power of the electronic unit, and the shutter speed

half-frame camera a camera used for making filmstrips that has an aperture one-half the size of a regular 35mm slide and oriented horizontally across the film width

halftone subjects printed illustrations consisting of uniformly spaced dots of varying size, which blend together and convey shades of gray

helical scan recording a method of recording video pictures diagonally on adjacent stripes across videotape

high-angle shot a scene photographed with the camera placed high, looking down at the subject

high-contrast film used to photograph paste-up or other line drawn or printed subjects to produce clear lines on opaque background (Kodalith film or equal)

high-contrast subject an illustration consisting solely of black lines or marks on white paper

hook-and-loop board a presentation board consisting of a cloth surface textured with minute nylon loops to which display materials backed with strips of tape having nylon hooks will intermesh and hold firmly

inaudible picture advance signal an electronic signal placed on tape and not heard, but which automatically activates a projector to change the picture

incident-light method the measurement of light falling on a scene by the use of an incident light meter held in the scene and aimed at the camera

individualized learning the procedure in which each student assumes responsibility for learning through independent use of appropriate materials and study at a preferred pace

input a receptacle or other connection through which an electronic signal is fed into an amplifier

input device a keyboard, graphics tablet, light pen, or other device used to put information into a computer

instructional-design plan procedure for instructional planning that involves application of a number of interrelated steps relating to objectives, instructional strategies, and evaluation of learning

instructional development process of designing an instructional program employing an objective, systematic procedure, such as an instructional-design plan

interactive video combining video images with computer programing, such that the sequence of the program will depend on user choices, answers to questions, and so forth

interface interconnection device between a microcomputer and video player, or other media equipment, which translates computer commands into signals that direct the media to start, stop, skip ahead, or back up

job aid (see *learning aid*)

jumper cord (see *patch cord*)

justify changing spacing between letters and words of printed lines comprising columns on a page so that each line is of the same length at both right and left margins

key light the brightest light source on a scene, forming a large portion of the total illumination

key word (see *operator*)

lamination applying a thin adhesive-backed clear acetate coating over a picture or other graphic material

layout a design on paper illustrating how all elements that comprise the visual, or a page, will appear

learning aid a checklist of steps or procedures, with or without line drawings and photographs, to be used by a person when assembling, operating, or maintaining equipment

legibilty the requirement for lettering size and style so that the farthest seated member of a potential audience can see and read projected or nonprojected material satisfactorily

leisure-time programming pressing keys on a microprocessor to store control signals for a multi-image presentation at one's own pace or convenience, and then playing back the commands for operating the program at regular showing pace

lens diaphragm the opening through which light enters a camera; its size is controlled by an adjustable diaphragm consisting of metal blades

lighting ratio the relationship between the intensity of the key light and the intensity of the fill light as measured with a light meter

line copy (see *high-contrast subject*)

line number first element put on paper when writing a computer program in BASIC

linear program information and instruction presented in a fixed sequence which is the same for all students

literal a string of characters or numbers, enclosed in quotation marks, to be translated literally by a computer

litho film   (see *high-contrast film*)

log sheet   a written record of all pictures taken, including scene numbers, takes, camera settings, and special remarks

long shot   a general view of a subject and its setting

low-angle shot   a scene photographed with the camera placed low, looking up at the subject

macro lens   one specially adapted for close-up and copy work that does not distort lines in a subject

magnetic chalkboard   a presentation board consisting of a metal sheet, covered with chalkboard paint, to which magnetic-backed objects will adhere and on which chalk marks can be made

main title   the name of the production, shown at the start of instructional media

matched action   the smooth continuation of action between two adjacent, related video scenes

mechanical   term applied to the completed paste-up ready to be reproduced on an offset press

medium shot   a view of a subject that eliminates unnecessary background and other details

menu   list of content or options within a computer program from which the user makes a selection

microcomputer   a self-contained computer including a keyboard and memory, and attached to a display screen or a printer

microprocessor programer   uses computer commands to control a number of slide projectors and dissolve functions for multi-image presentations

mixer   a control mechanism through which a number of sound-producing units can be fed in order to combine voice, music, or sound effects at desired recording levels onto a single audiotape

modem   a device that permits information to be transmitted between computers using telephone lines

module   a study unit or program designed for self-paced learning

monitor   (see *video monitor*)

monophonic recorder   a tape recorder with a single recording head, capable of only recording on one channel at a time

montage   a series of short scenes in a video recording used to condense time or distance or to show a series of related activities or places

multicamera technique   recordings of the same action with two or more cameras operated at the same time and located at different positions in relation to the subject

multi-image   simultaneous projection of two or more pictures on one or adjacent screens for group viewing

multitrack recorder   a recorder with 4, 8, 16 or more tracks, which allows a separate recording to be made on each track and played back in any combination

narration   the verbal comments made to accompany the visuals

negative opaque   a water-soluble carbon material brushed on high-contrast negatives to eliminate marks, spots, and areas

nondirectional microphone   one that picks up sounds coming from all directions

objective scene   a scene recorded with the camera aimed toward the subject, from a theater audience point of view

opaque projector   a projector that can enlarge information from paper, pages from a book or other nontranslucent or nontransparent materials

open reel recorder   a tape recorder that uses magnetic tape on separate reels

operand   tells the operator of a computer program what to act upon—literals, variables, expressions

operator   key word or action verb in a computer statement which tells the computer what to do

output   a connection on an electronic device at which the signal leaves the unit

output device   video screen, plotter, or printer to display the information provided by a computer

overhead projector   a projector that accepts transparent and translucent sheets and projects the information prepared on them onto a screen

overlay   one or more additional transparent sheets with lettering or other matter that can be placed over a base transparency or an opaque background

overprint   the superimposition of one scene over another; generally titles, captions, or labels over a background scene or a specially prepared background

page proof   a trial copy of final pages for checking before the press run

pan (panorama)   the movement of a video camera, while shooting, in a horizontal plane (sideways)

parallax   the difference between the vertical position of an object in a filmed scene as viewed through a viewfinder and as recorded on film through the camera lens

paste-up   the combination of illustrations and lettering, each unit of which is rubber-cemented in position on paper or cardboard

patch cord   an electrical wire used to connect together two pieces of sound equipment (such as tape recorders and record players) so that electrical impulses can be transferred between the units in order to make a recording

permanent mounting   the application of rubber cement to the back of a flat material to be mounted and also to the mount surface, then permitting the two surfaces to dry before adhering them together; or the use of dry-mount tissue placed between the material and the mount surface, with heat and pressure being applied to seal the layers together

photographic light meter   a device for measuring light levels, either incident upon or reflected from a scene

photomechanical transfer process (PMT)   a photographic method using high-contrast film and paper to prepare prints, projected materials, and plates for offset printing

photo modifier   a large-size camera used to enlarge or reduce art work

photostabilization   a two-step rapid processing method

for photographic paper

picas   a printing measurement in which 6 picas equals one inch of line length

picture transfer   the transfer of printing ink from a picture to a special acetate sheet after the two are sealed together with heat and/or pressure and then submerged in water to remove the paper

pinboard   (see *register board*)

pixels   rows of dots that comprise a computer-generated graphics image

planning board   a board with strips of acetate channels holding storyboard cards

points   a printing measurement in which 72 points equal one inch of line length

posterization   the process of converting a continuous-tone photograph to a high-contrast image through an application of the diffusion-transfer, or high-contrast, film process

process camera   a large-size, cut-film camera, used by a graphic artist or a printer to convert paste-up, or other original material, with necessary size change, to a high quality paper or film copy

programer   a control unit to activate one or more projectors to change images on screens according to scripted sequencing

progressive disclosure   exposing a series of items on a transparency or slide by moving an opaque mask during use or preparation

pumping in (pumping out)   the change of camera position from one video scene to the next without change of camera angle with relation to the same subject

radio microphone   (see *wireless microphone*)

rangefinder   a camera attachment that, upon proper setting, indicates the distance from camera to subject or sets the lens in focus at that range

readability formula   procedure to count the number of syllables and sentences for a certain number of words on a printed page in order to determine a grade level of difficulty for reading material

real-time programing   putting control signals on audiotape in actual clock time required to run the program when synchronizing a multi-image presentation

rear-screen projection   projecting on the back side of a translucent screen, the projector being behind the screen

reflected-light method   the measurement of light reflected from a scene by the use of a reflected light meter aimed at the subject

register board   a surface with two or more small vertical posts for holding paper, cardboard, or film materials all correctly aligned when more than one layer must be assembled for picture taking

register marks   cross symbols placed on a base sheet and also on each overlay sheet to register (match) colors when reproducing multi-color pages or transparencies

reinforcement   the result of success in learning which builds confidence in a student to continue learning

release form   the form used to obtain written permission

for use of pictures taken of persons or of their copyrighted materials

repro copy   corrected copy of typeset material to be used in preparing the paste-up

reproduction   making many copies (10 or more) of printed material

reversal film   film which, after exposure, is processed to produce a positive image on the same film

scene (shot)   the basic element that makes up the visuals of instructional media; each separate picture or video footage exposed when the release button is pushed

script   the specific directions for picture taking or art work in the form of a listing of scenes with accompanying narration or captions

self-paced learning   an instructional method whereby each student assumes responsibility for independent study at a preferred rate of study while using required materials

sequence   a section of instructional media, more or less complete in itself, and made up of a series of related scenes

shading film   textures and patterns printed on acetate sheets having adhesive backing for adhering to cardboard, paper, acetate, or film

shot   (see *scene*)

shotgun microphone   a unidirectional type that accepts only a narrow angle of sound and rejects most other noises

shutter speed   the interval between opening and closing of the shutter of a camera, measured in fractions of a second

simulation   a program which attempts to emulate a realistic situation

single-framing   exposing one frame at a time on motion-picture film, as opposed to continuous exposure

single-lens reflex camera   a compact camera employing a mirror or prism for accurate viewing directly through the camera lens

slant track recording   (see *helical scan recording*)

software   a program for use with a computer

sound effects   realistic sounds (street traffic, machinery, animals, and so on) that can be added to an audio or video program

special-effects generator   a device that allows the addition of various effects to visual images as they are selected and recorded on videotape

specifications   the framework and limits within which instructional media are produced; may include such factors as length, materials, special techniques, assistance required, completion date, and budget

spot meter   a reflected-type light meter that accepts only a narrow angle of light; used to measure accurately the light level from a small portion of a scene

statement   a line of instruction in a computer program

stop bath   a solution in which the chemicals stop the action of a developer on exposed film

storage device   tape recorder with audiocassette tape, disc drive with magnetic floppy or hard disc, or other device for storing computer information

storyboard   a series of sketches or pictures which visualize each topic or sequence in instructional media to be produced; generally prepared after the treatment and before the script

stencil lettering   lettering made by using a special pen to fill in the outline of a letter or number cut into a piece of plastic

stereophonic tape recorder   a recorder on which a recording is made on two tracks simultaneously and then played back together

subjective scene   a scene recorded with the camera placed in the subject's position and aimed at action being performed

subject specialist   a person having broad knowledge of the subject content for instructional media to be produced

switcher   a control unit that permits a recording to be made from any of a number of video cameras or studio equipment as selected by the program director

synchronized sound   sounds on videotape that correlate in proper relationship with pictures

take (of a scene)   one exposure among two or more for the same scene; successive takes of the same scene are numbered from 1 upward

tape cassette   (see *audiocassette recorder or videocassette recorder*)

technical staff   a person or persons responsible for the photography, graphic-art work, video and sound recording in producing instructional media

telephoto lens   a camera lens that permits a closer view of a subject than would be obtained by a normal lens from the same position

television receiver   receives signals through an antenna or cable connection which enters the circuits of the receiver and is converted into video and audio impulses

template lettering   performed by moving a scriber, one leg of which traces letters grooved in a template, while a pen in the scriber forms letters

temporary mounting   the application of rubber cement to the back of an illustration or lettering and immediate placement on a mount surface while the cement is still wet

thermal film   a film sensitive to infrared heat for producing overhead transparencies from carbon marks made on ordinary paper

thumbspot   a visible mark placed in the lower left corner of a slide to indicate the proper position for correctly viewing the slide

tilt   the movement of a video camera in a vertical plane

time-base corrector   a device used to ensure acceptable picture and sound qualiy when combining video-taped materials from various sources

time-code generator   electronic device that records eight digit numbers on each frame of a videotape and used during editing

time-compressed speech   method of increasing the word rate of recorded speech without distortion in vocal pitch (see also *variable speed recording*)

time-lapse photography   the exposing of individual motion-picture frames at a much slower rate than normal, for projection at normal speed; the method accelerates action that normally takes place too slowly for motion to be perceived

tone-control programer   uses audio frequency tones to control a number of slide projectors or different dissolve functions for a multi-image presentation

toner   a black powder used in the electrostatic duplicating process to deposit on the paper or film and provide the visible, opaque image

touch screen technique   user touches a finger or light pen to item on computer screen and computer responds accordingly

tracking control   a setting on videocassette recorders to adjust video head speed to overcome minor variations that affect picture quality of a tape during playback

transitional devices   the use of such techniques as fade-out, fade-in and dissolves to bridge space and time in video recordings

treatment   a descriptive synopsis of how the content of instructional media can be organized and presented

tutorial   a computer program that presents information followed by a question or problem, then based on user's answer, the next block of instruction is presented or remedial instruction is provided

unidirectional microphone   (see *cardioid microphone*)

variable   memory in a computer with a specific name, but the value of which can be changed

variable speed recording   speeding up or slowing down a recording while maintaining proper pitch and intelligibility

vellum   a high grade translucent paper or film on which drawings are prepared

VHS   a $\frac{1}{2}$ inch video format (differs from Beta format)

videocassette   a sealed, rectangular container holding reels with $\frac{3}{4}$ inch, $\frac{1}{2}$ inch, or 8mm size videotape

videocassette recorder (VCR)   a video unit that records and plays back visual images and sound as magnetic tape, on reels in a sealed container, passes by the record or playback head in the machine

videodisc   similar in appearance to a long-play audio record containing up to 30 minutes of video and sound or 54,000 still frames, in any combination, and played with a laser stylus

video monitor   receives video signals from a video player directly into video and audio circuits

videotape recorder (VTR)   a video unit that records and plays back visual images and sound as magnetic tape from one reel passes by the record or playback head in the machine and takes up on another reel

visual literacy   skills an individual develops in interpreting, judging, responding to, and using visual representations of reality

white balance   a video camera setting to fine-tune the camera for the particular light conditions under which the scene will be recorded

wide-angle lens   a camera lens that permits a wider view of a subject and its surroundings than would be obtained by a normal lens from the same position

wipe   a visual effect in video recordings in which a new scene seems to push the previous scene off the screen

wireless microphone   a microphone that acts as a radio transmitter, broadcasting the audio signal to a receiver

Zimdex   a method of placing index numbers on tape to facilitate locating specific parts of a recording on the other track

zoom lens   a camera lens of variable magnification that permits a smooth change of subject coverage between distance and close-up shots without changing the camera position

7½"

6½"

DATE DUE

9½" 8½"

MASK

The area enclosed wi
be used with an overhea

MASK FOR VIDEO AND MOTION-PICTURE FO

The area enclosed within the color rule is the area within which th
art work, and other visuals for a video recording and a motion pictur
should be composed (proportion 3 × 4).

Demco, Inc. 38-293